"十三五"国家重点图书出版物出版规划
经典建筑理论书系
加州大学伯克利分校环境结构中心系列

建筑模式语言

——城镇·建筑·构造

（上册）

A Pattern Language

—Towns · Buildings · Construction

［美］ C. 亚历山大　S. 伊希卡娃　M. 西尔沃斯坦

M. 雅各布逊　I. 菲克斯达尔－金　S. 安格尔　著

王听度　周序鸿　译

李道增　高亦兰　审校
关肇邺　刘鸿滨

知识产权出版社
全国百佳图书出版单位
—北京—

图书在版编目（CIP）数据

建筑模式语言：城镇·建筑·构造 / （美）C. 亚历山大等著；王昕度，周序鸿译 . —北京：知识产权出版社，2022.7

（经典建筑理论书系）

书名原文：A Pattern Language：Towns·Buildings·Construction

ISBN 978-7-5130-7406-3

Ⅰ . ①建… Ⅱ . ① C…②王…③周… Ⅲ . ①建筑设计－研究 Ⅳ . ① TU2

中国版本图书馆 CIP 数据核字（2021）第 013803 号

责任编辑：李 潇 刘 嚣　　　　　　责任校对：谷 洋
封面设计：红石榴文化·王英磊　　　　责任印制：刘译文

经典建筑理论书系

建筑模式语言——城镇·建筑·构造（上册）

A Pattern Language—Towns·Buildings·Construction

［美］ C. 亚历山大　S. 伊希卡娃　M. 西尔沃斯坦
M. 雅各布逊　I. 菲克斯达尔－金　S. 安格尔　　著

王昕度　周序鸿　译
李道增　高亦兰
关肇邺　刘鸿滨　审校

出版发行：知识产权出版社 有限责任公司		网　址：http://www.ipph.cn	
社　址：北京市海淀区气象路 50 号院		邮　编：100081	
责编电话：010-82000860 转 8119		责编邮箱：liuhe@cnipr.com	
发行电话：010-82000860 转 8101		发行传真：010-82000893/82005070	
印　刷：三河市国英印务有限公司		经　销：新华书店、各大网上书店及相关专业网点	
开　本：880mm×1230mm　1/32		总印张：69.5	
版　次：2022 年 7 月第 1 版		印　次：2022 年 7 月第 1 次印刷	
总字数：1730 千字		总定价：268.00 元（上、下册）	
ISBN 978-7-5130-7406-3			
京权图字：01-2016-8195			

译 序

克里斯托弗·亚力山大是美国杰出的建筑理论家。由他领衔撰写的《建筑模式语言》一出版就受到建筑界的广泛重视和高度赞誉，并对建筑业产生了深远的影响。

《建筑模式语言》别出心裁且有根有据地描述了城镇、邻里、住宅、花园和房间等共 253 个模式，提供了一幅幅设计、规划、施工等方面的崭新蓝图，构思新奇，妙想迭出，毫不流俗。

作者写道："我们深信无疑，本语言要比一本手册、一位教师、另一种可能的模式语言略胜一筹。这里的许多模式看来在今天和以后的 500 年间将成为人性的一部分，成为富有人情味的行动的一部分。"诚如美国《建筑设计》一则评论所说："这也许是 20 世纪出版的关于建筑设计的最重要的一本书了。"

本书的生命力在于"以人为本"。它是本书的主题思想，像一条鲜艳的红线贯穿始终。各模式的字里行间洋溢着浓浓的人情味和对人的无微不至的关爱，如何保护生态环境，如何绿化美化城镇和住宅，反对建筑风格的千篇一律，鼓励人际交往，强调人、社会和自然环境三者的和谐统一等。

本书旁征博引，内容丰富充实，但并不庞杂拖沓。前后连贯，遥相呼应，条理清晰，图文并茂。

《建筑模式语言》实在是一本不可多得的佳作，它充满着时代的气息和对未来美好的憧憬；它是建筑科学和艺术的完美结合，具有很强的生命力和无比的魅力。

广大读者，从建筑科学角度来看，可以把此书作为一种令人向往的城市建筑和规划的模式语言来阅读；从建筑艺术角度来看，也可以当成一部长篇叙事散文诗来欣赏：

其中既有对建筑的一般原理、人的行为心理的深入浅出的分析说明，又有对蓝天碧水、绿树鲜花、野草小虫、飞鸟游鱼，以及弥漫在旷野上的新鲜空气、沁人肺腑的芳香的抒情描绘，还有对世界各地千姿百态的建筑风貌、民间习俗和趣闻轶事的生动叙述……真可谓生花妙笔，异彩纷呈，使人赏心悦目，回味无穷。

我国建筑界享有声誉的资深学者李道增、高亦兰、关肇邺和刘鸿滨四位教授力促本书翻译出版，并拨冗审校译文，足见他们对本书的重视。知识产权出版社多位领导和编辑对本书给予了全力支持，付出了许多辛劳。在此，我们一并向他们表示诚挚的谢意。

本书《建筑》部分是周序鸿翻译的，其余部分均为王听度所译。我们历时数载，精心翻译，反复修改，力求译文正确和传神，唯恐不能如愿。

敬请读者批评指正。

王听度　　写于清华园
周序鸿

关于作者

C·亚历山大，美国建筑师协会颁发的最高研究勋章的获得者，是一位有实践经验的建筑师和营造师。他是加州大学伯克利分校建筑学教授，环境结构中心的负责人。

《建筑模式语言》是丛书的第二卷。这套丛书以崭新的观点来描述城镇规划和建筑，力图提供一整套可供选择的工作方法，以求逐步取代在建筑和规划方面的种种现行观念和实践。

about the author:

Christopher Alexander, winner of the first
medal for research ever awarded by the American
Institute of Architects, is a practising architect
and contractor, Professsor of Architecture at the
University of California, Berkeley, and Director of
the Center for Environmental Structure.

A Pattern Language is the second in a series of books
which describe an entirely new attitude to architecture and
planning. The books are intended to provide a complete
working alternative to our present ideas about architecture,
building, and planning—an alternative which will, we
hope, gradually replace current ideas and practices.

目 录
CONTENTS

CONTENTS
目　录

目　录
CONTENTS

USING THIS BOOK

本书使用说明

《建筑模式语言》

　　《建筑的永恒之道》（第一卷）和《建筑模式语言》（第二卷）是一部完整著作的两个组成部分。本书（第二卷）为城市建筑和规划提供一种模式语言；另一本书（第一卷）为使用该模式语言提供理论和说明。本书描述城镇、邻里、住宅、花园和房间的各种详尽的模式。另一本书解释原理，这种原理使利用这些模式来创造出新的建筑和城镇成为可能。本书是《建筑的永恒之道》的原始资料集；另一本书是它的实践和渊源。

　　这两本书互为渗透，平行发展。八年来，我们不断地耕耘，一方面努力去理解建筑过程的本质，另一方面又要尽量构思出切实可行的模式语言，在备尝甘苦的过程中，这两本书的内容也日臻完善。鉴于对种种实际问题的考虑，我们不得不将它们分别以单行本出版。但事实上，它们已成为不可分割的整体。分开来阅读这两本书也未尝不可，但要融会贯通地掌握这两本书中全部的新思想，对这两本书齐头并进地阅读是极为重要的。

　　《建筑的永恒之道》描述了建造各种城镇和建筑物的基本性质。书中已阐明，如果城镇和建筑不由社会全体成员共同参与建造，如果他们没有使用建筑上共同的模式语言，如果其模式语言本身不是富于生机的，那么城镇和建筑就不可能充满活力。

A PATTERN LANGUAGE

Volume 1,*The Timeless Way of Building*,and Volume 2,*A Pattern Language*,are two halves of a single work.This book provides a language,for building and planning;the other book provides the theory and instructions for the use of the language. This book describes the detailed patterns for towns and neigh borhoods,houses,gardens,and rooms.The other book explains the discipline which makes it possible to use these patterns to create a building or a town.This book is the sourcebook of the timeless way;the other is its practice and its origin.

The two books have evolved very much in parallel.They have been growing over the last eight years,as we have worked on the one hand to understand the nature of the building process,and on the other hand to construct an actual,possible pattern language.We have been forced by practical considerations,to publish these two books under separate covers;but in fact,they form an indivisible whole.It is possible to read them separately.But to gain the insight which we have tried to communicate in them,it is essential that you read them both.

The Timeless Way of Building describes the fundamental nature of the task of making towns and buildings.It is shown there,that towns and buildings will not be able to become alive,unless they are made by all the people in society,and

我们在本书中提供一种可行的模式语言，它是《建筑的永恒之道》所不可缺少的。这种语言切实可行。它是我们从过去八年内大量的建筑和规划的实践中精心提炼出来的。你可以利用它和你的邻居们合作，以改进你们的城镇和邻里。你可以利用它和你的家人一起齐心协力设计新的住宅；或和其他人合作设计出脱俗立新的办公室、车间或公共建筑——如学校。你还可把它作为实际施工中的指南。

本语言的要素都是称为模式的实体。每一模式描述我们周围环境中一再反复发生的某个问题，接着叙述解决这一问题的关键所在。这样，你就能千百次地重复利用这种解决问题的办法而又不会有老调重弹之感。

为了方便和明了起见，每一模式都具有统一的格式：第一，开头有一幅照片，表示该模式的原型实例。第二，在照片之后有一段引言，说明该模式承上启下、前后连贯的内容，并解释它是如何去协助完善一些较大的模式。第三，接下来是 ৪০৩ 符号，表示提出问题。符号的下面是粗体字的主题。它用三言两语点明问题的实质。继主题之后，就是问题的全部内容了。这是最长的一节。这一节描述该模式的经验性背景材料，为其可行性提供例证，并说明这一模式在建筑中可能被显示出来的各种途径等。第四，解决方案，它像标题一样，用粗体字印刷，这是该模式的核心部分。它表明为解决上文中提出的问题所需的各种空间实体关系和社会关系的范围。这种解决方案均以指示的形式出现。由此，你会确切地了解，为了使这一模式付诸实施，需要做些什么。第五，在答案之后是一张简图，它以图解的形式说明答案，其主要部分都由标出的文字一一注明。

unless these people share a common pattern language,within which to make these buildings,and unless this common pattern language is alive itself.

In this book,we present one possible pattern language,of the kind called for in *The Timeless Way*.This language is extremely practical.It is a language that we have distilled from our own building and planning efforts over the last eight years.You can use it to work with your neighbors,to improve your town and neighborhood.You can use it to design a house for yourself,with your family;or to work with other people to design an office or a workshop or a public building like a school.And you can use it to guide you in the actual process of construction.

The elements of this language are entities called patterns.Each pattern describes a problem which occurs over and over again in our environment,and then describes the core of the solution to that problem,in such a way that you can use this solution a million times over,without ever doing it the same way twice.

For convenience and clarity,each pattern has the same format.First,there is a picture,which shows an archetypal example of that pattern.Second,after the picture,each pattern has an introductory paragraph,which sets the context for the pattern,by explaining how it helps to complete certain larger patterns.Then there are three diamonds to mark the beginning of the problem.After the diamonds there is a headline,in bold type.This headline gives the essence of the problem in one or two sentences.After the headline comes the body of the problem.This is the longest section.It describes the empirical background of the pattern,the evidence for its validity,the range of different ways the pattern can be manifested in a building,and so on.Then,again in bold type,like the headline,is

简图的下面又是 ℛℭ 符号，表示该模式的主要内容已经叙述完毕。最后，在十字形符号之后还有一段结尾文字，它说明与本模式有联系的其他一些较小的模式。它们对于完善、修饰和充实该模式是必不可少的。

这种统一格式包含两大目的。第一，它使每一模式和其他模式联系起来，所以你可以把253个模式汇总作为一个整体、作为一种语言来掌握，从而能创造出无穷无尽、千变万化的组合。第二，它把每一模式的问题和解决方案统统摆在你的面前，你自己可以对它作出判断，并对它加以修正，但要抓住它的主要实质。

下面接着谈一下如何理解各模式间的相互关系。

模式是有顺序的，从最大的模式，如区域和城镇开始，接着便是邻里、住宅团组、住宅、房屋和凹室，最后是构造细部。

这种顺序是直线式的序列，它对发挥模式语言的作用确实非常重要。这一点将在下一节中做更加充分的介绍和说明。这种序列的最重要之处就在于它是以各模式间的相互关系作为基础的。本语言中的每一模式既和它前面的若干"较大的"模式相联系，又和它后面的若干"较小的"模式相联系。每一模式都有助于完善位于其前的较大模式，而它自身也为位于其后的较小模式所充实。

the solution—the heart of the pattern—which describes the field of physical and social relationships which are required to solve the stated problem,in the stated context.This solution is always stated in the form of an instruction-so that you know exactly what you need to do,to build the pattern.Then,after the solution,there is a diagram,which shows the solution in the form of a diagram,with labels to indicate its main components.

After the diagram,another three diamonds,to show that the main body of the pattern is finished.And finally,after the diamonds there is a paragraph which ties the pattern to all those smaller patterns in the language,which are needed to complete this pattern,to embellish it,to fill it out.

There are two essential purposes behind this format. First,to present each pattern connected to other patterns,so that you grasp the collection of all 253 patterns as a whole,as a language,within which you can create an infinite variety of combinations.Second,to present the problem and solution of each pattern in such a way that you can judge it for yourself,and modify it,without losing the essence that is central to it.

Let us next understand the nature of the connection between patterns.

The patterns are ordered,beginning with the very largest,for regions and towns,then working down through neighborhoods,clusters of buildings,buildings,rooms and alcoves,ending finally with details of construction.

This order,which is presented as a straight linear sequence,is essential to the way the language works.It is presented,and explained more fully,in the next section.What is most important about this sequence,is that it is based on the connections between the patterns.Each pattern is connected

例如，你可以把模式**近宅绿地**（60）作为实例。它和一些较大的模式有联系：**亚文化区边界**（13）、**易识别的邻里**（14）、**工作社区**（41）和**僻静区**（59）。这些模式都出现在它的第一页上。并且它也和一些较小的模式有联系：**户外正空间**（106）、**树荫空间**（171）和**花园围墙**（173）。这些模式都出现在它的最后一页上。

这意味着**易识别的邻里**、**亚文化区边界**、**工作社区**和**僻静区**都不是完整无缺的，除非它们都包含**近宅绿地**。而且，**近宅绿地**本身也不是完整无缺的，除非它包含**户外正空间**、**树荫空间**和**花园围墙**。

这还意味着，在利用实际的模式术语时，如果你想按照这一模式来布置一块绿地，你不仅要按说明该模式的各种指示行事，而且也必须竭力把这块绿地嵌入**易识别的邻里**或某一**亚文化区边界**之内，以利于形成**僻静区**；继而，你还应按**户外正空间**、**树荫空间**和**花园围墙**等模式不断经营这块绿地，使之逐步完善。

总之，没有任何一个模式是孤立存在的。每一个模式在世界上之所以能够存在，只因为在某种程度上为其他模式所支持：每一模式又都包含在较大的模式之中，大小相同的模式都环绕在它的周围，而较小的模式又为它所包含。

这就是我们对世界的基本观点。这就是说，当你建造一个东西的时候，你绝不能孤立地建造它，而是必须在它的内部和外界进行整理。这样，一个较大的空间就显得更加紧凑、更加完整。按你的意愿所建造的建筑在大自然的网络中就会适得其所。

to certain "larger" patterns which come above it in the language;and to certain "smaller" patterns which come below it in the language.The pattern helps to complete those larger patterns which are "above" it,and is itself completed by those smaller patterns which are "below" it.

Thus,for example,you will find that the pattern ACCESSIBLE GREEN(60),is connected first to certain larger patterns:SUBCULTURE BOUNDARY(13),IDENTIFIABLE NEIGHBORHOOD(14),WORK COMMUNITY (41),and QUIET BACKS(59).These appear on its first page.And it is also connected to certain smaller patterns:POSITIVE OUTDOOR SPACE (107),TREE PLACES(171),and GARDEN WALL(173). These appear on its last page.

What this means,is that IDENTIFIABLE NEIGHBORHOOD, SUBCULTURE BOUNDARY,WORK COMMUNITY,and QUIET BACKS are incomplete,unless they contain an ACCESSIBLE GREEN;and that an ACCESSIBLE GREEN is itself incomplete,unless it contains POSITIVE OUTDOOR SPACE,TREE PLACES,and a GARDEN WALL.

And what it means in practical terms is that,if you want to lay out a green according to this pattern,you must not only follow the instructions which describe the pattern itself,but must also try to embed the green within an IDENTIFIABLE NEIGHBORHOOD or in some SUBCULTURE BOUNDARY,and in a way that helps to form QUIET BACKS;and then you must work to complete the green by building in some POSITIVE OUTDOOR SPACE,TREE PLACES,and a GARDEN WALL.

In short,no pattern is an isolated entity.Each pattern can exist in the world,only to the extent that is supported by other patterns:the larger patterns in which it is embedded,the patterns of the same size

现在我们来解释各个单独模式中的问题和答案两者关系的性质。

每一答案都是如此叙述的：它指明为解决某一问题所不可缺的各种关系的基本方面，但极其概括抽象，因此你能够以你自己的方式，根据你自己的爱好，并根据当地的条件，解决你正在设法解决的问题。

为此，我们在写出的每一个解决方案中，力求避免有令人不悦之感。它只包括那些极其基本的要点，如果你真想解决问题，这些要点是不可避免的。从这种意义上来讲，我们要在每一个解决方案中竭力抓住在各种情况下都行得通的、共同的不变特性。

诚然，我们并不总是一帆风顺的。我们对于这些问题所给出的种种解决方案，在意义上有所不同。有一些解决方案较另一些解决方案更加切合实际、更加深刻、更加明确和肯定。为了明白无误地表明这一点，我们在每一个模式的标题后面，分别加上两个星号或一个星号，或是不加星号。

在有两个星号的模式中，我们认为，在叙述真正不变的特性方面已取得成功：简而言之，我们阐明的解决方案概括了解决问题的一切可能途径的共同特性。总而言之，我们认为，在有两个星号的情况下，如果没有按照我们所提出的模式以这种或那种方式来塑造环境，同时要想恰如其分地解决已涉及的问题几乎是不可能的；我们同样认为，在这些情况下，模式会对业已形成的良好环境的深刻性和不可避免性做出说明。

that surround it,and the smaller patterns which are embedded in it.

This is a fundamental view of the world.It says that when you build a thing you cannot merely build that thing in isolation,but must also repair the world around it,and within it,so that the larger world at that one place becomes more coherent,and more whole;and the thing which you make takes its place in the web of nature,as you make it.

Now we explain the nature of the relation between problems and solutions,within the individual patterns.

Each solution is stated in such a way that it gives the essential field of relationships needed to solve the problem,but in a very general and abstract way-so that you can solve the problem for yourself,in your own way,by adapting it to your preferences,and the local conditions at the place where you are making it.

For this reason,we have tried to write each solution in a way which imposes nothing on you.It contains only those essentials which cannot be avoided if you really want to solve the problem.In this sense,we have tried,in each solution,to capture the invariant property common to all places which succeed in solving the problem.

But of course,we have not always succeeded.The solutions we have given to these problems vary in significance.Some are more true,more profound,more certain,than others.To show this clearly we have marked every pattern,in the text itself,with two asterisks,or one asterisk,or no asterisks.

In the patterns marked with two asterisks,we believe that we have succeeded in stating a true invariant:in short,that the solution we have stated summarizes *aproperty* common to *all possible ways* of solving the stated problem.In these two-asterisk cases we believe,in short,that it is not possible to solve

在有一个星号的模式中，我们认为在识别不变的特性方面已取得若干进展：如果我们谨慎从事，肯定能使解决方案更趋完善。在这些情况下，我们认为应明智地对待这种模式，不必全盘照抄照搬。几乎可以肯定地说，因为在我们列举的解决方案中并未包括各种可能的解决途径，所以你可以从中找出其他的解决办法。

最后，在没有星号的模式中，我们确认，我们在说明不变的特性方面没有取得成功。但可以说，肯定存在着与我们的解决方案不同的其他解决途径。在这些情况下，我们仍然阐明了解决方案，无非是为了使它具体化——给读者至少提供一种解决方案——但尚未发现解决问题的各种可能的方案中的核心部分，即真正不变的特性。

当然，我们希望，阅读并利用这一语言的读者中的许多人将设法改进这些模式，以充沛的精力投入此项工作，以便发现更加真实的、更加深刻的不变特性，从而完成未竟的任务。我们还希望，这些更加真实的、被缓慢地发现的模式，随着时间的推移，将渐渐地进入我们大家都能使用的一种共同语言。

至此你就会明白，这些模式是富有生气的，不断发展的。事实上，如果你愿意，你可以把每一个模式都视为一种假设，如同科学中的假设一样。在这种意义上说，每一模式都代表着我们公认的一种最佳猜测：怎样安排总体的环境才有助于解决悬而未决的问题。人们根据切身的经验往往会提出种种疑问，归根结底会集中到一点：真会发生这种事吗？真会有我们所描述的那种感觉吗？而按照我们所建议的解决方案行事，难解的问题真会迎刃而解吗？但是，这些星号表示我们对这些假设的可靠性的信赖程度。无论星号含有何种意义，模式终归是模式。因此，253 个模式都是试验性的，在新的经验和观察研究的推动下，都会自由地向前发展。

the stated problem properly,without shaping the environment in one way or another according to the pattern that we have given—and that,in these cases,the pattern describes a deep and inescapable property of a well-formed environment.

In the patterns marked with one asterisk,we believe that we have made some progress towards identifying such an invariant:but that with careful work it will certainly be possible to improve on the solution.In these cases,we believe it would be wise for you to treat the pattern with a certain amount of disrespect—and that you seek out variants of the solution which we have given,since there are almost certainly possible ranges of solutions which are not covered by what we have written.

Finally,in the patterns without an asterisk,we are certain that we have not succeeded in defining a true invariant—that,on the contrary,there are certainly ways of solving the problem different from the one which we have given.In these cases we have still stated a solution,in order to be concrete—to provide the reader with at least one way of solving the problem—but the task of finding the true invariant,the true property which lies at the heart of all possible solutions to this problem,remains undone.

We hope,of course,that many of the people who read,and use this language,will try to improve these patterns—will put their energy to work,in this task of finding more true,more profound invariants and we hope that gradually these more true patterns,which are slowly discovered,as time goes on,will enter a common language,which all of us can share.

You see then that the patterns are very much alive and evolving.In fact,if you like,each pattern may be looked upon as a hypothesis like one of the hypotheses of science.In this sense,each pattern represents our current best guess as to what

最后，我们要解释一下本语言的情况，为何我们要取名为《建筑模式语言》，并且强调本语言不是唯一的和绝对的，以及我们又是怎样想象本语言会和其他的千千万万种语言发生关系的。我们期待着人们将来会为他们自己创造出无穷无尽的模式语言。

《建筑的永恒之道》一书指出，每一个充满活力的、完整的社会都有它自己独特而又清晰的模式语言；进而言之，在这样一个社会中的每个人都有一种独特的语言，部分地为他人所使用，但就整体性来说，它是该语言拥有者的心灵所独具的。从这点来说，在一个健全的社会中，有多少人就有多少种语言，即使这些语言为大家所共用，而且相似。

于是问题就发生了：这部作品的问世将起到什么作用？我们在此发表它，有什么思想条框？有什么意图呢？本语言作为一部著作而发表，这一事实意味着成千上万的人可以利用它。人们也许会依赖这本出版了的语言，不再通过他们自己的思索去发展他们自己的语言。这样一种危险难道真的不存在吗？

其实，这本书是抛砖引玉之作。它是人们在社会的广阔进程中将逐步认识并不断完善他们自己的模式语言的一种初步尝试。在《建筑的永恒之道》中，我们认为并且解释了人们今天所使用的语言是如此粗野且支离破碎，以致大多数人不再有任何语言可说了——他们现在使用的语言不是从人与自然出发考虑的。

我们历时数载，力图构成本语言。我们期待着使用者会感受到它的力量，并乐于使用它，结果他将重新认识到这样一种活生生的语言意味着什么。如果我们仅在这方面取得成功的话，那么很可能每个人都会再次着手构思和发展他自己的语言——或许会把印于本书的语言作为出发点。

arrangement of the physical environment will work to solve the problem presented.The empirical questions center on the problem—does it occur and is it felt in the way we have described it? And the solution—does the arrangement we propose in fact resolve the problem? And the as-terisks represent our degree of faith in these hypotheses.But of course,no matter what the asterisks say,the patterns are still hypotheses,all 253 of them—and are therefore all tentative,all free to evolve under the impact of new experience and observation.

Let us finally explain the status of this language,why we have called it "A Pattern Language" with the emphasis on the word "A," and how we imagine this pattern language might be related to the countless thousands of other languages we hope that people will make for themselves,in the future.

The Timeless Way of Building says that every society which is alive and whole,will have its own unique and distinct pattern language;and further,that every individual in such a society will have a unique language,shared in part,but which as a totality is unique to the mind of the person who has it.In this sense,in a healthy society there will be as many pattern languages as there are people—even though these languages are shared and similar.

The question then arises:What exactly is the status of this published language? In what frame of mind,and with what intention,are we publishing this language here? The fact that it is published as a book means that many thousands of people can use it.Is it not true that there is a danger that people might come to rely on this one printed language,instead of developing their own languages,in their own minds?

The fact is,that we have written this book as a first step in the society—wide process by which people will gradually

然而，我们深信不疑，本语言要比一本手册、一位教师，或者另一种可能的模式语言略胜一筹。这里的许多模式都是具有深刻含义的建筑原型，它们深深扎根于事物的本质之中。这些模式看来在今天和以后的 500 年间将成为人性的一部分，成为富有人情味的行动的一部分。我们很怀疑，是否有人能构思出一种行之有效的模式语言，而不包括，比如说，模式**拱廊**（119）或模式**凹室**（179）。

　　就这层意思来说，我们也已经竭尽全力尽可能地深入环境中事物的本质。我们希望，印于本书的模式语言大部分将成为任何人在他的脑海中能构想出来的任何一种切合实际的、富有人情味的模式语言的核心。就此而言，我们在这里提供的模式语言至少有一部分是一切可能的模式语言的原型核心，这就会使人们感到充满生机和富有人情味。

become conscious of their own pattern languages,and work to improve them.We believe,and have explained in *The Timeless Way of Building*,that the languages which people have today are so brutal,and so fragmented,that most people no longer have any language to speak of at all—and what they do have is not based on human,or natural considerations.

We have spent years trying to formulate this language,in the hope that when a person uses it,he will be so impressed by its power,and so joyful in its use,that he will understand again,what it means to have a living language of this kind.If we only succeed in that,it is possible that each person may once again embark on the construction and development of his own language—perhaps taking the language printed in this book,as a point of departure.

And yet,we do believe,of course,that this language which is printed here is something more than a manual,or a teacher,or a version of a possible pattern language.Many of the patterns here are archetypal—so deep,so deeply rooted in the nature of things,that it seems likely that they will be a part of human nature,and human action,as much in five hundred years,as they are today.We doubt very much whether anyone could construct a valid pattern language,in his own mind,which did not include the pattern ARCADES(119)for example,or the pattern ALCOVES(179).

In this sense,we have also tried to penetrate,as deep as we are able,into the nature of things in the environment:and hope that a great part of this language,which we print here,will be a core of any sensible human pattern language,which any person constructs for himself,in his own mind.In this sense,at least a part of the language we have presented here,is the archetypal core of all possible pattern languages,which can make people feei alive and human.

《建筑模式语言》概要

　　《建筑模式语言》具有一种网络性质。关于这一点在《建筑的永恒之道》一书中已做了充分的解释。可是，当我们使用这种语言网络时，我们始终把它作为一种序列，它贯穿全部模式，从较大的模式到较小的模式，从创造结构的模式到装饰结构的模式，再到美化装饰的模式……

　　既然本语言实际上是一种网络，那么任何序列都不能完全地把握住它。但是，下面的这种序列能把握住整个网络的绝大部分。为了实现这一点，本序列呈现为一条上下起伏的不规则的线，仿佛一根挑花针编织的缀锦。

　　模式的序列既是本语言的概要，同时又是模式的索引。如果你读一遍连接各组模式之间的词句，你就会对本语言的整体有一个通盘的了解。并且，一旦获得这种了解，你就能从中找到和你的设计有关的模式了。

　　最后，这种模式序列，正如我们将在下一节中解释的那样，也是一套"基本设计图汇编"，你可以从中选择对你最有用的模式，或多或少保持它们印在本书中的顺序，从而形成你自己的设计语言。

SUMMARY OF THE LANGUAGE

A pattern language has the structure of a network.This is explained fully in *The Timeless Way of Building*.However,when we use the network of a language,we always use it as a *sequence*,going through the patterns,moving always from the larger patterns to the smaller,always from the ones which create structures,to the ones which then embellish those structures,and then to those which embellish the embellishments...

Since the language is in truth a network,there is no one sequence which perfectly captures it.But the sequence which follows,captures the broad sweep of the full network;in doing so,it follows a line,dips down,dips up again,and follows an irregular course,a little like a needle following a tapestry.

The sequence of patterns is both a summary of the language,and at the same time,an index to the patterns.If you read through the sentences which connect the groups of patterns to one another,you will get an overview of the whole language. And once you get this overview,you will then be able to find the patterns which are relevant to your own project.

And finally,as we shall explain in the next section,this sequence of patterns is also the "base map," from which you can make a language for your own project,by choosing the

ഇൻ

现在我们从本语言说明城镇或社区的那一部分开始。这些模式绝不可能一下子"设计"或"建造"出来——而是经过缓慢的、逐步的发展过程，以如下的方式设计出来的：每个人的行动总是有助于创造或产生这些较大的全球性模式，并将缓慢而又确定不移地，在若干年之后，形成含有这些综合模式的社区。

1. 独立区域

在每一区域内都要按区域规定的保护土地和划定城市规模的各项政策行事；

2. 城镇分布

3. 指状城乡交错

4. 农业谷地

5. 乡村沿街建筑

6. 乡间小镇

7. 乡村

通过城市政策，鼓励逐步形成那些说明城市的主要结构；

8. 亚文化区的镶嵌

9. 分散的工作点

10. 城市的魅力

11. 地方交通区

通过基本上由两级自治社区控制的活动，在农业区建立起这些较大的城市模式，从总体上讲，它们所在的地方是容易识别的；

patterns which are most useful to you,and leaving them more or less in the order that you find them printed here.

80)03

We begin with that part of the language which defines a town or community.These patterns can never be"designed"or"built"in one fell swoop—but patient piecemeal growth,designed in such a way that every individual act is always helping to create or generate these larger global patterns,will,slowly and surely,over the years,make a community that has these global patterns in it.

1.INDEPENDENT REGIONS

within each region work toward those regional policies which will protect the land and mark the limits of the cities;

2.THE DISTRIBUTION OF TOWNS

3.CITY COUNTRY FINGERS

4.AGRICULTURAL VALLEYS

5.LACE OF COUNTRY STREETS

6.COUNTRY TOWNS

7.THE COUNTRYSIDE

through city policies,encourage the piecemeal formation of those major structures which define the city;

8.MOSAIC OF SUBCULTURES

9.SCATTERED WORK

10.MAGIC OF THE CITY

11.LOCAL TRANSPORT AREAS

build up these larger city patterns from the grass roots,through action essentially controlled by two levels of self-governing communities,which exist as physically identifiable places;

12. 7000 人的社区

13. 亚文化区边界

14. 易识别的邻里

15. 邻里边界

通过鼓励发展下列网络，将社区相互连接起来；

16. 公共交通网

17. 环路

18. 学习网

19. 商业网

20. 小公共汽车

制定社区和邻里政策，根据下列基本原理，控制地方环境的性质；

21. 不高于四层楼

22. 停车场不超过用地的 9%

23. 平行路

24. 珍贵的地方

25. 通往水域

26. 生命的周期

27. 男人和女人

在邻里和社区之内，以及它们之间的边界上，鼓励形成地方中心；

28. 偏心式核心区

29. 密度圈

30. 活动中心

31. 散步场所

32. 商业街

33. 夜生活

12.COMMUNITY OF 7000

13.SUBCULTURE BOUNDARY

14.IDENTIFIABLE NEIGHBORHOOD

15.NEIGHBORHOOD BOUNDARY

connect communities to one another by encouraging the growth of the following networks;

16.WEB OF PUBLIC TRANSPORTATION

17.RING ROADS

18.NETWORK OF LEARNING

19.WEB OF SHOPPING

20.MINI-BUSES

establish community and neighborhood policy to control the character of the local environment according to the following fundamental principles;

21.FOUR-STORY LIMIT

22.NINE PER CENT PARKING

23.PARALLEL ROADS

24.SACRED SITES

25.ACCESS TO WATER

26.LIFE CYCLE

27.MEN AND WOMEN

both in the neighborhoods and the communities,and in between them,in the boundaries,encourage the formation of local centers;

28.ECCENTRIC NUCLEUS

29.DENSITY RINGS

30.ACTIVITY NODES

31.PROMENADE

32.SHOPPING STREET

33.NIGHT LIFE

34. 换乘站

环绕这些中心，准备发展住宅团组，以面对面的人群为基础；

35. 户型混合

36. 公共性的程度

37. 住宅团组

38. 联排式住宅

39. 丘状住宅

40. 老人天地

在住宅团组之间，在这些中心的周围，尤其在邻里之间的边界内，鼓励形成工作社区；

41. 工作社区

42. 工业带

43. 像市场一样开放的大学

44. 地方市政厅

45. 项链状的社区行业

46. 综合商场

47. 保健中心

48. 住宅与其他建筑间杂

在住宅团组和工作社区之间，允许公路网络和小路网络逐渐地、不拘形式地发展；

49. 区内弯曲的道路

50. 丁字形交点

51. 绿茵街道

52. 小路网络和汽车

53. 主门道

54. 人行横道

55. 高出路面的便道

34.INTERCHANGE

around these centers,provide for the growth of housing in the form of clusters,based on face-to-face human groups;

35.HOUSEHOLD MIX

36.DEGREES OF PUBLICNESS

37.HOUSE CLUSTER

38.ROW HOUSES

39.HOUSING HILL

40.OLD PEOPLE EVERYWHERE

between the house clusters,around the centers,and especially in the boundaries between neighborhoods,encourage the formation of work communities;

41.WORK COMMUNITY

42.INDUSTRIAL RIBBON

43.UNIVERSITY AS A MARKETPLACE

44.LOCAL TOWN HALL

45.NECKLACE OF COMMUNITY PROJECTS

46.MARKET OF MANY SHOPS

47.HEALTH CENTER

48.HOUSING IN BETWEEN

between the house clusters and work communities,allow the local road and path network to grow informally,piecemeal;

49.LOOPED LOCAL ROADS

50.T JUNCTIONS

51.GREEN STREETS

52.NETWORK OF PATHS AND CARS

53.MAIN GATEWAYS

54.ROAD CROSSING

55.RAISED WALK

56. 自行车道和车架

57. 市区内的儿童

在社区和邻里，提供人们能休憩、与人交际、恢复精力和体力的宽敞的公共活动场所；

58. 狂欢节

59. 僻静区

60. 近宅绿地

61. 小广场

62. 眺远高地

63. 街头舞会

64. 水池和小溪

65. 分娩场所

66. 圣地

在每一住宅团组和工作社区之间，提供若干块较小的公共土地，以备地方上共同需要的各种活动之用；

67. 公共用地

68. 相互沟通的游戏场所

69. 户外亭榭

70. 墓地

71. 池塘

72. 地方性运动场地

73. 冒险性的游戏场地

74. 动物

在公共用地、房屋团组和工作社区的范围内要鼓励最小的独立的社会组织：家庭、工作小组和聚会地的演变。包括家庭的各种组成形式；

56.BIKE PATHS AND RACKS

57.CHILDREN IN THE CITY

in the communities and neighborhoods,provide public open land where people can relax,rub shoulders and renew themselves;

58.CARNIVAL

59.QUIET BACKS

60.ACCESSIBLE GREEN

61.SMALL PUBLIC SQUARES

62.HIGH PLACES

63.DANCING IN THE STREET

64.POOLS AND STREAMS

65.BIRTH PLACES

66.HOLY GROUND

in each house cluster and work community,provide the smaller bits of common land,to provide for local versions of the same needs;

67.COMMON LAND

68.CONNECTED PLAY

69.PUBLIC OUTDOOR ROOM

70.GRAVE SITES

71.STILL WATER

72.LOCAL SPORTS

73.ADVENTURE PLAYGROUND

74.ANIMALS

within the framework of the common land,the clusters,and the work communities encourage transformation of the smallest independent social institutions:the families,workgroups,and gathering places.The family,in all its forms;

75. 家庭

76. 小家庭住宅

77. 夫妻住宅

78. 单人住宅

79. 自己的家

工作小组包括各种车间、办公室，甚至儿童学习小组；

80. 自治工作间和办公室

81. 小服务行业

82. 办公室之间的联系

83. 师徒情谊

84. 青少年协会

85. 店面小学

86. 儿童之家

也包括地方上的商店和聚会场所。

87. 个体商店

88. 临街咖啡座

89. 街角杂货店

90. 啤酒馆

91. 旅游客栈

92. 公共汽车站

93. 饮食商亭

94. 在公共场所打盹

至此，定义城镇或社区的综合模式已完备。现在我们开始叙述模式语言的另一部分，它要从三维空间上赋予许多建筑群和个体建筑物以不同的形状。这些都是能够被"设计"或"建造"出来的模式，它们说明个体建筑及各建筑物之间的空间；现在我们第一次涉及由个人或小团体控制的模式，这些模式是能够由他们立即建立的。

75.THE FAMILY

76.HOUSE FOR A SMALL FAMILY

77.HOUSE FOR A COUPLE

78.HOUSE FOR ONE PERSON

79.YOUR OWN HOME

the workgroups,including all kinds of workshops and offices and even children's learning groups;

80.SELF-GOVERNING WORKSHOPS AND OFFICES

81.SMALL SERVICES WITHOUT RED TAPE

82.OFFICE CONNECTIONS

83.MASTER AND APPRENTICES

84.TEENAGE SOCIETY

85.SHOPFRONT SCHOOLS

86.CHILDREN'S HOME

the local shops and gathering places.

87.INDIVIDUALLY OWNED SHOPS

88.STREET CAFE

89.CORNER GROCERY

90.BEER HALL

91.TRAVELER'S INN

92.BUS STOP

93.FOOD STANDS

94.SLEEPING IN PUBLIC

This completes the global patterns which define a town or a community.We now start that part of the language which gives shape to groups of buildings,and individual buildings,on the land,in three dimensions.These are the patterns which can be "designed"or"built"-the patterns which define the individual buildings and the space between buildings;where we are dealing for the first time with patterns that are under the

下面第一组模式有助于建筑群的总体布置：这些建筑物的高度和数量，通往基地的入口，主要停车区以及穿过建筑群体的交通线路；

95. 建筑群体

96. 楼层数

97. 有屏蔽的停车场

98. 内部交通领域

99. 主要建筑

100. 步行街

101. 有顶街道

102. 各种入口

103. 小停车场

根据基地的性质、树木和阳光等因素，在建筑群体内部的基地上对个体建筑逐个确定位置：这是本语言中最重要的组成部分之一；

104. 基地修整

105. 朝南的户外空间

106. 户外正空间

107. 有天然采光的翼楼

108. 鳞次栉比的建筑

109. 狭长形住宅

在房屋的各个侧面,布置入口、花园、庭院、屋顶和露台,从而同时形成房屋的体量和房屋之间的空间——切记,室内空间和室外空间、阴和阳总是一气呵成；

control of individuals or small groups of individuals,who are able to build the patterns all at once.

The first group of patterns helps to lay out the overall arrangement of a group of buildings:the height and number of these buildings,the entrances to the site,main parking areas,and lines of movement through the complex;

95.BUILDING COMPLEX
96.NUMBER OF STORIES
97.SHIELDED PARKING
98.CIRCULATION REALMS
99.MAIN BUILDING
100.PEDESTRIAN STREET
101.BUILDING THOROUGHFARE
102.FAMILY OF ENTRANCES
103.SMALL PARKING LOTS

fix the position of individual buildings on the site,within the complex,one by one,according to the nature of the site,the trees,the sun:this is one of the most important moments in the language;

104.SITE REPAIR
105.SOUTH FACING OUTDOORS
106.POSITIVE OUTDOOR SPACE
107.WINGS OF LIGHT
108.CONNECTED BUILDINGS
109.LONG THIN HOUSE

within the buildings'wings,lay out the entrances,the gardens,courtyards,roofs,and terraces:shape both the volume of the buildings and the volume of the space between the buildings at the same time-remembering that indoor space and outdoor space,yin and yang,must always get their shape together;

110. 主入口

111. 半隐蔽花园

112. 入口的过渡空间

113. 与车位的联系

114. 外部空间的层次

115. 有生气的庭院

116. 重叠交错的屋顶

117. 带阁楼的坡屋顶

118. 屋顶花园

当房屋的主要部分和室外各区已具轮廓后要及时对房屋之间的小路和广场给予更细致的考虑和注意；

119. 拱廊

120. 小路和标志物

121. 小路的形状

122. 建筑物正面

123. 行人密度

124. 袋形活动场地

125. 能坐的台阶

126. 空间中心有景物

在确定了道路之后，现在再把话题回到房屋上来，在任何一幢房屋的各个不同的侧翼内，要精心设计出空间的基本梯度，并决定室内交通将如何把各种梯度的空间连接起来；

127. 私密性层次

128. 室内阳光

129. 中心公共区

130. 入口空间

131. 穿越空间

110.MAIN ENTRANCE

111.HALF-HIDDEN GARDEN

112.ENTRANCE TRANSITION

113.CAR CONNECTION

114.HIERARCHY OF OPEN SPACE

115.COURTYARDS WHICH LIVE

116.CASCADE OF ROOFS

117.SHELTERING ROOF

118.ROOF GARDEN

when the major parts of buildings and the outdoor areas have been given their rough shape,it is the right time to give more detailed attention to the paths and squares between the buildings;

119.ARCADES

120.PATHS AND GOALS

121.PATH SHAPE

122.BUILDING FRONTS

123.PEDESTRIAN DENSITY

124.ACTIVITY POCKETS

125.STAIR SEATS

126.SOMETHING ROUGHLY IN THE MIDDLE

now,with the paths fixed,we come back to the buildings:within the various wings of any one building,work out the fundamental gradients of space,and decide how the movement will connect the spaces in the gradients;

127.INTIMACY GRADIENT

128.INDOOR SUNLIGHT

129.COMMON AREAS AT THE HEART

130.ENTRANCE ROOM

131.THE FLOW THROUGH ROOMS

132. 短过道

133. 有舞台感的楼梯

134. 禅宗观景

135. 明暗交织

在房屋的各个侧翼及其内部空间梯度的范围内，并根据室内交通的情况，来说明最重要的区和房间。首先，是针对住宅说的；

136. 夫妻的领域

137. 儿童的领域

138. 朝东的卧室

139. 农家厨房

140. 私家的沿街露台

141. 个人居室

142. 起居空间的序列

143. 多床龛卧室

144. 浴室

145. 大储藏室

其次，同样是针对办公室、车间和公共建筑物而言的；

146. 灵活办公空间

147. 共同进餐

148. 工作小组

149. 宾至如归

150. 等候场所

151. 小会议室

152. 半私密办公室

增设一些小的、稍稍独立于主结构的附加建筑，并将它们设置在从上面的楼层通往街道和花园的必经的途中；

153. 出租房间

132.SHORT PASSAGES

133.STAIRCASE AS A STAGE

134.ZEN VIEW

135.TAPESTRY OF LIGHT AND DARK

within the framework of the wings and their internal gradients of space and movement,define the most important areas and rooms.

First,for a house;

136.COUPLE'S REALM

137.CHILDREN'S REALM

138.SLEEPING TO THE EAST

139.FARMHOUSE KITCHEN

140.PRIVATE TERRACE ON THE STREET

141.A ROOM OF ONE'S OWN

142.SEQUENCE OF SITTING SPACES

143.BED CLUSTER

144.BATHING ROOM

145.BULK STORAGE

then the same for offices,workshops,and public buildings;

146.FLEXIBLE OFFICE SPACE

147.COMMUNAL EATING

148.SMALL WORK GROUPS

149.RECEPTION WELCOMES YOU

150.A PLACE TO WAIT

151.SMALL MEETING ROOMS

152.HALF-PRIVATE OFFICE

add those small outbuildings which must be slightly independent from the main structure,and put in the access from the upper stories to the street and gardens;

153.ROOMS TO RENT

将房屋的内部和外部有机地结合起来，使两者之间的边缘处于恰到好处的位置，并形成富有人情味的建筑细节；

对花园和园中各处的布置做出决定；

154.TEENAGER'S COTTAGE

155.OLD AGE COTTAGE

156.SETTLED WORK

157.HOME WORKSHOP

158.OPEN STAIRS

prepare to knit the inside of the building to the outside,by treating the edge between the two as a place in its own right,and making human details there;

159.LIGHT ON TWO SIDES OF EVERY ROOM

160.BUILDING EDGE

161.SUNNY PLACE

162.NORTH FACE

163.OUTDOOR ROOM

164.STREET WINDOWS

165.OPENING TO THE STREET

166.GALLERY SURROUND

167.SIX-FOOT BALCONY

168.CONNECTION TO THE EARTH

decide on the arrangement of the gardens,and the places in the gardens;

169.TERRACED SLOPE

170.FRUIT TREES

171.TREE PLACES

172.GARDEN GROWING WILD

173.GARDEN WALL

174.TRELLISED WALK

175.GREENHOUSE

176.GARDEN SEAT

177.VEGETABLE GARDEN

178.COMPOST

现在回来谈住宅的内部。为了完善主要房间，附设必要的次要房间和凹室；

179. 凹室

180. 窗前空间

181. 炉火熊熊

182. 进餐气氛

183. 工作空间的围隔

184. 厨房布置

185. 坐位圈

186. 共宿

187. 夫妻用床

188. 床龛

189. 更衣室

精细调整房间和凹室的形状和尺寸，使其结构准确且易于建造；

190. 天花板高度变化

191. 室内空间形状

192. 俯视外界生活之窗

193. 半敞开墙

194. 内窗

195. 楼梯体量

196. 墙角的房门

要使所有的墙都具有一定的纵深，不管在什么地方都得如此，以便布置凹室、窗、搁架、壁橱或坐位；

197. 厚墙

198. 居室间的壁橱

199. 有阳光的厨房工作台

200. 敞开的搁架

201. 半人高的搁架

go back to the inside of the building and attach the necessary minor rooms and alcoves to complete the main rooms;

179.ALCOVES

180.WINDOW PLACE

181.THE FIRE

182.EATING ATMOSPHERE

183.WORKSPACE ENCLOSURE

184.COOKING LAYOUT

185.SITTING CIRCLE

186.COMMUNAL SLEEPING

187.MARRIAGE BED

188.BED ALCOVE

189.DRESSING ROOM

fine tune the shape and size of rooms and alcoves to make them precise and buildable;

190.CEILING HEIGHT VARIETY

191.THE SHAPE OF INDOOR SPACE

192.WINDOWS OVERLOOKING LIFE

193.HALF-OPEN WALL

194.INTERIOR WINDOWS

195.STAIRCASE VOLUME

196.CORNER DOORS

give all the walls some depth,wherever there are to be alco ves,windows,shelves,closets,or seats;

197.THICK WALLS

198.CLOSETS BETWEEN ROOMS

199.SUNNY COUNTER

200.OPEN SHELVES

201.WAIST-HIGH SHELF

202. 嵌墙坐位

203. 儿童猫耳洞

204. 密室

在这一阶段，你已具备了完整的个体建筑的设计方案。如果遵循了提供的模式，那么你就会有一幅空间草图，或是用桩在地面上标出的，或是在纸上标出的，其精确性几乎分毫不差。你知道了房间的高度、门和窗的大致尺寸和位置，你也粗略地知道了房屋的屋顶是怎么一回事，以及花园是如何布置的。

下面是本语言的最后一部分，将叙述如何按照这幅大致精确的空间草图直接建造容易建造的房屋，并将详细地告诉你如何去建造它。

在你布置结构细节之前，要确立结构的基本原理，以便使你的住宅平面图和构想能直接付诸实施；

205. 结构服从社会空间的需要

206. 有效结构

207. 好材料

208. 逐步加固

运用结构的这一原理，并根据你所作的平面图，精心设计出完美无缺的结构布置图；这是在你实际开始建造之前，要在纸上做的最后一件事情了；

209. 屋顶布置

210. 楼面和天花板布置

211. 加厚外墙

212. 角柱

213. 柱的最后分布

在建筑基地上打桩以标志柱的位置，并依照这些桩的布局，着手安装房屋的主构架；

202.BUILT-IN SEATS

203.CHILD CAVES

204.SECRET PLACE

At this stage,you have a complete design for an individual building.If you have followed the patterns given,you have a scheme of spaces,either marked on the ground,with stakes,or on a piece of paper,accurate to the nearest foot or so.You know the height of rooms,the rough size and position of windows and doors,and you know roughly how the roofs of the building,and the gardens are laid out.

The next,and last part of the language,tells how to make a buildable building directly from this rough scheme of spaces,and tells you how to build it,in detail.

Before you lay out structural details,establish a philosophy of structure which will let the structure grow directly from your plans and your conception of the buildings;

205.STRUCTURE FOLLOWS SOCIAL SPACES

206.EFFICIENT STRUCTURE

207.GOOD MATERIALS

208.GRADUAL STIFFENING

within this philosophy of structure,on the basis of the plans which you have made,work out the complete structural layout;this is the last thing you do on paper,before you actually start to build;

209.ROOF LAYOUT

210.FLOOR AND CEILING LAYOUT

211.THICKENING THE OUTER WALLS

212.COLUMNS AT THE CORNERS

213.FINAL COLUMN DISTRIBUTION

put stakes in the ground to mark the columns on the site,and start erecting the main frame of the building according

214. 柱基

215. 底层地面

216. 箱形柱

217. 圈梁

218. 墙体

219. 楼面天花拱结构

220. 拱式屋顶

在房屋的主构架内,选定门、窗开口的正确位置,并给开口装框;

221. 借景的门窗

222. 矮窗台

223. 深窗洞

224. 低门道

225. 门窗边缘加厚

当你建造主构架和开口时,采用下列附属模式,并使它们各得其所;

226. 柱旁空间

227. 柱的连接

228. 楼梯拱

229. 管道空间

230. 辐射热

231. 老虎窗

232. 屋顶顶尖

修饰表面和室内细节;

233. 地面面层

234. 鱼鳞板墙

235. 有柔和感的墙内表面

236. 大敞口窗户

to the layout of these stakes;

214.ROOT FOUNDATIONS

215.GROUND FLOOR SLAB

216.BOX COLUMNS

217.PERIMETER BEAMS

218.WALL MEMBRANES

219.FLOOR-CEILING VAULTS

220.ROOF VAULTS

within the main frame of the building,fix the exact positions for openings-the doors and windows-and frame these openings;

221.NATURAL DOORS AND WINDOWS

222.LOW SILL

223.DEEP REVEALS

224.LOW DOORWAY

225.FRAMES AS THICKENED EDGES

as you build the main frame and its openings,put in the following subsidiary patterns where they are appropriate;

226.COLUMN PLACE

227.COLUMN CONNECTION

228.STAIR VAULT

229.DUCT SPACE

230.RADIANT HEAT

231.DORMER WINDOWS

232.ROOF CAPS

put in the surfaces and indoor details;

233.FLOOR SURFACE

234.LAPPED OUTSIDE WALLS

235.SOFT INSIDE WALLS

236.WINDOWS WHICH OPEN WIDE

237. 镶玻璃板门

238. 过滤光线

239. 小窗格

240. 半英寸宽的压缝条

建造室外细节以修饰室外空间布置，如同室内空间布置一样完美无缺；

241. 户外设座位置

242. 大门外的条凳

243. 可坐矮墙

244. 帆布顶篷

245. 高花台

246. 攀援植物

247. 留缝的石铺地

248. 软质面砖和软质砖

利用各种装饰、灯光、颜色和生活中的纪念品来完善你的房屋；

249. 装饰

250. 暖色

251. 各式坐椅

252. 投光区域

253. 生活中的纪念品

237.SOLID DOORS WITH GLASS

238.FILTERED LIGHT

239.SMALL PANES

240.HALF-INCH TRIM

build outdoor details to finish the outdoors as fully as the indoor spaces;

241.SEAT SPOTS

242.FRONT DOOR BENCH

243.SITTING WALL

244.CANVAS ROOFS

245.RAISED FLOWERS

246.CLIMBING PLANTS

247.PAVING WITH CRACKS BETWEEN THE STONES

248.SOFT TILE AND BRICK

complete the building with ornament and light and color and your own things;

249.ORNAMENT

250.WARM COLORS

251.DIFFERENT CHAIRS

252.POOLS OF LIGHT

253.THINGS FROM YOUR LIFE

为你的设计选用模式语言

253 个模式汇编成了建筑模式语言。它们创造了一幅有关一个完整区域的有条有理的蓝图，它们有能力以千千万万种形式创造出这样的区域，并以各种姿态呈现在所有的建筑细节上。

本语言中的任何一个小的模式序列本身就是一小部分环境的语言。这也是千真万确的。这张小小的模式表能够产生出以百万计的公园、小路、住宅、车间或花园。

例如，试考虑下列 10 个模式：

私家的沿街露台（140）

有阳光的地方（161）

有围合的户外小空间（163）

六英尺深的阳台（167）

小路和标志物（120）

天花板高度变化（190）

角柱（212）

大门外的条凳（242）

高花台（245）

各式坐椅（251）

这一短短的模式表本身就是一种语言——它是关于住宅正面的门廊的上千种可能的语言之一。我们选择这种小语言去建造通往住宅正面的门廊。本语言及其模式就是以这种方式来帮助建成此门廊的。

CHOOSING A LANGUAGE FOR YOUR PROJECT

All 253 patterns together form a language.They create a coherent picture of an entire region,with the power to generate such regions in a million forms,with infinite variety in all the details.

It is also true that any small sequence of patterns from this language is itself a language for a smaller part of the environment;and this small list of patterns is then capable of generating a million parks,paths,houses,workshops,or gardens.

For example,consider the following ten patterns:
PRIVATE TERRACE ON THE STREET(140)
SUNNY PLACE(161)
OUTDOOR ROOM(163)
SIX-FOOT BALCONY(167)
PATHS AND GOALS(120)
CEILING HEIGHT VARIETY(190)
COLUMNS AT THE CORNERS(212)
FRONT DOOR BENCH(242)
RAISED FLOWERS(245)
DIFFERENT CHAIRS(251)

This short list of patterns is itself a language:it is one of a thousand possible languages for a porch,at the front of a house.One of us chose this small language,to build a porch onto the front of his house.This is the way the language,and its

我曾从**私家的沿街露台**（140）开始。这模式需要一个露台。它缓缓升高，并同住宅连接。它位于街道的一侧。**有阳光的地方**（161）模式中建议，在庭院中有阳光一侧的某个特定的地点应当得到加强并把这个地点做成庭院、阳台和有围合的户外小空间。我曾使用这两个模式在住宅的南侧给一个加高的平台选定了位置。

　　为了把这一平台做成**有围合的户外小空间**（163），我把它的一半置于现有的悬挑屋顶的下面，在它的正中央种了一簇玫瑰，开花时节，玫瑰花散发出的清香阵阵袭人。她那临空婆娑的叶子增添了屋顶下面空间围合感。我还在平台的西侧放置了一个镶嵌玻璃的防风屏，结果使它的空间围合感更加强了。

　　我利用了**六英尺深的阳台**（167）定出平台的尺寸。但要审慎，切不可盲目采用这一模式。其理由是采用这一模式后，务必保证人们坐得舒服，以及确保人们围着靠墙的小几交谈时所必需的最小空间。因为我想要的空间至少有两个可供聊天的地方：一个在屋顶下，供热天和雨天用；另一个在室外的蓝天下，供做日光浴用；这样阳台的面积应当为 $12 \times 12 \text{ft}^2$。

　　现在来谈**小路和标志物**（120）：通常，这一模式涉及邻里中的便道，它在语言中出现得较早。但我结合这里的特点使用这模式。就是说，人们在土地上步行自然而然地踩踏出来的小路应当保留并加以整修。一条通往我家大门的小路恰好穿过我原来计划要布置平台的那块土地的一角，所以我到平台去就抄近路了。

patterns,helped to generate this porch.

I started with PRIVATE TERRACE ON THE STREET (140).That pattern calls for a terrace,slightly raised,connected to the house,and on the street side.SUNNY PLACE(161) suggests that a special place on the sunny side of the yard should be intensified and made into a place by the use of a patio,balcony,outdoor room,etc.I used these two patterns to locate a raised platform on the south side of the house.

To make this platform into an OUTDOOR ROOM(163),I put it half under the existing roof overhang,and kept a mature pyracanthus tree right smack in the middle of the platform.The overhead foliage of the tree added to the roof-like enclosure of the space.I put a wind screen of fixed glass on the west side of the platform too,to give it even more enclosure.

I used SIX-FOOT BALCONY(167)to determine the size of the platform.But this pattern had to be used judiciously and not blindly—the reasoning for the pattern has to do with the minimum space required for people to sit comfortably and carry on a discussion around a small side—table.Since I wanted space for at least two of these conversation areas—one under the roof for very hot or rainy days,and one out under the sky for days when you wanted to be full in the sun,the balcony had to be made 12×12 feet square.

Now PATHS AND GOALS(120):Usually,this pattern deals with large paths in a neighborhood,and comes much earlier in a language.But I used it in a special way.It says that the paths which naturally get formed by people's walking,on the land,should be preserved and intensified.Since the path to our front door cut right across the corner of the place where I had planned to put the platform,I cut the corner of the platform off.

平台离地面的高度取决于**天花板高度变化**（190）。平台要造得高于地平线约 1ft，天花被覆盖部分的高度为 6 ～ 7ft。这恰好相当于平台大小的空间。由于离地面的高度大约正是适合坐人的高度，所以模式**大门外的条凳**（242）自然而然地满足了这一要求。

有三根立柱支承着旧门廊上的屋顶。它们过去不得不待在它们现在所待的地方，因为要支撑屋顶。但是，根据**角柱**（212），平台与角柱的相对位置作精心的调整——结果，角柱有助于限定其两侧的社会空间。

最后，我们在"大门外的条凳"附近放置了几个花箱——当你坐在那里时，你可以闻到一阵阵扑鼻的花香，真是好极了——这是根据**高花台**（245）而进行安排的。而且你在门廊内能看到的旧椅子是**各式坐椅**（251）。

从这一简短的例子中，你能理解到这一模式语言是多么简洁有力。而现在，当你为自己、为自己的设计而构思模式语言时，或许会意识到你必须是何等的小心翼翼。

经过修饰的门廊
The finished porch

The height of the platform above the ground was determined by CEILING HEIGHT VARIETY(190).By building the platform approximately one foot above the ground line,the ceiling height of the covered portion came out at between 6 and 7 feet—just right for a space as small as this.Since this height above the ground level is just about right for sitting,the pattern FRONT DOOR BENCH (242)was automatically satisfied.

There were three columns standing,supporting the roof over the old porch.They had to stay where they are,because they hold the roof up.But,following COLUMNS AT THE CORNERS(212),the platform was very carefully tailored to their positions—so that the columns help define the social spaces on either side of them.

Finally,we put a couple of flower boxes next to the "front door bench" -it's nice to smell them when you sit there—according to RAISED FLOWERS (245).And the old chairs you can see in the porch are DIFFERENT CHAIRS(251).

You can see,from this short example,how powerful and simple a pattern language is.And you are now,perhaps ready to appreciate how careful you must be,when you construct a language for yourself and your own project.

The character of the porch is given by the ten patterns in this short language.In just this way,each part of the environment is given its character by the collection of patterns which we choose to build into it.The character of what you build,will be given to it by the language of patterns you use,to generate it.

For this reason,of course,the task of choosing a language for your project is fundamental.The pattern language we have given here contains 253 patterns.You can therefore use it to

这一门廊的特点是由上述 10 个模式构成的简短语言所提供的。环境每一部分的特性，正是以这种方式由我们所选择的一组模式赋予的。你造的建筑物所具有的特性是由你使用的模式语言形成的。

由此可见，为设计选择模式语言是极其重要的。我们在此介绍的模式语言共包含 253 个模式。你可利用本语言创造出几乎多得无法想象的各种可能的较小模式语言，或可用它来为各种不同的设计服务，而你所要做的只是从本语言中挑选模式而已。

现在我们要来描述一下你为自己的设计而挑选模式语言的大致程序：你可先从我们记载于此的语言中选取模式，然后再补充你自己的模式。

1. 首先，复制一份模式语言的总目录表，并将其中形成你的设计语言所需的模式标上小记号。如果你无法弄到一台复印机，可从本书所印的目录表中标出模式，夹上纸条，标明页数；利用书签——你喜欢的随便什么东西，在它们上面写下你自己的模式语言。现在万事俱备，为了清楚说明起见，我们暂且假定在你的面前放着一份模式语言目录表。

2. 先大略地浏览一下这份目录表，选一个你最中意的、能描绘你的设计全貌的模式。这是你的设计的起始模式。用记号标出。（如果有两三个可能中选的模式，不必担心——从中挑选看来是最佳的一个，其余的待以后选用。）

3. 现在就参考本书的这一起始模式将它通读一遍。注意你正在阅读的这一模式的开头和结尾处指名提及的其他模式，都是你的设计语言可能选中的候选模式。在开头的那些模式总是比你的设计"要大"。先不必包括它们，除非你有能力在你所设计的小天地内，至少小规模地能实现这些模式。但在结尾处的那些模式是"较小的"。它们几乎全部都是重要的。在你的目录表上将它们一一用记号标出。除非你有某种特殊的理由，不把它们包括进你的设计之中。

generate an almost unimaginably large number of possible different smaller languages,for all the different projects you may choose to do,simply by picking patterns from it.

We shall now describe a rough procedure by which you can choose a language for your own project,first by taking patterns from this language we have printed here,and then by adding patterns of your own.

1.First of all,make a copy of the master sequence(pages xix-xxxiv)on which you can tick off the patterns which will form the language for your project.If you don't have access to a copying machine,you can tick off patterns in the list printed in the book,use paper clips to mark pages,write your own list,use paper markers—whatever you like.But just for now,to explain it clearly,we shall assume that you have a copy of the list in front of you.

2.Scan down the list,and find the pattern which best describes the overall scope of the project you have in mind.This is the starting pattern for your project.Tick it.(If there are two or three possible candidates,don't worry:just pick the one which seems best:the others will fall in place as you move forward.)

3.Turn to the starting pattern itself,in the book,and read it through.Notice that the other patterns mentioned by name at the beginning and at the end,of the pattern you are reading,are also possible candidates for your language.The ones at the beginning will tend to be "larger" than your project.Don't include them,unless you have the power to help create these patterns,at least in a small way,in the world around your project.The ones at the end are "smaller." Almost all of them will be important.Tick all of them,on your list,unless you have some special reason for not wanting toinclude them.

4. 现在你的目录表上记号多了一些。参考目录表上的下一个位于最高处且标有记号的模式，并且把书翻到这个模式的页面。这一模式会再一次把你引向另外一些模式。而你再一次将这些有关的模式标上记号——尤其是末尾的那些"较小的"模式。照例，对那些"较大的"模式通常用不着打记号，除非你能利用它们一些合理的因素说得具体一些就是用在你的设计中。

5. 当你对某一模式产生怀疑时，就不要涵盖它。否则，你的目录表就很容易变得冗长不堪了——若果真如此，这目录表就会使人混淆不清而无所适从了。即使仅仅包括你所偏爱的那些模式，它也会长得可观。

6. 照此办理，直到你把你的设计所需的模式一一标出记号为止。

7. 再加上你自己的语言材料以调整目录表的顺序。如果想在你的设计中包括一些东西，但是还不能找出与之相应的模式，那么，就将它们写入目录表顺序中某一合适的地方，靠近那些大小和重要性大致相同的模式。例如，没有桑拿浴这一模式。如果你想把它纳入你的设计，就在你的目录表中靠近**浴室**（144）旁边添加上桑拿浴。

8. 当然，你想改动任何模式，就改动好了。常常会有这样的情况，你对模式具有个人的一些看法，更确切地说，模式和你的关系已更为密切。在此情况下，你完全有"权"来支配模式语言，并最有效地使它成为你自己的语言，如果你在本书合适的地方写下你所做的种种变动的话。而且，如果你还变更模式的名称，它将成为最具体的语言了——这样，它就更明确地反映了你的改动。

4.Now your list has some more ticks on it.Turn to the next highest pattern on the list which is ticked,and open the book to that pattern.Once again,it will lead you to other patterns.Once again,tick those which are relevant—especially the ones which are "smaller" that come at the end.As a general rule,do not tick the ones which are "larger" unless you can do something about them,concretely,in your own project.

5.When in doubt about a pattern,don't include it.Your list can easily get too long:and if it does,it will become confusing. The list will be quite long enough,even if you only include the patterns you especially like.

6.Keep going like this,until you have ticked all the patterns you want for your project.

7.Now,adjust the sequence by adding your own material. If there are things you want to include in your project,but you have not been able to find patterns which correspond to them,then write them in,at an appropriate point in the sequence,near other patterns which are of about the same size and importance.For example,there is no pattern for a sauna.If you want to include one,write it in somewhere near BATHING ROOM(144)in your sequence.

8.And of course,if you want to change any patterns,change them.There are often cases where you may have a personal version of a pattern,which is more true,or more relevant for you.In this case,you will get the most "power" over the language,and make it your own most effectively,if you write the changes in,at the appropriate places in the book.And,it will be most concrete of all,if you change the name of the pattern too- so that it captures your own changes clearly.

∞ 𝒞𝒢

假设现在你已取得了你自己的设计语言。使用语言的方式很大程度上取决于语言的规模。涉及城镇的模式只能通过农业区的活动逐步加以实施；关于建筑的模式可以在你头脑中形成和在地面上标示出来；至于构造施工模式必须在工地上总体建成。为此，我们对这三种不同的情况，分别给出了三种说明。有关城镇部分，参阅第 67 页；有关建筑部分，参阅第 899 页；有关构造部分，参阅第 1767 页。

在《建筑的永恒之道》一书中，对这三种不同规模的模式的程序，在适当的章节，利用广泛的实例逐一进行了较充分的描述。关于城镇部分，参阅该书第 24 章和第 25 章；关于个体建筑部分，参阅第 20 章、第 21 章和第 22 章；关于描述一幢房屋实际建造起来的方法——施工过程，参阅第 23 章。

Suppose now that you have a language for your project. The way to use the language depends very much on its scale.Patterns dealing with towns can only be implemented gradually,by grass roots action;patterns for a building can be built up in your mind,and marked out on the ground;patterns for construction must be built physically,on the site.For this reason we have given three separate instructions,for these three different scales.For towns,see page 66;for buildings,see page 980;for construction,see page 1876.

The procedures for each of these three scales are described in much more detail with extensive examples,in the appropriate chapters of The Timeless Way of Building.For the town—see chapters 24 and 25;for an individual building-see chapters 20,21,and 22;and for the process of construction which describes the way a building is actually built see chapter 23.

诗意盎然的模式语言

最后，还有一点请注意。本模式语言正如英语一样，既可用于散文也可用于诗。散文和诗两者的差异不是所用的语言不同，而是同一种语言以不同的方式表达罢了。在一个普普通通的英语句子中，每个词都有一个意义，而这个句子本身也有一个简单的意义。在一首诗中的含义却多得多、密集得多。每个词具有好几种含义；而作为整体的这一句子具有大量密集的相互交错的含义，共同表达全诗的内容。

模式语言也同样如此。以相当松散的方式把模式串联起来而构成建筑物是可能的。类似这样的建筑物就是模式的组合，也不是密集性的，也不是深奥的。但是，在同一的总体空间内，把许许多多模式重叠在一起，也是可能的。以这种方式构成的建筑物是具有高度密集性的；它在这小小的空间内具有多层含义。并且，正是通过这种密集性而成为深奥的。

在诗中，这种密集性往往是通过我们以前并不理解的词和意义之间的同一性表达出来的。在《哦，玫瑰花，你，病态的美》一诗中，玫瑰花是和任何一朵玫瑰花都比不了的，它和更大的、更具有私人性质的许多东西等同了。于是，由于这种联系，这首诗阐明了人和玫瑰花的关联。这种联系不仅阐明了词义，而且也阐明了我们真实的生活。

THE POETRY OF THE LANGUAGE

Finally,a note of caution.This language,like English,can be a medium for prose,or a medium for poetry.The difference between prose and poetry is not that different languages are used,but that the same language is used,differently.In an ordinary English sentence,each word has one meaning,and the sentence too,has one simple meaning.In a poem,the meaning is far more dense.Each word carries several meanings;and the sentence as a whole carries an enormous density of interlocking meanings,which together illuminate the whole.

The same is true for pattern languages.It is possible to make buildings by stringing together patterns,in a rather loose way.A building made like this,is an assembly of patterns. It is not dense.It is not profound.But it is also possible to put patterns together in such a way that many many patterns overlap in the same physical space:the building is very dense;it has many meanings captured in a small space;and through this density,it becomes profound.

In a poem,this kind of density,creates illumination,by making identities between words,and meanings,whose identity we have not understood before.In "O Rose thou art sick," the rose is identified with many greater,and more personal things than any rose—and the poem illuminates the person,and

哦，玫瑰花，你，病态的美。

那肉眼看不见的小小昆虫，

在一片茫茫的漆黑之夜，

飞舞在呼啸的暴风雨中：

它已经寻觅到了你那美丽

而又呈深红色的欢乐窝：

它的爱慕之情珍藏在心底，

会在黑暗中使你枯萎凋谢。

威廉·布莱克

（译者注：William Blake，1757—1827，英国诗人兼艺术家。他在这首诗中刻画了一只不显眼的小昆虫在玫瑰花中找到了安乐窝，但它那邪恶而隐秘的爱情却毁了美丽的玫瑰花的生命）。

　　同样的情况也发生在建筑中。例如，考虑这样两个模式——**浴室**（144）和**池塘**（71）。浴室是住宅的一部分，你可以在那里慢条斯理地洗澡，洗个痛快，或许还有同伴一起洗。浴室是你伸展四肢轻松休息的地方。池塘是邻里中一块有水面的地方，可以注目凝视或在其中游泳，孩子们也可在水面上划船嬉水，使水花四溅。水塘里的水滋润着我们需要水的地方，水是我们宇宙的伟大元素之一。

the rose,because of this connection.The connection not only illuminates the words,but also illuminates our actual lives.

O Rose thou art sick.

The invisible worm,

That flies in the night

In the howling storm:

Has found out thy bed

Of crimson joy:

And his dark secret love

Does thy life destroy.

WILLIAM BLAKE

The same exactly,happens in a building.Consider,for example,the two patterns BATHING ROOM(144)and STILL WATER(71).One defines a part of a house where you can bathe yourself slowly,with pleasure,perhaps in company;a place to rest your limbs,and to relax.The other is a place in a neighborhood,where this is water to gaze into,perhaps to swim in,where children can sail boats,and splash about,which nourishes those parts of ourselves which rely on water as one of the great elements of the unconscious.

Suppose now,that we make a complex of buildings where individual bathing rooms are somehow connected to a common pond,or lake,or pool—where the bathing room merges with this common place;where there is no sharp distinction between the individual and family processes of the bathing room,and the common pleasure of the common pool.In this place,these two patterns exist in the same space;they are identified;there is a compression of the two,which requires less space,and which is more profound than in a place where they are merely side by side.The compression illuminates each of the patterns,sheds

现在假设我们建造一个建筑群体，在那里许多私人浴室以某种方式和公共池塘、湖泊或水池相连接。浴室与这些公用地点合并到一起。在私人和家庭浴室洗澡与在公共浴池内共同欢乐地洗澡两者之间没有什么明显的区别。在这里，这两个模式存在于同一空间，它们是同一的；两者压缩所需的空间较小，而比它们仅仅在一处并列更为深奥。这种压缩阐明每一个模式，并使其意义明朗化。当我们对我们内心世界需要的种种联系得理解多一点的时候，它也阐明我们的生活。

但是，这种压缩不只是富有诗意的、情趣深奥的。它不仅仅是诗的要素以及奇特、华丽而动人的故事的要素，在某种程度上，也是每个英语句子的要素。通常，在我们的谈吐中，每一个单独的词都有或多或少的压缩性，因为每个词都恰恰具有与之联系的其他词汇未表露尽的含义。甚至像"弗里达，请把奶酪端过来"这样一句话也有某种压缩，因为句中的这些言词与它前面的全部言辞有着种种联系，含有言外之意和弦外之音。

我们每个人同自己的朋友或家人谈话时，都利用语言中的词语相互之间的联系而形成种种压缩。我们越是感到了语言中的全部联系，我们平常谈论的东西的内容就越丰富、越微妙。

再重复一遍，建筑中的情况也同样如此。把许多模式压缩到某一单独的空间内，使之成为艺术作品的特殊建筑并不具有诗意，也无妙处可言。这只是最普通的节省空间而已。很可能，一幢房屋的全部模式会以某种方式重叠在一所简易的单室小屋之中。模式不需要串连成一列，也不需要相互分离。每幢楼房、每个房间、每座花园所需要的模式都要尽量压缩，才能更加完美。建筑的造价将会降低，而其中的含义将更加密集，耐人寻味。

light on its meaning;and also illuminates our lives,as we understand a little more about the connections of our inner needs.

But this kind of compression is not only poetic and profound.It is not only the stuff of poems and exotic statements,but to some degree,the stuff of every English sentence.To some degree,there is compression in every single word we utter,just because each word carries the whisper of the meanings of the words it is connected to.Even "Please pass the butter,Fred" has some compression in it,because it carries overtones that lie in the connections of these words to all the words which came before it.

Each of us,talking to our friends,or to our families,makes use of these compressions,which are drawn out from the connections between words which are given by the language.The more we can feel all the connections in the language,the more rich and subtle are the things we say at the most ordinary times.

And once again,the same is true in building.The compression of patterns into a single space,is not a poetic and exotic thing,kept for special buildings which are works of art.It is the most ordinary economy of space.It is quite possible that all the patterns for a house might,in some form be present,and overlapping,in a simple one-room cabin.The patterns do not need to be strung out,and kept separate.Every building,every room,every garden is better,when all the patterns which it needs are compressed as far as it is possible for them to be.The building will be cheaper;and the meanings in it will be denser.

It is essential then,once you have learned to use the language,that you pay attention to the possibility of compressing the many patterns which you put together,in

一旦你学会使用本语言，那重点就是你要把注意力集中到在尽可能小的空间内压缩你放置在一起的、为数众多的模式的可能性上。你可以认为压缩模式是一种方法，能使你的建筑物的造价尽量降低。同样，也是使建筑富有诗意的唯一途径。

the smallest possible space.You may think of this process of compressing patterns,as a way to make the cheapest possible building which has the necessary patterns in it.It is,also,the only way of using a pattern language to make buildings which are poems.

TOWNS

城

镇

现在我们就从描述城镇或社区的这一部分模式语言开始吧。这些模式绝不可能一下子"被设计"或"被建造"出来，但是，它们会逐渐地发展起来，并以下列方式被设计出来：每个人的行为都是朝着有助于创造或产生这些较大的综合模式的方向前进，经过若干年后，终将缓慢而又肯定地形成包含这些综合模式的社区。

₧⊳

　　第一批 94 个模式是处理环境中的大尺度结构的：城镇和乡村的发展、各种道路的布置、工作和家庭之间的关系、邻里内合适的公共机构的建立、支持这些公共机构所需的各种公共空间。

　　我们认为，在本篇中所提供的模式最好通过渐进的方式加以实现。每一工程项目的建成或每一规划决定的作出都要由社区批准，都要取决于它是否协助完成一些大尺度的模式。我们并不认为，给城镇和邻里带来如此多结构的这些大模式可以由集权的当局、法律或城市总平面图创造出来。相反，我们认为，如果每一个大大小小的建筑行为对自己逐步形成的"世界之一角"承担责任，使这些较大的模式出现，那么它们就能逐步地、有机地、几乎是自然而然地应运而生。

We begin with that part of the language which defines a town or a community. These patterns can never be "designed" or "built" in one fell swoop-but patient piecemeal growth, designed in such a way that every individual act is always helping to create or generate these larger global patterns, will, slowly and surely, over the years, make a community that has these global patterns in it.

⏳⏳

The first 94 patterns deal with the large-scale structure of the environment:the growth of town and country,the layout of roads and paths,the relationship between work and family,the formation of suitable public institutions for a neighborhood,the kinds of public space required to support these institutions.

We believe that the patterns presented in this section can be implemented best by piecemeal processes,where each project built or each planning decision made is sanctioned by the community according as it does or does not help to form certain large-scale patterns. *We do not believe that these large patterns,which give so much structure to a town or of a neighborhood,can be created by central-ized authority,or by laws,or by master plans.* We believe instead that they can emerge gradually and organically,almost of their own accord,if every act of building,large or small,takes on the responsibility for gradually shaping its small corner of the world to make

在下面的寥寥数页中,我们将描述规划过程,我们认为,这一过程是与循序渐进法相符合的。

1. 我们建议规划过程的核心如下：区域是由社会和政治团体的各阶层构成的，即从最小的和最具有地方性的团体——家庭、邻里和工作小组——直至最大的团体——市议会和区域会议。

例如，试想象一下一个大都会区域由下列团体构成，每一个团体都是一个相应的政治实体：

A. 区域: 8000000 人。

B. 主要城市: 500000 人。

C. 社区和小城镇: 各为 5 ～ 10000 人。

D. 邻里: 每一邻里为 500 ～ 1000 人。

E. 住宅团组和工作社区: 各为 30 ～ 50 人。

F. 家庭和工作小组: 各为 1 ～ 15 人。

2. 每个团体都对其共同使用的环境作出自己的决定。理想的情况是，每个团体实际上都在它自己的"级别"上拥有公共土地。较高一级的团体并不拥有或控制属于低一级的团体的土地——它们只拥有并控制位于它们之间的公共土地，并为更高一级的团体服务。比如，7000 人的社区可以拥有位于构成社区的各邻里之间的公共土地，但并不占有邻里本身。具有合作性质的住宅团组拥有住宅之间的公共土地，但不占有住宅本身。

3. 每一个团体都要对与其自身内部结构有关的那些模式承担责任。

因此，我们设想，例如，我们已经提及的各种不同的团体会选用下列模式：

these larger patterns appear there.

In the next few pages we shall describe a planning process which we believe is compatible with this piecemeal approach.

1.The core of the planning process we propose is this:The region is made up of a hierarchy of social and political groups,from the smallest and most local groups— families,neighborhoods,and work groups—to the largest groups—city councils,regional assemblies.

Imagine for example a metropolitan region composed very roughly of the following groups,each group a coherent political entity:

A.The region:8,000,000 people.

B.The major city:500,000 people.

C.Communities and small towns:5-10,000 people each.

D.Neighborhoods:500-1000 people each.

E.House clusters and work communities:30-50 people eacho

F.Families and work groups:1-15 people each.

2.*Each group makes its own decisions about the environment it uses in common.*Ideally,each group actually owns the common land at its "level." And higher groups do not own or control the land belonging to lower groups-they only own and control the common land that lies *between* them,and which serves the higher group.For instance,a community of 7000 might own the public land lying between its component neighborhoods,but not the neighborhoods themselves.A cooperative house cluster would own the common land between the houses,but not the houses themselves.

3.Each of these groups takes responsibility for those patterns relevant to its own internal structure.

Thus,we imagine,for example,that the various groups we have named might choose to adopt the following patterns:

A. 区域：独立区域

　　　　　城镇分布

　　　　　指状城乡交错……

B. 城市：亚文化区的镶嵌

　　　　　分散的工作点

　　　　　城市的魅力……

C. 社区：7000 人的社区

　　　　　亚文化区边界

4. 每一邻里、社区或城市都可以自由施展才能，通过各种途径去说服各自的成员，即不同的团体和个人去逐步实现这些模式。

在任何情况下，这一点将取决于某种动机。可是，所选的实际动机，就其力量和实现的可能性来说，都会有极大的不同。一些模式，如**指状城乡交错**或许是区域法律范围内的事——因为没有什么东西能阻止爱财如命、贪得无厌的投资者到处大兴土木。另一些模式，如**主门道、分娩场所、池塘**或许是纯粹自发性的。还有其他一些模式或许具有各式各样的动机，它们介乎上述两种极端例子之间。

例如，**小路网络和汽车、近宅绿地**和其他一些模式是可以实现的，办法是：对这些发展项目实行有助于它们实现的减税措施。

5. 实现模式要尽量采取松散、自愿的方式，要以社会责任感为基础，而不是以立法和强制为基础。

例如，假设有一个全市性的决定：在市内某些地区增加工业设施。就此而言,通过划分地区或征用权或其他活动，这个城市不能要求它的邻里负责人推行这一政策。城市的决策人可以建议说，这项决定十分重要，哪个邻里愿意协助实现这一较大的模式，哪个邻里就会得到源源不断的金钱。总之，他们是办得到的，如果他们发现地方上的邻里

A.Region: INDEPENDENT REGIONS

               DISTRIBUTION OF TOWNS

               CITY COUNTRY FINGERS...

B.City: MOSAIC OF SUBCULTURES

               SCATTERED WORK

               THE MAGIC OF THE CITY...

C.Community: COMMUNITY OF 7000

               SUBCULTURE BOUNDARY...

4.Each neighborhood,community,or city is then free to find various ways of persuading its constituent groups and individuals to implement these patterns gradually.

In every case this will hinge on some kind of incentive. However,the actual incentives chosen might vary greatly,in their power,and degree of enforcement.Some patterns,like CITY COUNTRY FINGERS,might be made a matter of regional law—since nothing less can deter money—hungry developers from building everywhere.Other patterns,like MAIN GATEWAY,BIRTH PLACES,STILL WATER,might be purely voluntary.And other patterns might have various kinds of incentives,intermediate between these extremes.

For example,NETWORK OF PATHS AND CARS, ACCESSIBLE GREENS,and others might be formulated so that tax breaks will be given to those development projects which help to bring them into existence.

5.As far as possible,implementation should be loose and voluntary,based on social responsibility,and not on legislation or coercion.

Suppose,for example,that there is a citywide decision to increase industrial uses in certain areas.Within the process here defined,the city could not implement this policy over the heads

愿意看到他们自己美好的前途，愿意改变他们自己的环境使这一模式在当地出现。他们一旦找到这样的邻里，这一模式一定会在若干年后逐步实现，因为地方上的邻里会见机行事的。

6. 一旦这一过程进展顺利，例如，已经采用模式**保健中心**的社区可以邀请一批医生来商议并建造这一中心。设计诊所的用户团队就会从**保健中心**以及社区模式语言的其他有关模式开始工作。他们会竭力把社区已采用的任何比较高级的模式，如**停车场不超过用地的9%、地方性运动场地、小路网络和汽车、近宅绿地**等纳入他们自己的设计。

7. 当然，对于个人的建筑行为来说，人们按照自己的方式来着手对待这些较大的社会模式是可能的，甚至在邻里、社区和地区团体形成之前。

因此，例如一群设法摆脱他们房前声音嘈杂的、危险的交通状况的人们可做出决定：撬开那里的沥青路面，重建**绿茵街道**。他们会根据这一模式中所提供的论据及对现存的沥青路面的街道模式所做的分析，向交通部门提出令人信服的改建理由。

另一群想要在目前已划定好的只供建住宅用的邻里内建造一个小型公共车间的人们，会根据**分散的工作点、固定的工作点**等模式为自己争辩，并有可能使城市的或划区的部门对地区的划分进行一些调整，以此朝引进模式的方向缓慢开展工作，有时一些模式就在现行的规章制度允许的范围之内和划分的地区之内。

我们现已精心设计出了在俄勒冈大学的尤金校园内实施这一过程的部分方案。《俄勒冈实验》一书已对此做了论述。但是一所大学和一个城镇是截然不同的，因为大学只有一个集权的占有者，只有一种资金来源。因此，不可避免的是，这一过程只有通过个人的通力协作，共同努力，形成较大的完整模式而不受上述规划的限制，才有可能部

of the neighborhoods,by zoning or the power of eminent domain or any other actions.They can suggest that it is important,and can increase the flow of money to any neighborhoods willing to help implement this larger pattern.They can implement it,in short,if they can find local neighborhoods willing to see their own future in these terms,and willing to modify their own enviromnent to help make it happen locally.As they find such neighborhoods,then it will happen gradually,over a period of years,as the local neighborhoods respond to the incentives.

6.Once such a process is rolling,a community,having adopted the pattern HEALTH CENTER,for example,might invite a group of doctors to come and build such a place. The team of users,designing the clinic would work from the HEALTH CENTER pattern,and all the other relevant patterns that are part of the community's language.They would try to build into their project any higher patterns that the community has adopted—NINE PER CENT PARKING,LOCA LSPORTS,NETWORK OF PATHS AND CARS,ACCESSIBLE GREEN,etc.

7.It is of course possible for individual acts of building to begin working their way toward these larger communal patterns,even before the neighborhood,community,and regional groups are formed.

Thus,for example,a group of people seeking to get rid of noisy and dangerous traffic in front of their houses might decide to tear up the asphalt,and build a GREEN STREET there instead.They would present their case to the traffic department based on the arguments presented in the pattern,and on an analysis of the existing street pattern.

Another group wanting to build a small communal

分地在那里实现。

我们在此解释大模式是如何从较小的模式逐步发展起来的理论，详见《建筑的永恒之道》第 24 章和第 25 章。

我们希望，将来有朝一日再续写一卷来解释为了在城镇内彻底实现这一过程所需的政治和经济上的法律手续。

你要为建立一个拥有 1000 个独立区域而不是国家的世界政府而全力以赴。

1. 独立区域

workshop,in a neighborhood currently zoned for residential use only,can argue their case based on SCATIERED WORK,SETTLED WORK,etc.,and possibly get the city or zoning department to change the zoning regulation on this matter,and thereby slowly work toward introducing patterns,one at a time within the current framework of codes and zoning.

We have worked out a partial version of this process at the Eugene campus of the University of Oregon.That work is described in Volume 3,*The Oregon Experiment*.But a university is quite dif-ferent from a town,because it has a single centralized owner,and a single source of funds.It is inevitable,therefore,that the process by which individual acts can work together to form larger wholes without restrictive planning from above,can only partly be put into practice there.

The theory which explains bow large patterns can be built piecemeal from smaller ones,is given in Chapters 24 and 25 of The Timeless Way of Building.

At some time in the future,we hope to write another volume,which explains the political and economic processes needed to implement this process fully,in a town.

Do what you can to establish a world government,with a thousand independent regions,instead of countries;

1.INDEPENDENT REGIONS

模式1 独立区域**

    大都会各区只有在它的每一区都成为小的自治区,并足以成为独立的文化区时才会趋于平衡。

    有4个独立的论据使我们得出了这一结论:1.人类政府的本质和局限性;2.世界社会中各区域间的平等权利;3.区域规划的种种设想;4.鼓励人类文化的鲜明性和多样性。

    1.在规模上有着天然局限性的群体,能以富有人情味的方式来治理自己。生物学家霍尔丹在他的论文《论合适的规模》中对此做了评论:

1   INDEPENDENT REGIONS**

Metropolitan regions will not come to balance until each one is small and autonomous enough to be an independent sphere of culture.

There are four separate arguments which have led us to this conclusion:1.The nature and limits of human government. 2.Equity among regions in a world community.3.Regional planning considerations.4.Support for the intensity and diversity of human cultures.

1.There are natural limits to the size of groups that can govern themselves in a human way.The biologist J.B.S.Haldane has remarked on this in his paper, "On Being the Right Size":

...just as there is a best size for every animal,so the same is true for every human institution.In the Greek type of democracy all the citizens could listen to a series of orators and vote directly on questions of legislation.Hence their philosophers held that a small city was the largest possible democratic state...(J.B.S Haldane, "On Being the Right Size," *The World of Mathematics*,Vol.II,J.R.Newman,ed.New York:Simon and Schuster,1956,pp.962-967.)

It is not hard to see why the government of a region becomes less and less manageable with size.In a population of N persons,there are of the order of N2 person-to-person links needed to keep channels of communication open.Naturally,when N goes beyond a certain limit,the channels of communication needed for democracy and justice and information are simply too clogged,and too complex;bureaucracy overwhelms human processes.

……正如每一种动物都有一个最佳的规模一样，每一个人类社会也是如此。在希腊式的民主中，全体公民都能聆听到一大批演说家的雄辩演说，并且都能对立法问题进行直接投票。因此，他们的哲学家认为一个小城市可能是最民主的国家了……（J.B.S.Haldane，"On Being the Right Size，" *The World of Mathematics*，Vol.II，J.R.Newman，ed.New York：Simon and Schuster，1956，pp.962～967.）

不难理解，为什么一个区域政府随其规模的扩大而变得越来越不便管理了。设某区的人口数量为 N，为了使交际渠道畅通无阻，人际接触联系的次数则大致为 N^2。很自然，当 N 超过一定限度，为民主、正义和信息所需的交际渠道就简直会堵塞不堪、非常错综复杂；官僚政治就会阻碍人类的进程。

毫无疑问，当 N 增长时，政府的各级官僚机构也会相应增加。在丹麦等小国，这样层层重叠的机构甚少，所以任何公民都能亲自拜访教育部长。但是，这种直接地到访，在英、美等较大的国家里是完全不可能的。

我们认为，某一区域的人口数量达到 200 万～1000 万，就算达到饱和状态了。若超过这个规模，人民同日理万机的政府之间的关系就会日益趋向于疏远。就近代史的观点来看，我们的估计似乎是不同寻常的：民族国家强大有力，其政府就能管辖几千万乃至几亿人民。但是，这些大国不能声称已具有天然合理的规模。它们不能声称在城镇、社区和作为一个整体的世界社会之间的需要已经达到了平衡。的确，大国的倾向一贯是忽视地方的需要，并压抑地方的文化，同时却把它们的权势扩大到它们从未涉足过的、一般人几乎难以想象的范围。

2. 如果一个区域不拥有数百万居民，就不会强大到足以在世界政府内占一席位，从而也将不能取代现在的民族国家的权力和威望。

And,of course,as N grows the number of levels in the hierarchy of government increases too.In small countries like Denmark there are so few levels,that any private citizen can have access to the Minister of Education.But this kind of direct access is quite impossible in larger countries like England or the United States.

We believe the limits are reached when the population of a region reaches some 2 to 10 million.Beyond this size,people become remote from the large-scale processes of goverranent.Our estimate may seem extraordinary in the light of modern history:the nation-states have grown mightily and their governments hold power over tens of millions,sometimes hundreds of millions,of people.But these huge powers cannot claim to have a natural size. They cannot claim to have struck the balance between the needs of towns and communities,and the needs of the world community as a whole.Indeed,their tendency has been to override local needs and repress local culture,and at the same time aggrandize themselves to the point where they are out of reach,their power barely conceivable to the average citizen.

2.Unless a region has at least several million people in it,it will not be large enough to have a seat in a world government,and will therefore not be able to supplant the power and authority of present nationstates.

We found this point expressed by Lord Weymouth of Warminster,England,in a letter to the New York Times,March 15,1973:

我们从英格兰华敏斯特的魏玛诗勋爵 1973 年 3 月 15 日致《纽约时报》的一封信中找到了这种观点：

世界联盟：一千个州

……以民主为基础的世界联盟最重要的基石是在集权政府的管辖范围内实行区域化。……这一论点所依据的观点是：如果每个代表不是代表着世界人口中大致相等的部分，世界政府在道义上就缺乏权威。到 2000 年，估计世界人口数量可达 100 亿大关。现在言归正传，我提议，我们应当通过一种理想的区域性的州来进行构思，这种区域性的州的人口数量约为 1000 万，或为 500 万～1500 万，以便给予它更大的灵活性。这将为联合国大会提供一个相当于 1000 个区域代表的组织：一个宣称自己真正代表世界人口的机构。

魏玛诗坚信，西欧会表现出某种首创精神去传播世界政府这一概念。他希望区域自治运动以便约束设于斯特拉斯堡的欧洲议会，并希望权力能够从威斯敏斯特、巴黎、波恩等地逐步转移到斯特拉斯堡联邦的地方议会。

我正在建议，在未来的欧洲，我们会看到英格兰本土将划分为肯特区、威塞克斯区、莫西亚区、盎格里亚区和诺森伯里亚区，当然，还有独立的英格兰区、威尔斯区和爱尔兰区。欧洲的其他例子将包括布列塔尼区、巴伐里亚区和卡拉勃里亚区。我们当代欧洲的民族一致性将丧失其政治意义。

3. 如果各个区域没有自己管理自己的本领，那么它们将对解决各自的环境问题无能为力。州与州、国家与国家间的边界往往穿越自然的区域边界，所以，人们要解决区域问题，通过直接的和人为有效的途径几乎是不可能的。

法国经济学家格拉维埃针对这种看法已做了详尽无遗的分析。他在自己的一系列著作和论文中提出把欧洲划分

WORLD FEDERATION:A THOUSAND STATES

...the essential foundation stone for world federation on a democratic basis consists of regionalization within centralized government...This argument rests on the idea that world government is lacking in moral authority unless each delegate represents an approximately equal portion of the world's population.Working backward from an estimate of the global population in the year 2000,which is anticipated to rise to the 10,000 million mark,I suggest that we should be thinking in terms of an ideal regional state at something around ten million,or between five and fifteen million,to give greater flexibility.This would furnish the U.N.with an assembly of equals of 1000 regional representatives:a body that would be justified in claiming to be truly representative of the world's population.

Weymouth believes that Western Europe could take some of the initiative for triggering this conception of world government.He looks for the movement for regional autonomy to take hold in the European Parliament at Strasbourg;and hopes that power can gradually be transferred from Westminister, Paris,Bonn,etc.,to regional councils,federated in Strasbourg.

I am suggesting that in the Europe of the future we shall see England split down into Kent,Wessex,Mercia,Anglia and Northumbria,with an independent Scotland,Wales and Ireland,of course.Other European examples will include Brittany,Bavaria and Calabria.The national identities of our contemporary Europe will have lost their political significance.

3.Unless the regions have the power to be self-governing, they will not be able to solve their own environmental problems.The arbitrary lines of states and countries, which

成许多区域的概念，也即围绕着穿越现存的国界和次国界的划分区域问题，提出了分散的、重新组织的欧洲的概念。（例如，巴塞尔—斯特拉斯堡区域包括法国、德国、瑞士的部分国土；利物浦区域包括英格兰和威尔士部分地区）。（See Jean-Francois Gravier, "L'Europe des regions," in 1965 Internationale Regio Planertagung, Schriften der Regio 3, Regio, Basel, 1965, pp.211 ~ 222; and in the same Volume see also Emrys Jones, "The Conflict of City Re-gions and Administrative Units in Britain," pp.223 ~ 235.)

4. 最后，如果今天的一些伟大国家不把其权力极大地分散，那么世界上各国人民的迥然不同的美妙的语言、文化、风俗和生活方式——所有这一切对我们赖以生存的这个行星的健康气氛来说是极端重要的——都将一一消失。总之，我们认为，独立区域是语言、文化、风俗、经济和法律的天然容器，每个区域应当是分离的、独立的，使之足以保持其文化的力量和勃勃生机。

在**亚文化区的镶嵌**（8）中，我们要详细论述如下事实：在一个城市的范围内，人类的文化只有在至少是部分地与其相邻的文化分离时才能繁荣昌盛。我们认为，同样的论点也适用于独立区域：即世界上的各个区域，为了要作为文化区而生存下去，它们相互之间必须保持一定距离和各自独特的风格。

在中世纪的黄金时代，城市就起到了这种作用。它们曾经为各种不同的文化影响和经济交流提供永久性的、充满热情的活动场所；这些场所就是大的公社，它们的公民是共同成员，每个人对城市的命运都有某种发言权。我们认为独立区域能够成为为人类实体提供文化、语言、法律、劳务、经济交流、杂耍等活动场所的现代城邦——一种新的公社，而在过去这些活动场所是由有城墙的城市或城邦提供给它的成员的。

often cut across natural regional boundaries, make it all but impossible for people to solve regional problems in a direct and humanly efficient way.

An extensive and detailed analysis of this idea has been given by the French economist Gravier,who has proposed,in a series of books and papers,the concept of a Europe of the Regions,a Europe decentralized and reorganized around regions which cross present national and subnational boundaries.(For example,the Basel-Strasbourg Region includes parts of France,Germany,and Switzerland;the Liverpool Region includes parts of England and parts of Wales).See Jean-François Gravier, "L'Europe des regions," in 1965 Internationale Regio Planertagung,Schriften der Regio 3,Regio,Basel,1965,pp.211-222;and in the same volume see also Emrys Jones, "The Conflict of City Regions and Administrative Units in Britain," pp.223-235.

4.Finally,unless the present-day great nations have their power greatly decentralized,the beautiful and differentiated languages,cultures,customs,and ways of life of the earth's people,vital to the health of the planet,will vanish.In short,we believe that independent regions are the natural receptacles for language,culture,customs,economy,and laws and that each region should be separate and independent enough to maintain the strength and vigor of its culture.

The fact that human cultures within a city can only flourish when they are at least partly separated from neighboring cultures is discussed in great detail in MOSAIC OF SUBCULTURES(8).We are suggesting here that the same argument also applies to regions—that the regions of the earth must also keep their distance and their dignity in order to survive as cultures.

因此：

无论你在什么地方，都要竭尽全力去促进世界独立区域的发展；每个独立区域的人口数量为 **200 万 ~ 1000 万**；每个独立区域都有自己天然的和地理的边界；每个独立区域都有自己的经济；每个独立区域都是自治的；每个独立区域都在世界政府内占有一个席位，不受较大的州或国家的干预。

1000个独立区域

每个独立区域200万~1000万人口

⊱⟡⟡⟢

在每一独立区域内，鼓励人口分布尽可能地遍及全区——**城镇分布**（2）……

在每一独立区域内制定出那些保护土地并指明城市的极限的区域政策：

2. 城镇分布

3. 指状城乡交错

4. 农业谷地

5. 乡村沿街建筑

6. 乡间小镇

7. 乡村

In the best of medieval times,the cities performed this function.They provided permanent and intense spheres of cultural influence,variety,and economic exchange;they were great communes,whose citizens were comembers,each with some say in the city's destiny.We believe that the independent region can become the modern polis—the new commune—that human entity which provides the sphere of culture,language, laws,services,economic exchange,variety,which the old walled city or the polis provided for its members.

Therefore:

Wherever possible,work toward the evolution of independent regions in the world;each with a population between 2 and 10 million;each with its own natural and geographic boundaries;each with its own economy;each one autonomous and self-governing;each with a seat in a world government,without the intervening power of larger states or countries.

ᛒᚥᚷ

Within each region encourage the population to distribute itself as widely as possible across the region-THE DISTRIBUTION OF TOWNS (2)...

within each region work toward those regional policies which will protect the land and mark the limits of the cities:

2.THE DISTRIBUTION OF TOWNS

3.CITY COUNTRY FINGERS

4.AGRICULTURAL VALLEYS

5.LACE OF COUNTRY STREETS

6.COUNTRY TOWNS

7.THE COUNTRYSIDE

模式2 城镇分布

　　……现在考虑独立区域内住区的特性：村庄、城镇和城市要与区域的独立性——**独立区域**（1）保持一种怎样的平衡呢？

୨୦୯୨

　　如果一个区域的人口过度集中在小村庄，现代文明就绝不会产生；但是，如果一个区域的人口过度集中于大城市，那么地球定将走向毁灭，因为它会不堪重负。

2   THE DISTRIBUTION OF TOWNS

...consider now the character of settlements within the region:what balance of villages,towns,and cities is in keeping with the independence of the region—INDEPENDENT REGIONS(I)?

⊱⊰

If the population of a region is weighted too far toward small villages,modern civilization can never emerge;but if the population is weighted too far toward big cities,the earth will go to ruin because the population isn't where it needs to be,to take care of it.

Two different necessities govern the distribution of population in a region.On the one hand,people are drawn to cities:they are drawn by the growth of civilization,jobs,education, economic growth,information.On the other hand,the region as a social and ecological whole will not be properly maintained unless the people of the region are fairly well spread out across it,living in many different kinds of settlements-farms,villages,towns,and cities-with each settlement taking care of the land around it.Industrial society has so far been following only the first of these necessities.People leave the farms and towns and villages and pack into the cities,leaving vast parts of the region depopulated and undermaintained.

In order to establish a reasonable distribution of population within a region,we must fix two separate features of the distribution:its statistical character and its spatial character.

两种不同的需要支配着一个区域的人口分布。一方面，人们迁往城市：因为城市文明的发展、职业、教育、经济增长、信息渠道等都具有极大的吸引力；另一方面，区域作为一个社会的和生态的整体，除非居民相当合理地分布在全区，生活在许多不同的聚居地——农场、村庄、城镇和城市——每个聚居地都要照管好它四周的土地，否则将不可能妥善地维持下去。迄今为止，工业社会只关注上述必需品中的第一项。而人们却离乡背井，涌入城市，致使区域内的大片沃土荒芜、人烟稀少。

为了在区域内建立一个合理的人口布局，我们必须注意人口分布的两个独立的特性：即人口分布的统计学特性和空间特性。

第一，我们必须确保城镇的统计学分布在规模上是合适的：有许许多多小城镇和很少几个大城镇。第二，我们必须确保区域内城镇的空间分布是合适的：特定规模的城镇均匀地分布在全区，而并非高度集中于一处。

实际上，统计学上的分布会自行处理。大量的研究表明，在城市发展和人口流动中一直起作用的自然人口统计过程以及政治和经济过程，将创造出这样一种城镇布局：区域内有许许多多小城镇和少数几个大城镇；的确，这种分布的性质与我们在本模式中提出的对数分布大致符合。克里斯托勒、席泼、赫伯特·西蒙等给出了种种解释；布赖恩·贝利和威廉·加里森都在他们的著作中进行了概括。(See "Alternate Explanations of Urban Rank-Size Relationships," *Annals of the Association of American Geographers*, Vol.48, March 1958, No.1, pp.83 ～ 91.)

下面让我们接着来假设，城镇具有合理的分布规模。但是，各城镇是相互紧挨在一起的还是分散在各处的？如果一个区域内的所有大、中、小城镇都统统拥挤在一个连续不断的城市区域内，那么大小城镇兼有这一事实虽在政

First,we must be sure that the statistical distribution of towns,by size,is appropriate:we must be sure that there are many small towns and few large ones.Second,we must then be sure that the spatial distribution of towns within the region is appropriate:we must be sure that the towns in any given size category are evenly spread out across the region,not highly concentrated.

In practice,the statistical distribution will take care of itself.A large number of studies has shown that the natural demographic and political and economic processes at work in city growth and population movement will create a distribution of towns with many small towns and few large ones;and indeed,the nature of this distribution does correspond,roughly,to the logarithmic distribution that we propose in this pattern.Various explanations have been given by Christaller,Zipf,Herbert Simon,and others;they are summarized in Brian Berry and William Garrison, "Alternate Explanations of Urban Rank-Size Relationships," *Annals of the Association of American Geographers*,Vol.48,March 1958,No.1,pp.83-91.

Let us assume,then,that towns will have the right distribution of sizes.But are they adjacent to one another,or are they spread out?If all the towns in a region,large,medium,and small,were crammed together in one continuous urban area,the fact that some are large and some are small,though interesting politically,would have no ecological meaning whatsoever.As far as the ecology of the region is concerned,it is the spatial distribution of the towns which matters,not the statistics of political boundaries within the urban sprawl.

Two arguments have led us to propose that the towns in any one size category should be uniformly distributed across the region:an economic argument and an ecological argument.

治上会令人感兴趣，但却毫无生态学方面的意义。至于说到区域内的生态平衡，至关紧要的是城镇的空间分布，而不是在市区扩展的范围内政治边界的统计数字。

经济学的和生态学的两种论据已使我们提出了下列建议：任何规模的城镇都应当在全区域内均匀分布。

经济学的论据：全世界不发达地区的经济正濒临崩溃。因为工作机会和人口都纷纷涌往最大的城市，因为大城市的经济对他们具有巨大的吸引力。瑞典、苏格兰、以色列和墨西哥都是例子。那里的人口流入斯德哥尔摩、格拉斯哥、特拉维夫和墨西哥城。只要这种人口流动继续下去，城市就会创造出层出不穷的新产业，于是，就会有更多的人奔向城市谋求职业。农村和城市之间的不平衡状态就会日趋严重。城市变得越来越富足，而远离城市中心的地区则每况愈下，越发贫困。最后，该地区城市中心的生活水准可能是世界上最高的，而在只不过相距数英里的城市边缘地区，人们可能饥肠辘辘，正在挨饿。

这种贫富悬殊的情况只有通过保证全区域平等分享资源和发展经济的各项政策才能加以阻止。例如，在以色列，早有过某种尝试：把政府能够用来补贴经济发展的有限资源，源源不断地输往经济上最落后的地区。(See "Urban Growth Policies in Six European Countries," Urban Growth Policy Study Group, Office of International Affairs, HUD, Washington, D.C., 1972.)

生态学的论据：在空间上过分集中的人口会给区域的整个生态系统造成巨大的负担。随着大城市的发展，人口的流入加剧了城市的空气污染、交通堵塞、水资源短缺、住房短缺，以及居住密度超出合乎常情的合理限度。在一些大城市中心区，生态平衡已岌岌可危，行将彻底崩坏。相反，如果人口比较均匀地分布于全区域，对环境的生态平衡的冲击就会减少到最小程度，在这种情况下，人们会更加谨慎处世和

*Economic.*All over the world,underdeveloped areas are facing economic ruin because the jobs,and then the people,move toward the largest cities,under the influence of their economic gravity.Sweden,Scotland,Israel,and Mexico are all examples.The population moves toward Stockholm,Glasgow,Tel Aviv,Mexico City—as it does so,new jobs get created in the city,and then even more people have to come to the city in search of jobs.Gradually the imbalance between city and country becomes severe.The city becomes richer,the outlying areas continuously poorer.In the end the region may have the highest standard of living in the world at its center,yet only a few miles away,at its periphery,people may be starving.

This can only be halted by policies which guarantee an equal sharing of resources,and economic development,across the entire region.In Israel,for example,there has been some attempt to pour the limited resources with which the government can subsidize economic growth into those areas which are most backward economically.(See "Urban Growth Policies in Six European Countries," Urban Growth Policy Study Group,Office of International Affairs, HUD, Washington,D.C.,1972.)

*Ecological.*An overconcentrated population,in space,puts a huge burden on the region's overall ecosystem.As the big cities grow,the population movement overburdens these areas with air pollution,strangled transportation,water shortages,housing shortages,and living densities which go beyond the realm of human reasonableness.In some metropolitan centers,the ecology is perilously close to cracking.By contrast,a population that is spread more evenly over its region minimizes its impact

更加周到地照料自己的土地，减少浪费且更加人性化：

这是因为当城镇的规模超过某一限度时，每个居民实际所需的城市大体量建筑急剧增加。例如，按人口计算，高层公寓的造价比普通房屋的造价高得多；道路和其他运输线路的造价随着上下班的人的增多而提高。同样，按人口计算，用于其他设施，如分配粮食和清除废物等设施的费用在城市比在小乡镇和村庄高得多。因此，假如人人都住在村子里，污水处理工厂的需求就会在某种程度上减少，而在完全城市化的社会中，污水处理厂是不可缺少的，而且处理污水的费用十分昂贵。一般来说，只有通过分散这一途径，我们才能提高自给自足的能力——如果我们想要最大限度地减轻各种社会系统对它们赖以维持的相应的生态系统的负担，自给自足是极其重要的。(The Ecologist, *Blueprint for Survival*, England : Penguin, 1972, pp. 52～53.)

因此：

鼓励区域内城镇的诞生与衰落消亡过程会逐步产生如下效果：

1. 人口按不同的规模均匀地分布：例如，有 100 万人的 1 个城镇，各有 10 万人的 10 个城镇，各有 1 万人的 100 个城镇，各有 100 人的 1000 个城镇。

2. 这些城镇在空间上是这样分布的：各种不同规模的城镇均匀地分布于全区域。

这一过程是能够实现的，办法是依靠区域内的土地分区政策、土地转让证书和各种刺激措施，这将鼓励工业按分布条令进行选址。

on the ecology of the environment,and finds that it can take care of itself and the land more prudently,with less waste and more humanity:

This is because the actual urban superstructure required per inhabitant goes up radically as the size of the town increases beyond a certain point.For example,the per capita cost of high rise flats is much greater than that of ordinary houses;and the cost of roads and other transportation routes increases with the number of commuters carried.Similarly,the per capita expenditure on other facilities such as those for distributing food and removing wastes is much higher in cities than in small towns and villages.Thus,if everybody lived in villages the need for sewage treatment plants would be somewhat reduced,while in an entirely urban society they are essential,and the cost of treatment is high.Broadly speaking,it is only by decentralization that we can increase self-sufficiency and self-sufficiency is vital if we are to minimize the burden of social systems on the ecosystems that support them.(The Ecologist,Blueprint for Sur vival,England:Penguin,1972,pp.52-53.)

Therefore:

Encourage a birth and death process for towns within the region,which gradually has these effects:

1.The population is evenly distributed in terms of different sizes-for example,one town with 1,000,000 people,10 towns with 100,000 people each,100 towns with 10,000 people each,and 1000 towns with 100 people each.

2.These towns are distributed in space in such a way that within each size category the towns are homogeneously distributed all across the region.

This process can be implemented by regional zoning

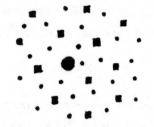

100万人口的城镇——相距250mi（1mi=1.61km）

10万人口的城镇——相距80mi

1万人口的城镇——相距25mi

1000人口的城镇——相距8mi

8OCB

随着分布的开展，保护好可耕的沃土良田——**农业谷地**（4）；保护好较小的偏僻的城镇，在其四周建立乡村地带，并且分散工业，这会使这些城镇经济稳定——**乡间小镇**（6）。在较大的、较中心的城区制定土地政策，使城市地带之间保持空旷的乡村地带——**指状城乡交错**（3）……

policies,land grants,and incentives which encourage industries to locate according to the dictates of the distribution.

⁛⁜

As the distribution evolves,protect the prime agricultural land for farming—AGRICULTURAL VALLEYS(4);protect the smaller outlying towns,by establishing belts of countryside around them and by decentralizing industry,so that the towns are economically stable—COUNTRY TOWNS(6).In the larger more central urban areas work toward land policies which maintain open belts of countryside between the belts of city—CITY COUNTRY FINGERS(3)...

模式3 指状城乡交错**

......城镇的分配需要一个平衡的区域——**城镇分布**（2）——在城镇和城市范围内，这可以通过控制城市土地和广阔的乡村之间的平衡得到进一步的完善。

※※※

城市化连续不断地蔓延扩大会破坏居民的生活，使城市不堪重负。但是，城市的绝对规模是有价值的和强有力的。

当人们去郊野踏青，观赏一望无际的田园风光，置身于野生植物和飞禽走兽之间时会顿觉心旷神怡。为了让人们能接近大自然，城市中的各个点附近都必须有与乡村相衔接的边界。同时，只有当一个城市包含大量的人与工作之间的密切的联系和接触，以及多样的生活方式时，它对

3   CITY COUNTRY FINGERS**

...the distribution of towns required to make a balanced region-DISTRIBUTION OF TOWNS(2)—can be further helped by controlling the balance of urban land and open countryside within the towns and cities themselves.

৪৩০৪৫

Continuous sprawling urbanization destroys life,and makes cities unbearable.But the sheer size of cities is also valuable and potent.

People feel comfortable when they have access to the countryside,experience of open fields,and agriculture; access to wild plants and birds and animals.For this access,cities must have boundaries with the countryside near every point.At the same time,a city becomes good for life only when it contains a great density of interactions among people and work,and different ways of life.For the sake of this interaction,the city must be continuous—not broken up.In this pattern we shall try to bring these two facts to balance.

Let us begin with the fact that people living in cities need contact with true rural land to maintain their roots with the land that supports them.A 1972 Gallup poll gives very strong evidence for this fact.The poll asked the question: "If you could live anywhere,would you prefer a city,suburban area,small town,or farm?" and received the following answers from 1465 Americans:

生活才是有益的。为了这种相互接触，城市必须是连续的，而不是分裂的。在本模式中我们试图把这两种实际情况加以平衡。

让我们从下面这一实际情况开始吧：生活在城市里的人需要同真正的农田阡陌密切接触，与养育他们的土地融为一体。1972 年的一次盖洛普民意测验为这一事实提供了非常有力的证据。民意测验中提出了这样一个问题："假如你能在任何地方生活，你愿意生活在城市、市郊、小城镇，还是农场？"从 1465 名美国人中获得如下的答案：

城市	13％
郊区	13％
小城镇	32％
农场	23％

这种对城市的不满情绪愈演愈烈。在 1966 年，22％的人表示他们宁可生活在城市里，而在 1972 年，仅仅事隔 6 年，这一数字竟然下降到 13％。（"Most don't want to live in a city," George Gallup, *San Francisco Chronicle*, *Monday*, December 18, 1972, p.12.）

不难理解为什么城市居民渴望同乡村保持接触。仅仅 100 年前，85％的美国人生活在农村；而今，70％的美国人生活在城市里。很明显，我们不能完完全全生活在城市里——至少是我们迄今所建的城市里——我们同乡村保持密切接触的需要实在太深刻了，这是一种生理上的需要：

虽然我们可以认为我们人类是无与伦比的，可是，从遗传学上说，我们多半与其他哺乳动物一样，偏爱天然的栖息地：那里风景如画，苍松翠柏，绿草如茵，空气新鲜，沁人心脾。为了使我们自己得到轻松的休息并感到身心健康，一般来说，就是让我们的躯体，以在多少万年的进化过程中所养成的方式对外界做出各种反应。从生物学和遗传学角度来看，我们似乎最适应热带大草原的生态环境，

City	13%
Suburb	13%
Small town	32%
Farm	23%

And this dissatisfaction with cities is getting worse. In 1966,22 percent said they preferred the city-in 1972,only six years later,this figure dropped to 13 percent.("Most don't want to live in a city," George Gallup,*San Francisco Chronicle*,Monday,December 18,1972,p.12.)

It is easy to understand why city people long for contact with the countryside.Only 100 years ago 85 percent of the Americans lived on rural land;today 70 percent live in cities.Apparently we cannot live entirely within cities—at least the kinds of cities we have built so far—our need for contact with the countryside runs too deep,it is a biological necessity:

Unique as we may think we are,we are nevertheless as likely to be genetically programmed to a natural habitat of clean air and a varied green landscape as any other mammal.To be relaxed and feel healthy usually means simply allowing our bodies to react in the way for which one hundred millions of years of evolution has equipped us.Physically and genetically,we appear best adapted to a tropical savanna,but as a cultural animal we utilize learned adaptations to cities and towns.For thousands of years we have tried in our houses to imitate not only the climate,but the setting of our evolutionary past:warm,humid air,green plants,and even animal companions.Today,if we can afford it,we may even build a greenhouse or swimming pool next to our living room,buy a place in the country,or at least

但是，我们作为文化动物，利用后天学到的应变能力来适应城市和城镇的生活环境。千百年来，我们不仅一直试图在室内模拟气候，而且还有我们过去的进化环境：温暖而又湿润的空气、绿色的植物、甚至动物伴侣。今天，如果我们力所能及，甚至可以在我们的起居室附近建造温室或游泳池，在乡村购置一块土地，或者至少带着我们自己的子女去海滨度假。我们面对着大自然的美丽和多样性，它的千姿百态和绚丽色彩（尤其是绿色），对其他动物的优美动作和悦耳动听的声音（如鸟儿的飞翔跳跃和清脆的啼鸣），都会做出特殊的生理反应，至今我们还未透彻理解其中的奥妙。但是，很显然，大自然在我们的日常生活中应当被认为是生物学需要的一部分。这一点在我们讨论为人类造福的资源政策时是不能忽视的。(H.H.Iltis，P.Andres，and O.L.Loucks，in *Population Resources Environment*：*Issues in Human Ecology*，P.R.Ehrlich and A.H.Ehrlich，San Francisco：Freeman and Co.，1970，p.204.)

但是，现在城市居民想要接触乡村生活正变得越发困难。在旧金山海湾区，每年有 $21mi^2$（$1mi^2=2.59km^2$）的空地流失。(Gerald D.Adams，"The Open Space Explosion，" Cry California，Fall 1970，pp.27 ～ 32.) 随着城市变得越来越大，农田就离得越来越远。

由于城市居民和乡村的联系已破裂，城市正在变成监狱。农场度假、城市儿童在农场度假一年和老人退休后回乡村安度晚年——这一切现在都已一一被耗费巨大的避暑胜地、夏令营和退休村所取代。大多数人眼下与乡村的唯一接触，就是每逢周末成群结队地离城去乡间度假，结果造成公路交通堵塞，并使仅有的几个正规娱乐中心游人爆满。许多周末度假者在星期天夜里返城时，感到他们的神经比离城时更加疲惫不堪。

take our children vacationing on the seashore.The specific physiological reactions to natural beauty and diversity,to the shapes and colors of nature(especially to green),to the motions and sounds of other animals,such as birds,we as yet do not comprehend.But it is evident that nature in our daily life should be thought of as a part of the biological need. It cannot be neglected in the discussions of resource policy for man.(H.H.Iltis,P.Andres,and O.L.Loucks,in *Population Resources Environment:Issues in Human Ecology*,P. R.Ehrlich and A.H.Ehrlich,San Francisco:Freeman and Co.,1970,p.204.)

But it is becoming increasingly difficult for city dwellers to come into contact with rural life.In the San Francisco Bay Region 21 square miles of open space is lost each year(Gerald D.Adams, "The Open Space Explosion," *Cry California*,Fall 1970,pp.27-32.)As cities get bigger the rural land is farther and farther away.

With the breakdown of contact between city dwellers and the countryside,the cities become prisons.Farm vacations,a year on the farm for city children,and retirement to the country for old people are replaced by expensive resorts,summer camps,and retirement villages.And for most,the only contact remaining is the weekend exodus from the city,choking the highways and the few organized recreation centers.Many weekenders return to the city on Sunday night with their nerves more shattered than when they left.

If we wish to reestablish and maintain the proper connection between city and country,and yet maintain the density of urban interactions,it will be necessary to stretch out the urbanized area into long sinuous fingers which extend into

当乡村远离时，城市就变成一座监狱
When the countryside is far away the city becomes a prison

如果我们希望重建并保持城乡之间的适当衔接，并且还希望保持市区内十分频繁的人际接触交流，那么，扩展市区将是必要的：把市区扩大成蜿蜒曲折的、向农田延伸的指形地带，如下一张图所示。不仅城市，而且与之毗连的农田都将呈现为狭长的指状带。

城市指状带的最大宽度取决于从市中心至乡村的最大容许距离。据我们测算，每人应当在10min之内能够步行到乡村。这将为城市指状带规定1mi的最大宽度。

农田指状带的最小宽度取决于典型农场的最小容许规模。因为90％的农场拥有的土地面积仍为500acre（1acre=4046.86m²）或以下，而且没有相当重要的证据足以说明大农场比小农场更有效率（Leon H.Keyserling, *Agriculture and the Public Interest*, Conference on Economic Progress, Washington, D.C., February 1965），所以，农田指状带的宽度以不超过1mi为宜。

落实这一模式需要三种不同的新政策。第一，关于农田，一定要有这样一种政策：鼓励小农场重建，使小农场符合1mi宽的农田带。第二，一定要有鼓励城市向各个方向扩散的政策。第三，乡村必须是真正公共的，这样，人们甚至可以同私人耕种的那部分土地建立联系。

请想一想，这模式将如何改造城市的生活。每个城市居民都能去乡间游览；从城市商业区到旷野，骑半小时自

the farmland,shown in the diagram below.Not only will the city be in the form of harrow fingers,but so will the farmlands adjacent to it.

The maximum width of the city fingers is determined by the maximum acceptable distance from the heart of the city to the countryside.We reckon that everyone should be within 10 minutes'walk of the countryside.This would set a maximum width of 1 mile for the city fingers.

The minimum for any farmland finger is determined by the minimum acceptable dimensions for typical working farms.Since 90 percent of all farms are still 500 acres or less and there is no respectable evidence that the giant farm is more efficient(Leon H.Keyserling, *Agriculture and the Public Interest*,Conference on Economic Progress,Washington,D. C.,February 1965),these fingers of farmland need be no more than 1 mile wide.

The implementation of this pattern requires new policies of three different kinds.With respect to the farmland,there must be policies encouraging the reconstruction of small farms,farms that fit the one-mile bands of country land.Second,there must be policies which contain the cities'tendency to scatter in every direction.And third,the countryside must be truly public,so that people can establish contact with even those parts of the land that are under private cultivation.

Imagine how this one pattern would transform life in cities.Every city dweller would have access to the countryside;the open country would be a half-hour bicycle ride from downtown.

行车即可到达。

因此:

务使农田和城市土地均呈指状交错，即使是在大都市的中心。城市指状带的宽度应不超过 **1mi**，而农田指状带的宽度不应小于 **1mi**。

城市指状带至多1mi宽

❧❦

在多丘陵的地方，无论何时，务使乡村指状带位于山谷，而城市指状带位于山岗顶部的斜坡上——**农业谷地**（4）。把城市指状带划分为数百个有鲜明特色的、自治的亚文化区——**亚文化区的镶嵌**（8），并使主要公路和铁路都通往这些城市指状带的中心区——**公共交通网**（16），**环路**（17）……

Therefore:

Keep interlocking fingers of farmland and urban land, even at the center of the metropolis.The urban fingers should never be more than 1 mile wide,while the farmland fingers should never be less than 1 mile wide.

৪০০৪

Whenever land is hilly,keep the country fingers in the valleys and the city fingers on the upper slopes of hillsides—AGRICULTURAL VALLEYS(4).Break the city fingers into hundreds of distinct self-governing subcultures—MOSAIC OF SUBCULTURES(8),and run the major roads and railways down the middle of these city fingers—WEB OF PUBLIC TRANSPORTATION(16),RING ROADS(17)...

模式4　农业谷地*

　　……本模式有助于维持**独立区域**（1），使其在农业上更加自给自足；并且，本模式将在市区内保留农业土地，从而几乎是自动地创造出**指状城乡交错**（3）。但是，究竟什么样的土地应当加以保护？什么样的土地应当用于建造？

⟡⟡⟡

　　最适合于农业的土地也是最适合于建造的。但是，土地是有限的——一旦遭到毁坏，多少个世纪都无法恢复。

　　在最近几年内城市郊区的发展遍及农业和非农业的全部土地。这种发展侵吞着有限的资源，更加糟糕的是，彻底破坏了耕种城市附近农田的可能性。但是，我们从**指状城乡交错**（3）的论据中认识到，重要的是在靠近人们居住的地方，要有开阔的农田。因为可以用来耕作的耕地主要

4   AGRICULTURAL VALLEYS*

...this pattern helps maintain the INDEPENDENT REGIONS(1)by making regions more self-sufficient agriculturally;and it will create CITY COUNTRY FINGERS(3) almost automatically by preserving agricultural land in urban areas.But just exactly which land ought to be preserved,and which land built upon?

৪০৫৪

The land which is best for agriculture happens to be best for building too.But it is limited—and once destroyed,it cannot be regained for centuries.

In the last few years,suburban growth has been spreading over all land,agricultural or not.It eats up this limited resource and,worse still,destroys the possibility of farming close to cities once and for all.But we know,from the arguments of CITY COUNTRY FINGERS (3),that it is important to have open farmland near the places where people live.Since the arable land which can be used for farming lies mainly in the valleys,it is essential that the valley floors within our urban regions be left untouched and kept for farming.

The most complete study of this problem that we know,comes from Ian McHarg(*Design With Nature*,New York:Natural History Press,1969).In his "Plan for the Valleys" (Wallace-McHarg Associates,Philadelphia,1963), he shows how town development can be diverted to the hillsides and plateaus,leaving the valleys clear.The

位于山谷中，所以，在我们市区范围内的山谷底部保持其原始状态和耕耘状态十分必要。

据我们所知，伊恩·麦哈格对此问题做了最完整的研究（*Design With Nature*，New York：Natural History Press，1969）。他在《谷地的规划》"Plan for the Valleys"（Wallace-McHargAssociates，Philadelphia，1963）一书中指明，城镇发展怎样才能向山坡和高地转移，并使谷地成为一览无余的田野。本模式也得到下列事实的支持：完成这一任务有数种可能的、切合实际的方法（McHarg，pp.79～93）。

因此：

保护作为农田的全部农业谷地，防止这片土地遭受因任何发展所造成的危害，或是破坏或扼杀它的土壤的罕有的肥力。即使目前谷地荒芜，未加开垦，也要妥善保护：将谷地留给农场、公园和荒野。

务使城镇和城市沿着小山顶和山坡发展——**指状城乡交错**（3）。在谷地，把土地所有权作为一种管理权（包括担负起生态平衡的责任）来加以处理——**乡村**（7）……

pattern is supported,also,by the fact that there are several possible practical approaches to the task of implementation (McHarg,pp.79-93).

Therefore:

Preserve all agricultural valleys as farmland and protect this land from any development which would destroy or lock up the unique fertility of the soil.Even when valleys are not cultivated now,protect them:keep them for farms and parks and wilds.

8003

Keep town and city development along the hilltops and hillsides—CITY COUNTRY FINGERS(3).And in the valleys,treat the ownership of the land as a form of stewardship,embracing basic ecological responsibilities—THE COUNTRYSIDE(7)...

模式5　乡村沿街建筑

　　……根据模式指**状城乡交错**（3），城市土地和农业土地界线分明。但是，在展现出乡村指状带的城市指状带的终端需要一种附加结构。这种结构在传统上一直是市郊。可是……

❧❧❧

　　市郊是人类聚居地过时的、充满着矛盾的形式。

　　许多人向往去农村居住，同时又想靠近大城市。但是，从主要的市中心几分钟内就能到达的郊区，从几何空间上来讲，要建立起成千上万个小农场是不可能的。

　　为了在乡村生活得美满愉快，你必须有一片适当的土地——它大得足以放牧牛、马等家畜并能饲养家禽，还有一个果园——并且你必须能立即到达你能看见的那连绵不

5   LACE OF COUNTRY STREETS

...according to the pattern CITY COUNTRY FINGERS (3),there is a rather sharp division between city land and rural land.But at the ends of city fingers,where the country fingers open out,there is a need for an additional kind of structure.This structure has traditionally been the suburbs.But...

❧❧❧

The suburb is an obsolete and contradictory form of human settlement.

Many people want to live in the country;and they also want to be close to a large city.But it is geometrically impossible to have thousands of small farms,within a few minutes of a major city center.

To live well in the country,you must have a reasonable piece of land of your own—large enough for horses, cows,chickens,an orchard-and you must have immediate access to continuous open countryside,as far as the eye can see.To have quick access to the city,you must live on a road,within a few minutes'drive from city centers,and with a bus line outside your door.

It is possible to have both,by arranging country roads around large open squares of countryside or farmland,with houses closely packed along the road,but only one house deep.Lionel March lends support to this pattern in his paper, "Homes beyond the Fringe" (Land Use and Built Form Studies,Cambridge,England,1968).March shows that such a

断的旷野。为了能迅速抵达城市，你必须居住在公路一侧，在你的门外有一条公共汽车线路，从市中心驱车直达你的家门只需几分钟。

两者兼得也是可能的，办法是：在旷野或农场的四周修筑乡村公路，在公路沿线安排密集的住宅，但它们离公路只有一幢住宅的间距。莱昂内尔·马奇在他的《郊外的住宅》（*Land Use and Built Form Studies*，Cambridge，England，1968）这篇论文中对这一模式表示支持。马奇指明，这样一种已充分发展了的模式可以为数百万人服务，甚至为国土面积小而人口密度大的英国的居民服务。

"乡村沿街建筑"包括如下内容：若干平方英里的旷野，高速公路从城市延伸到旷野的边缘，住宅密集在公路两侧，从市区通往乡村的小路纵横交错，四通八达。

1. 若干平方英里的旷野：我们认为，1mi² 是最小的一片开阔地，它依然保持着乡村的完整性。这一数字来自小农场的需求，而我们在论证指**状城乡交错**（3）中已提出来了。

2. 公路：为了使乡村免遭城市的蚕食鲸吞，通往乡村的公路数量上要大幅度削减。一种相互沟通的松散的公路网即已足够。各公路间相隔的距离为 1mi，不宜提倡直达交通。

3. 宅基地：家园、住宅和别墅坐落在乡村公路两侧，离公路有一块或两块宅基地的间距。后面是一片开阔地。家园的最小占地面积必须是半英亩左右，以便容许基本耕作。可是，一些住宅可能要成排或成组地建造，这样，人们可以集体耕种宅后的土地。假定半英亩上下的宅基地星罗棋布地分散在 1mi² 的开阔地周围，就会有 400 户人家居住在此。按每户 4 口计算，每平方英里为 1600 人。这与一个人口密度低的普通的市郊相差无几。

4. 小路：市民们可从城市边缘处的小路和田间崎岖的

pattern,fully developed,could work for millions of people even in a country as small and densely populated as England.

A "lace of country streets" contains square miles of open countryside,fast roads from the city at the edge of these square miles,houses clustered along the roads,and footpaths stretching out from the city,crisscrossing the countryside.

1.Square miles of open countryside.We believe that onesquare mile is the smallest piece of open land which still maintains the integrity of the countryside.This figure is derived from the requirements of small farms,presented in the argument for CITY COUNTRY FINGERS(3).

2.Roads.To protect the countryside from suburban encroachment,the roads running out into the countryside must be vastly reduced in number.A loose network of interconnected roads,at one-mile intervals with little encouragement for through traffic to pass through them,is quite enough.

3.Lots.Situate homesteads,houses,and cottages along these country roads one or two lots deep,always setting them off the road with the open land behind them.The *minimum* land for a home-stead must be approximately one-half acre to allow for basic farming.However,some of the housing could be in rows or clusters,with people working the land behind collectively. Assuming one-half acre lots around a one mile square of open land,we can have 400 households to the square mile.With four people per household,that is 1600 people per square mile;not very different from an ordinary low density suburb.

4.Footpaths.The countryside can be made accessible to city people by means of footpaths and trails running from the edge of the city and from the country roads into the countryside,across the squares of open land.

小道通往郊野，也可从乡村公路穿越一片片广阔的田地通往郊野。

因此：

在城乡交会的地区，各乡村公路至少相距 1mi，以便公路环绕的一片片原野和农田至少有 1mi² 的面积。在公路沿线建设家园，离公路有一宅基地的间距，宅基地至少有半英亩面积，在住宅后面有 1mi² 的旷野和农田。

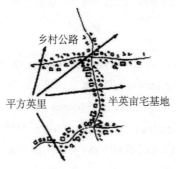

务使每一平方英里的旷野，包括农场和公园，都向公众开放——**乡村**（7）；将这半英亩宅基地整理形成住宅团组和邻里，即使它们相当分散——**易识别的邻里**（14），**住宅团组**（37）……

Therefore:

In the zone where city and country meet,place country roads at least a mile apart,so that they enclose squares of countryside and farmland at least one square mile in area.Build homesteads along these roads,one lot deep,on lots of at least half an acre,with the square mile of open countryside or farmland behind the houses.

✿✿✿

Make each square mile of countryside,both farm and park,open to the public—THE COUNTRYSDE(7);arrange the half acre lots to form clusters of houses and neighborhoods,even when they are rather spread out—IDENTIFIABLE NEIGHBORHOOD(14),HOUSE CLUSTER(37)...

模式6 乡间小镇*

……本模式构成**城镇分布**（2）的主要部分，它需要数以百计的较小型村镇来支持区域内的大型村镇和大城市。

୫୦୫

大城市犹如一块磁铁。面对中心城市的不断发展，乡间小镇要继续生存下去并保持自己的繁荣是极端艰难的。

近三十年间，3000 万美国农业人口被迫离开他们的农场和小城镇，迁入拥挤不堪的大城市。这种被迫的人口迁移，以每年 80 万人的速度在继续进行着。留在农村而未迁出的家庭对生活忧心忡忡，前途未卜：其中约有半数家庭的年收入低于 3000 美元。

人们不只是为寻找职业才离开小城镇而涌入大城市的，他们也想获取信息和接触流行的文化。例如，在爱尔兰和印度，精力充沛的人们远离农村故土，并非因为那里无事

6   COUNTRY TOWNS*

...this pattern forms the backbone of the DISTRIBUTION OF TOWNS (2),which requires that scores of smaller country towns support the larger towns and cities of the region.

❧❀❧

The big city is a magnet.It is terribly hard for small towns to stay alive and healthy in the face of central urban growth.

During the last 30 years,30 million rural Americans have been forced to leave their farms and small towns and migrate to crowded cities.This forced migration continues at the rate of 800,000 people a year.The families that are left behind are not able to count on a future living in the country:about half of them live on less than $3000 a year.

And it is not purely the search for jobs that has led people away from small towns to the cities.It is also a search for information,for connection to the popular culture. In Ireland and India,for example,lively people leave the villages where there is some work,and some little food,and they go to the city,looking for action,for better work,for a better life.

Unless steps are taken to recharge the life of country towns,the cities will swamp those towns which lie the nearest to them;and will rob those which lie furthest out of their most vigorous inhabitants.What are the possibilities?

可做，缺衣少食，他们进城是为了有所作为，谋求更好的职业和更好的生活。

如果不采取措施给乡间小镇注入新的活力，那些大城市将会吞没掉离它们最近的一些城镇；并将掠夺离它们最远的那些城镇中的活力。

那么，可能的出路在哪里呢？

1. 经济重建：刺激工商业分散到小城镇去。激励小城镇居民开展基础商业并进行生产创业活动。(See, for example, the bill introduced by Joe Evins in the House of Representatives, *Congressional Record*-House, October 3, 1967, 27687.)

2. 分区：采取分区的政策来保护小城镇及其四周的乡村。早在 20 世纪初埃比尼泽·霍华德就阐明了绿化带分区的意义，而美国各级政府现在才认真对待此事。

3. 社会服务：通过社会服务的形式密切小城镇和大城市之间的种种联系，这些联系是不可替代的：市民参观访问小城镇，去农场过周末或度假；城里的儿童去乡村上学和野营；不喜欢城市生活节奏的老年人去小城镇隐居。让大城市吸引小城镇提供一些服务，作为基础创业模式，而城市或私人团体将支付服务费用。

因此：

保护现存的乡间小镇；并鼓励发展人口达 500 ~ 10000 人的新型的自给自足的城镇，城镇周围是开放式的乡村原野，与邻近的城镇至少相隔 10mi。拨给每个城镇为建立其地方工业基地所需的款项——这要成为全区域的人民集体关心的大事。从而，这些城镇就不再是外地工作人员的集体宿舍，而是名副其实的能够自力更生的城镇了。

1.Economic reconstruction.Incentives to business and industry to decentralize and locate in small towns.Incentives to the inhabitants of small towns to begin grassroots business and production ventures.(See,for example,the bill introduced by Joe Evins in the House of Representatives,*Congressional Record-House*,October 3,1967,27687.)

2.Zoning.Zoning policy to protect small towns and the countryside around them.Greenbelt zoning was defined by Ebenezer Howard at the turn of the century and has yet to be taken seriously by American governments.

3.Social services.There are connections between small towns and cities that take the form of social services,that are irreplaceable:small town visits,farm weekends and vacations for city dwellers,schools and camps in the countryside for city children,small town retirement for old people who do not like the pace of city life.Let the city invite small towns to provide these services,as grassroots ventures,and the city,or private groups,will pay for the cost of the service.

Therefore:

Preserve country towns where they exist;and encourage the growth of new self-contained towns,with populations between 500 and 10,000,entirely surrounded by open countryside and at least 10 miles from neighboring towns.Make it the region's collective concern to give each town the wherewithal it needs to build a base of local industry,so that these towns are not dormitories for people who work in other places,but real towns—able to sustain the whole of life.

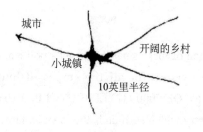

城市

小城镇

开阔的乡村

10英里半径

⟊⟊⟊

把每个小城镇作为政治社区来对待，并充分保障人生各个阶段所需的生活必需品——**7000 人的社区**（12），**生命周期**（26）。把小城镇周围开阔的乡村地带作为农田来处理。农田属于人民所有，人民可以自由地使用土地——**乡村**（7）……

Treat each of these small towns as a political community, with full provision for all the stages of human life—COMMUNITY OF 7000 (12),LIFE CYCLE(26).Treat the belt of open country which surrounds the town as farm land which belongs to the people and can be freely used by them—THE COUNTRYSIDE(7)...

模式7 乡村*

……在每一区域内,在各城镇之间,有广袤的原野——农田、公园、森林、沙漠、牧场、湖泊和河流。这片原野的立法性质和生态性质是这一地区平衡的关键所在。如果处理得当,本模式有助于完善下列模式:**城镇分布** (2)、**指状城乡交错** (3)、**农业谷地** (4)、**乡村沿街建筑** (5) 和**乡间小镇** (6)。

∞∞

我认为,土地的使用权应当属于一个大家庭的全体成员:其中许多人已经死去,少数人还活着,无数人尚未降临人世。

——引自一位尼日利亚部落成员

7 THE COUNTRYSIDE*

...within each region,in between the towns,there are vast areas of countryside-farmland,parkland,forests,deserts ,grazing meadows,lakes,and rivers.The legal and ecological character of this countryside is crucial to the balance of the region.When properly done,this pattern will help to complete THE DISTRIBUTION OF TOWNS(2),CITY COUNTRY FINGERS(3),AGRICULTURAL VALLEYS(4),LACE OF COUNTRY STREETS(5)and COUNTRY TOWNS(6).

෫෬

I conceive that land belongs for use to a vast family of which many are dead,few are living,and countless members are still unborn.

—a Nigerian tribesman

Parks are dead and artificial.Farms,when treated as private property,rob the people of their natural biological heritage—the countryside from which they came.

In Norway,England,Austria,it is commonly understood that people have a right to picnic in farmland,and walk and play—provided they respect the animals and crops.And the reverse is also true—there is no wilderness which is abandoned to its own processes—even the mountainsides are terraced,mown,and grazed and cared for.

We may summarize these ideas by saying that there is only one kind of nonurban land—*the countryside*.There are

公园是呆板的和矫揉造作的。作为私人财产的农场剥夺了大自然赐予人的生物的遗产——他们的故土乡村。

土地的所有权已被农场窃取

Property is theft

在挪威、英国和奥地利，众所周知，人民有权在农田进行野餐、散步和嬉戏——如果他们珍惜动物和庄稼。反之亦然——在发展农田的同时已经没有荒地了——甚至连山麓也被修成了梯田，辟为草场，人们割草放牧，照料土地。

我们把这些思想概括为：只有一种非城市的土地类型——乡村。那里没有公园，没有农场，没有未经勘察和绘图的荒地。每一片土地都有管理人员看管。如果它是可耕地，他们有权耕种它；如果它是荒地，他们有义务来看管它；每块土地对一般人都是开放的，只要他们能保护那里有机的生态平衡。

这种土地理论贯串一个中心思想，是奥尔道·利奥波尔特在他的论文《土地伦理学》（*A Sand County Almanac*, New York：Oxford University Press，1949）中提出的。利奥波尔特认为，我们与土地的关系的理论将为人类社会下一步大规模的伦理改造提供可遵循的原则：

直到现在，只有哲学家才研究过伦理学的这种延伸，实际上是一种生态进化的过程。这一延伸均可用生态学和哲学术语来描述。伦理，从生态学方面讲，就是限制人在生存斗争中的行动自由；从哲学方面讲，就是区别合乎社会的行为

no parks;no farms;no uncharted wilderness.Every piece of countryside has keepers who have the right to farm it,if it is arable;or the obligation to look after it,if it is wild;and every piece of land is open to the people at large,provided they respect the organic processes which are going on there.

The central conception behind this view of the land is given by Aldo Leopold in his essay, "The Land Ethic" (*A Sand County Almanac*,New York:Oxford University Press,1949);Leopold be-lieves that our relationship to the land will provide the framework for the next great ethical transformation in the human community:

This extension of ethics,so far studied only by philosophers,is actually a process in ecological evolution. Its sequences may be described in ecological as well as in philosophical terms.An ethic,ecologically,is a limitation on freedom of action in the struggle for existence.An ethic, philosophically,is a differentiation of social from antisocial conduct.These are two definitions of one thing.The thing has its origin in the tendency of interdependent individuals or groups to evolve modes of cooperation.The ecologist calls these symbioses.Politics and economies are advanced symbioses in which the original free-for-all competition has been replaced,in part,by cooperative mechanisms with an ethical content....

All ethics so far evolved rest upon a single premise: that the individual is a member of a community of interdependent parts.His instincts prompt him to compete for his place in that community,but his ethics prompt him also to cooperate....

The land ethic simply enlarges the boundaries of the community to include soils,waters,plants,and animals,or

和反社会的行为。这是一件事物的两种定义。此事的根源就在于这样一种倾向：即相互依赖的个人和团体都想共同发展各种合作。生态学家称为共生现象。政治和经济是高级的共生现象，在此现象内原始的完全自由的竞争部分地已为含有一定伦理内容的合作机构所代替了……

全部伦理学发展到今天，它所依据的唯一前提是：个人是由各个相互依赖的部分所组成的社会的一名成员。他的本能驱使他为他所处的那个社会中的地位而竞争，但是他的道德观念也促使他与其他人相互合作……

土地伦理学简直扩大着社会的边界，把土壤、水、植物和动物都包括进去了，或一言以蔽之：整片土地……

根据这种伦理学的原则来看，不考虑土地本身的固有价值，公园和野营地一向被认为是人们娱乐的"天然场所"，是死的东西与不道德的行为。被农场主"所占有的"、为他们自己谋取高额利润的农场也被如此看待。如果我们继续把土地当作享乐的工具和经济利润的源泉，则我们的公园和野营地将变得更人工感、塑料感，更像迪士尼乐园了。而我们的农场就将变得越来越像工厂了。土地伦理学以独一无二的"乡村"这一概念取代了公园和公共野营地。

支持这一概念的一个实例是《生存的蓝图》一书。该书建议给传统的社区对一些河口三角洲和沼泽地以服务管理权。这些低洼地是鱼类、贝壳类动物的产卵地。这些产卵地形成占全部海洋捕获量 60％ 的食物链基地。只有珍惜这些产卵地并把它们视为生命链中的一环的人们才能恰如其分地进行管理。(The Ecologist, England: Penguin, 1972, p.41.)

日本护林人员居住的森林区为这一概念提供了另一例证。村落沿森林边缘发展蔓延；村民管理森林。合理采伐是他们的职责之一。凡是想去游览这片森林区的人都可以去，并可分享村民的欢乐：

collectively:the land....

Within the framework of this ethic,parks and campgrounds conceived as "pieces of nature" for people's recreation,without regard for the intrinsic value of the land itself,are dead things and immoral.So also are farms conceived as areas "owned" by the farmers for their own exclusive profit.If we continue to treat the land as an instrument for our enjoyment,and as a source of economic profit,our parks and camps will become more artificial,more plastic,more like Disneyland.And our farms will become more and more like factories.The land ethic replaces the idea of public parks and public campgrounds with the concept of a single countryside.

One example of support for this idea lies in the Blueprint for Survival,and the proposal there to give traditional communities stewardship over certain estuaries and marshes. These wetlands are the spawning grounds for the fish and shellfish which form the base of the food chain for 60 per cent of the entire ocean harvest,and they can only be properly managed by a group who respects them as a cooperating part in the chain of life.(The Ecologist,England:Penguin,1972,p.41.)

The residential forests of Japan provide another example. A village grows up along the edge of a forest;the villagers tend the forest.Thinning it properly is one of their responsibilities. The forest is available to anyone who wants to come there and partake in the process:

The farmhouses of Kurume-machi stand in a row along the main road for about a mile.Each house is surrounded by a belt of trees of similar species,giving the aspect of a single large forest.The main trees are located so as to produce a shelter-belt.In addition,these small forests are homes for birds,a device

在久留米町，在离主要公路约 1mi 处有一排农舍。每一农舍的四周环绕着一条相同树种的林带，举目远眺，林海茫茫，景色迷人。主要的树种形成防护林带。还有许多小片森林，它们是鸟儿的故乡，也是蓄水容器、炭薪和木材的源泉。由于有选择地进行砍伐，它们也是调节气候的手段。在有村民居住的森林中，气温冬暖夏凉。

应当指出，300 多年前建立起来的这些住人的森林区，由于护林村民小心翼翼地、有选择地合理采伐和轮番更新森林的结果，至今这些林区依然生机盎然，完好无损。(John L.Creech, "Japan—Like a National Park," *Yearbook of Agriclture* 1963, U.S.Department of Agriculture, pp.525 ~ 528.)

因此：

把所有的农场定义为公园。公众有权去农场；同时务使全区域内的公园都变成可耕耘的农场。

向小组、家庭和合作社授予乡村管理权，分片包干，各自负责。向管理者颁发土地租借证书，管理者可以自由地管理土地，并制定土地使用法规——如何使用小农场、森林、沼泽地、沙漠等。公众可以自由地去乡村观赏田园风光，徒步旅行，举行野餐，寻奇探幽，泛舟游览，只要他们遵守土地法规。据此方案，夏季，在靠近城市的农场田野里，每天都会有络绎不绝的野餐者。

for conserving water,a source of firewood and timber,which is selectively cut,and a means of climate control,since the temperature inside the residential forest is cooler in summer and warmer in winter.

It should be noted that these residential forests,established more than 300 years ago,are still intact as a result of the careful selective cutting and replacement program followed by the residents.(John L.Creech, "Japan—Like a National Park," *Yearbook of Agriculture* 1963,U.S.Department of Agriculture,pp.525-28.)

Therefore:

Define all farms as parks,where the public has a right to be;and make all regional parks into working farms.

Create stewardships among groups of people,families and cooperatives,with each stewardship responsible for one part of the countryside.The stewards are given a lease for the land,and they are free to tend the land and set ground rules for its use—as a small farm,a forest,marshland,desert,and so forth.The public is free to visit the land,hike there,picnic,explore,boat,so long as they conform to the ground rules.With such a setup,a farm near a city might have picnickers in its fields every day during the summer.

在每一个自然保护区内，我们设想有一定数量的住宅，**住宅团组**（37），可通往未铺砌路面的乡村小径——**绿茵街道**（51）……

通过各项城市政策，鼓励逐步形成下列界定城市特征的主要结构：

8. 亚文化区的镶嵌

9. 分散的工作点

10. 城市的魅力

11. 地方交通区域

Within each natural preserve,we imagine a limited number of houses—HOUSE CLUSTER(37)—with access on unpaved country lanes—GREEN STREETS(51)...

Through city policies,encourage the piecemeal formation of those major structures which define the city:

8. MOSAIC OF SUBCULTURES

9. SCATTERED WORK

10. MAGIC OF THE CITY

11. LOCAL TRANSPORT AREAS

模式8 亚文化区的镶嵌**

……城市最基本的结构是由城市土地与辽阔的乡村之间的关系给出的——**指状城乡交错**（3）。在大片的城市土地上，最重要的结构必须来自能在那里共存的各种不同的人群和亚文化区。

⊱⊰

现代城市的大同小异和千篇一律的特征，扼杀了丰富多彩的生活方式，并抑制着人的个性发展。

试比较一下人们分布于全市的三种可能的选择途径：

1.在一个庞杂的城市里，人们混居在一起而不必顾及他们各自的生活方式或文化。这似乎是丰富多彩的，其实不然。这种情况压制了各种有意义的多样性，可能存在的千差万别大多显示不出来，而却鼓励千人一面、千篇一律

8   MOSAIC OF SUBCULTURES**

...the most basic structure of a city is given by the relation of urban land to open country—CITY COUNTRY FINGERS(3).Within the swaths of urban land the most important structure must come from the great variety of human groups and subcultures which can coexist there.

⟡⟡⟡

The homogeneous and undifferentiated character of modern cities kills all variety of life styles and arrests the growth of individual character.

Compare three possible alternative ways in which people may be distributed throughout the city:

1.In the heterogeneous city,people are mixed together, irrespective of their life style or culture.This seems rich. Actually it dampens all significant variety,arrests most of the possibilities for differentiation,and encourages conformity. It tends to reduce all life styles to a common denominator. What appears heterogeneous turns out to be homogeneous and dull.

2.In a city made up of ghettos,people have the support of the most basic and banal forms of differentiation race or economic status.The ghettos are still homogeneous internally,and do not allow a significant variety of life styles to emerge.People in the

的风格。这倾向于把一切生活方式都简化成为一个公分母。而呈现出的庞杂性原来是一些单调划一和枯燥无味的东西。

庞杂的城市
The heterogeneous city

2. 在由民族聚居区构成的城市里，人们的立脚点是以平庸的种族或经济状况为基础的分片划块。这些民族聚居区在内部仍然是千篇一律的，不允许任何多样化的生活方式诞生。民族聚居区的居民通常被迫住在那里而无法脱身，他们与社会的其余部分隔绝孤立，不能发展自己的生活方式，而且常常不能容忍异己的生活方式。

由民族聚居区构成的城市
City of ghettos

3. 在一个由大量的规模较小的亚文化区构成的城市里，每一个亚文化区都占有一块容易识别的地方，并以一条无居民居住的地带作为边界与其他亚文化区分隔开来，这样，新的生活方式才能发展。人们可以选择他们愿意居住的亚文化区，还可以体验许多异己的生活方式。因为每一个亚文化区环境都在培养一种强烈的相互支持的共同的价值观念，所以，个人才能发展。

亚文化区的镶嵌
Mosaic of subcultures

ghetto are usually forced to live there,isolated from the rest of society,unable to evolve their way of life,and often intolerant of ways of life different from their own.

3.In a city made of a large number of subcultures relatively small in size,each occupying an identifiable place and separated from other subcultures by a boundary of nonresidential land,new ways of life can develop.People can choose the kind of subculture they wish to live in,and can still experience many ways of life different from their own.Since each environment fosters mutual support and a strong sense of shared values,individuals can grow.

This pattern for a mosaic of subcultures was originally proposed by Frank Hendricks.His latest paper dealing with it is "Concepts of environmental quality standards based on life styles," with Malcolm MacNair(Pittsburg,Pennsylvania :University of Pittsburgh,February 1969).The psychological needs which underlie this pattern and which make it necessary for subcultures to be spatially separated in order to thrive have been described by Christopher Alexander, "Mosaic of Subcultures," Center for Environmental Structure, Berkeley, 1968.The following statement is an excerpt from that paper.

I.

We are the hollow men,

We are the stuffed men.

Leaning together

Headpiece filled with straw.Alas.

....

亚文化区的镶嵌这一模式最初是由弗兰克·亨德里克斯提出的。他和马尔科姆·麦克内厄合写的一篇最新论文《关于以生活方式为基础的环境质量标准的概念》（Pittsburg，Pennsylvania：University of Pittsburgh，February，1969）也涉及了这一模式。克里斯托弗·亚力山大说明了构成这一模式基础的心理需求，他并认为，心理需求使得亚文化区必然不断繁荣且在空间上分散。（"Mosaic of Subcultures，"Center for Environmental Structure，Berkeley，1968.）下面是这篇论文的摘录。

Ⅰ.

我们的精神空虚，

但却都自命不凡。

人云亦云岂应该？

个个都是木脑袋。呜呼！

....

无形之骸，无色之影，

四肢无力，脸无表情；

……

T.S. 埃里奥特

（译注：T.S.Eliot；1888—1965，英国诗人兼评论家）

生活在大城市的许多人性格脆弱。实际上，大都市区似乎都有这一特征：城市居民与处在比较简陋和坎坷的环境中的人相比，性格上更加脆弱。无独有偶，这种性格脆弱性是大都市区的另一种更加明显的特征——居民生活的千篇一律和缺乏多样性——的对应物。当然，性格的脆弱和缺乏多样性是相辅相成的，正如一个硬币的两面：相对而言，这就是形成人们个性无差别的社会条件。性格只能产生于自我，它具有鲜明的特色和完整性：根据定义，千人一面的社会产生不出具有鲜明性格的个人的社会。

Shape without form,shade without color,

Paralyzed force,gesture without motion;

...

—T.S.Eliot

Many of the people who live in metropolitan areas have a weak character.In fact,metropolitan areas seem almost marked by the fact that the people in them have markedly weak character,compared with the character which develops in simpler and more rugged situations.This weakness of character is the counterpart of another,far more visible feature of metropolitan areas:the homogeneity and lack of variety among the people who live there.Of course,weakness of character and lack of variety,are simply two sides of the same coin:a condition in which people have relatively undifferentiated selves.Character can only occur in a self which is strongly differentiated and whole:by definition,a society where people are relatively homogeneous,is one where individual selves are not strongly differentiated.

Let us begin with the problem of variety.The idea of men as millions of faceless nameless cogs pervades 20th century literature.The nature of modern housiug reflects this image and sustains it.The vast majority of housing built today has the touch of mass-production.Adjacent apartments are identical. Adjacent houses are identical.The most devastating image of all was a photograph published in *Life* magazine several years ago as an advertisement for a timber company:The photograph showed a huge roomful of people;all of them had exactly the same face.The caption underneath explained:In honor of the chairman's birthday,the shareholders of the corporation are wearing masks made from his face.

These are no more than images and indications.... But

现在让我们先从多样性问题谈起。把世界上千千万万个名字不详或无名字的人视为芸芸众生的这样一种观念充斥于 20 世纪的文学。现代的住宅建筑在本质上反映并保持这种形象。今天业已建成的大多数住宅建筑与工厂化的成批生产结下了不解之缘。相邻的公寓一模一样，相邻的住宅一模一样，千篇一律，毫无特色。数年前，某一木材公司在《生活》杂志上刊登了一幅广告照片，其中的人物形象含有最辛辣的挖苦味道。从照片中可以看出：高朋满座，可是他们的面孔恍若一人。照片下面的解释语是：为喜庆木材公司董事长寿辰，各股东个个头戴假面具，假面具的脸形即董事长的脸形。

这些至多不过是照片中的形象和说明而已……但是，令人可怕的芸芸众生的千人一面的脸谱究竟从何而来？为什么卡夫卡、凯默思和萨特尔言恳词切的话都句句说到了我们的心坎上？

许多作者已经详尽地答复了这一问题（David Riesman in The Lonely Crowd；Kurt Goldstein in The Organism；Max Wertheimer inThe Story of Three Days；Abraham Maslow in Motivation and Personality：Rollo May in Man's Search for Himself, etc.）。他们的回答都集中在下面的一个要点上：虽然一个人可能和他的邻居具有不同的综合人格，但直到他有了一个强有力的中心，他的独特性已经凝聚而又坚强有力时，才和他的邻居真正有所区别。目前，在大都会内，看来情况并非如此。尽管人们在细微末节方面有所不同，可是他们却永远相互依赖，总是竭力避免惹人不悦，害怕按自己真实的个性为人。

人们以一种"如此办事才能办得成"的世故的处世方式取代了"相信如此办才对"的行事方式。妥协，就是和他人和睦相处，与委员会的精神不背道而驰，以及妥协所意味的一切——在大都会区，这些性格都体现在成熟的、

where do all the frightening images of sameness,human digits,and human cogs,come from?Why have Kafka and Camus and Sartre spoken to our hearts?

Many writers have answered this question in detail— [David Riesman in The *Lonely Crowd*;Kurt Goldstein in *The Organism*;Max Wertheimer in *The Story of Three Days*;Abraham Maslow in *Motivation and Personality*;Rollo May in *Man's Search for Himself*,etc.].Their answers all converge on the following essential point:Although a person may have a different mixture of attributes from his neighbour,he is not truly different,until he has a strong center,until his uniqueness is integrated and forceful.At present,in metropolitan areas,this seems not to be the case.Different though they are in detail,people are forever leaning on one another,trying to be whatever will not displease the others,afraid of being themselves.

People do things a certain way "because that's the way to get them done" instead of "because we believe them right." Compromise,going along with the others,the spirit of committees and all that it implies—in metropolitan areas,these characteristics have been made to appear adult,mature,well-adjusted.But euphemisms do little to disguise the fact that people who do things because that's the way to get along with others,instead of doing what they believe in,do it because it avoids coming to terms with their own self,and standing on it,and confronting others with it.It is easy to defend this weakness of character on the grounds of expediency.But however many excuses are made for it,in the end weakness of character destroys a person;no one weak in character can love himself.The self-hate that it creates is not a condition in which

随机应变的成年人身上。可是，委婉已无助于掩饰这样一个事实：人们所采取的处世行事的方式是因为这种方式便于自己与他人相处，而不是他们相信自己的所作所为是正确的。他们这样做的目的是避免发生瓜葛，明哲保身而已。立足于权术，为这种性格上的弱点辩解是很容易的。虽然有种种借口，毕竟性格脆弱会把一个人毁灭掉。没有一个性格脆弱的人会珍爱自己。性格脆弱所造成的自我怨恨并不是使人成为一个完人的条件。

相反，一个完人的性格总是显示出他自己的本质，清晰可见的、外向的、引人注目的、开朗的、人人都可以理解的。他不害怕他自己的个性；他坚持他自己原来的个性；他就是他而非别人，他为自己的个性而骄傲；他承认自己的缺点，并努力加以改正。他为自己的个性而感到自豪和自信。

但是，要让你在外表的掩盖下显现出自我绝非易事。而要你按照别人的意志生活，使你的真正自我屈从于社会的习惯势力，使你在不会自我完善的违心的需求中掩饰自己倒是轻而易举的。

因此，显而易见的是，多样性、性格和发现你自己的自我，这三者是相互紧密地交织在一起的。在人们能够发现其自我的社会中，他们的性格一定是千差万别且十分坚强的。在人们发现其自我而有重重困难的社会里，他们一定会呈现出千人一面的模样，缺乏多样性，他们的性格必定十分脆弱。

如果今天，在大都市内人们的性格脆弱确是事实，并且我们想要为此做点补救工作，那么，我们必须做的第一件事情就是要弄清楚大城市是如何造成这种后果的。

Ⅱ.

大都市是怎样创造出人们难以发现自我个性的那些条

a person can become whole.

By contrast,the person who becomes whole,states his own nature,visibly,and outwardly,loud and clear,for everyone to see. He is not afraid of his own self;he stands up for what he is;he is himself,proud of himself,recognising his shortcomings,trying to change them,but still proud of himself and glad to be himself.

But it is hard to allow that you which lurks beneath the surface to come out and show itself.It is so much easier to live according to the ideas of life which have been laid down by others,to bend your true self to the wheel of custom,to hide yourself in demands which are not yours,and which do not leave you full.

It seems clear,then,that variety,character,and finding your own self,are closely interwoven.In a society where a man can find his own self,there will be ample variety of character,and character will be strong.In a society where people have trouble finding their own selves,people will seem homogeneous,there will be less variety,and character will be weak.

If it is true that character is weak in metropolitan areas today,and we want to do something about it,the first thing we must do,is to understand how the metropolis has this effect.

Ⅱ.

How does a metropolis create conditions in which people find it hard to find themselves?

We know that the individual forms his own self out of the values,habits and beliefs,and attitudes which his society presents him with.[George Herbert Mead,Mind,*Self and Society*.] In a metropolis the individual is confronted by a vast tableau of different values,habits and beliefs and attitudes.Whereas,in a primitive society,he had merely to integrate the traditional beliefs(in a sense,there was a self already there for the asking),in

件的呢？

我们都知道，任何一个人形成自己的个性是由于社会向他提供种种价值观念、习惯、信仰和观点的结果（George Herbert Mead，mind，*Self and Society*）。在大都市每个人都面临着不同的价值观念、习惯、信仰和观点所交织成的广泛的人世画面。就原始社会而论，每个人只需把自己同传统的信仰结合在一起（在某种意义上说，如果你愿意要个性的话，那时已存在个性了）；在现代社会里，每个人由于他周围的价值观念极度混乱，不得不为他自己捏造出一个自我来。

如果你每天不是无所事事，你就会遇到一些背景略为不同的人，他们对你的所作所为反应是各不相同的，甚至当你的行为始终如一时反应也不会相同，那么，这种情况就会越发令人迷惑不解。你有可能成为你自身的佼佼者，无忧无虑；你确信你现在的一切，你确信你现在正在做的一切，可是这种可能性会迅速化为泡影。人们永恒地面对着变幻莫测的社会，不再依靠自己产生力量；他们越来越仰人鼻息，察言观色，见机行事，投他人之所好，避他人之所恶。见人眉开眼笑，就说个滔滔不绝；见人不理不睬，就闭口不谈。在那样一个人世间，任何人要孕育一种内在的力量是难以实现的。

一旦我们接受如下的思想——个性的形成是一个社会过程，那么，不言而喻，一个坚强有力的社会个性的形成取决于周围社会秩序的力量。当各种观点、价值观念、信仰和习惯在大都市中广为扩散和混合掺杂在一起时，在这些条件中成长起来的人们也将扩散和混杂在一起，这一点几乎是不可避免的。脆弱的性格是现在大都市社会的直接产物。

这一论据已由玛格丽特·米德以令人信服的措辞进行了概括（*Culture，Change and Character Structure*）。一

modern society each person has literally to fabricate a self,for himself,out of the chaos of values which surrounds him.

If,every day you do something,you meet someone with a slightly different background,and each of these people's response to what you do is different even when your actions are the same,the situation becomes more and more confusing.The possibility that you can become secure and strong in yourself, certain of what you are,and certain of what you are doing,goes down radically.Faced constantly with an unpredictable changing social world,people no longer generate the strength to draw on themselves;they draw more and more on the approval of others;they look to see whether people are smiling when they say something,and if they are,they go on saying it,and if not,they shut up.In a world like that,it is very hard for anyone to establish any sort of inner strength.

Once we accept the idea that the formation of the self is a social process,it becomes clear that the formation of a strong social self depends on the strength of the surrounding social order.When attitudes,values,beliefs and habits are highly diffuse and mixed up as they are in a metropolis,it is almost inevitable that the person who grows up in these conditions will be diffuse and mixed up too.Weak character is a direct product of the present metropolitan society.

This argument has been summarized in devastating terms by Margaret Mead [*Culture,Change and Character Structure*].A number of writers have supported this view empirically:Hartshorne,H.and May,M.A.,*Studies in the Nature of Character*,New York,Macmillan,1929;and "A Summary of the Work of the Character Education Inquiry," *Religious Education*,1930,Vol.25,607-619 and 754-762. "Contradictory

些作者以自己的切身经验表示支持这一观点（Hartshorne，H.and May，M.A.，*Studies in the Nature of Character*，New York，Macmillan，1929；and "A Summary of the Work of the Character Education Inquiry，" *Religious Education*，1930，Vol.25，607～619 and 754～762）。"对于处在变化多端的环境中的、只对成人负责的孩子提出自相矛盾的要求，不仅妨碍他前后一贯的性格的形成，而且实际上，为了克服这种性格上的前后矛盾性，他就要以自己内心的平静和自尊作为代价。"……

但事情到此并未结束。迄今为止，我们已经看到大都会的这种扩散是如何制造出脆弱性格的。但当这种扩散变得十分明显时，就会形成一种特殊的表面一致性。当许多种颜色在密密麻麻的、微小的、凹凸不平的点和面上混合时，其总效应呈灰色。这种灰色以其独特的方式促使人形成脆弱性格。

在一个有许多声音和许多价值观念的社会里，人们绝不肯放弃他们共有的那些为数不多的东西。因此，玛格丽特·米德引用了一段话："现在存在着一种倾向，它把一切价值观念都简化成为美元、学校等级之类简单的衡量尺度或某种其他的简易定量测量的东西了。因此，在许多分门别类的文化价值观念中，根本不能用同一尺度来衡量的东西可以容易地——虽说只是表面地——被调和起来了。"约瑟夫·T. 克拉珀（*The Effects of Mass Communication*，Free Press，1960）也提道：

"芸芸众生的社会不仅造成了使人难以发现自己个性的困惑而迷茫的环境，而且造成了一团混乱，在这种混乱中人们所面临的情况是无法实现多样性——多样性已成为一句充满愚妄和伤感的空洞辞藻了，这一点是最明显不过的。"

demands made upon the child in the varied situations in which he is responsible to adults,not only prevent the organisation of a consistent character,but actually compel inconsistency as the price of peace and self-respect."...

But this is not the end of the story.So far we have seen how the diffusion of a metropolis creates weak character.But diffusion,when it becomes pronounced,creates a special kind of superficial uniformity.When many colors are mixed,in many tiny scrambled bits and pieces,the overall effect is grey.This greyness helps to create weak character in its own way.

In a society where there are many voices,and many values,people cling to those few things which they all have in common.Thus Margaret Mead(op.cit.): "There is a tendency to reduce all values to simple scales of dollars,school grades,or some other simple quantitative measure,whereby the extreme incommensurables of many different sets of cultural values can be easily,though superficially,reconciled." And Joseph T.Klapper[*The Effects of Mass Communication*,Free Press,1960]:

"Mass society not only creates a confusing situation in which people find it hard to find themselves—it also...creates chaos,in which people are confronted by impossible variety— the variety becomes a slush,which then concentrates merely on the most obvious."

...It seems then,that the metropolis creates weak character in two almost opposite ways;first,because people are exposed to a chaos of values;second,because they cling to the superficial uniformity common to all these values.A nondescript mixture of values will tend to produce nondescript people.

……于是，看起来似乎是大都市以两种几乎对立的方式形成人的脆弱性格；第一，因为人们暴露在价值观念的混乱之中；第二，因为他们执着于这些价值观念所共有的表面一致性。价值观念不可名状的混合将趋向于产生无特征可言的人。

Ⅲ.

很明显，解决这一问题有许多途径。其中一些解决办法会是个体性质的；另一些解决办法将涉及许多不同的社会过程，至少包括教育、工作、娱乐和家庭。现在我来描述一种特殊的解决办法，它涉及大都市大规模的社会组织。

这一解决方案说明如下：大都市必须包含大量的彼此不同的亚文化区。每个亚文化区都是非常有条不紊、井然有序的，以其自身的价值观念勾勒出鲜明的特征，并与其他的亚文化区泾渭分明地一一区别开来。虽然这些亚文化区必须是轮廓清晰的、各具特色的、分散的，但一定不是封闭式的；它们必须是相互容易沟通的，所以任何人都可以十分方便地从一个亚文化区迁往另一个亚文化区，并在他认为最适合他的亚文化区定居下来。

这种解决方案所依据的是下列两种假设：

1. 如果一个人处在这样一种社会环境里：他能从周围的人们和价值观念中为他的个性发展获得支持的力量，那么，他一定能发现自己的个性，从而成为一个性格坚强的人。

2. 为了发现他自己的个性，他也需要生活在另一种社会环境里：在那里许多可能彼此不同的价值观念体系会明确地得到承认和尊重。他尤其需要许多不同的选择，以使他不会把自己的本性引入歧途，使他能理解存在着各种不同的人，并能从中发现与自己的价值观念和信仰极为相符的一些人。

……一种心理机制潜伏在人们的需要之中：即他们需要有像他们自己那样的外界文化。马斯洛已经指出，自我

Ⅲ.

There are obviously many ways of solving the problem. Some of these solutions will be private.Others will involve a variety of social processes including,certainly,education, work,play,and family.I shall now describe one particular solution,which involves the large scale social organisation of the metropolis.

The solution is this. *The metropolis must contain a large number of different subcultures,each one strongly articulated,with its own values sharply detineated,and sharply distinguished from the others.But though these subcultures must be sharp and distinct and separate,they must not be closed;they must be readily accessible to one another,so that a person can move easily from one to another,and can settle in the one which suits him best.*

This solution is based on two assumptions:

1.A person will only be able to find his own self,and therefore to develop a strong character,if he is in a situation where he receives support for his idiosyncrasies from the people and values which surround him.

2.In order to find his own self,he also needs to live in a milieu where the possibility of many different value systems is explicitly recognized and honored.More specifically,he needs a great variety of choices,so that he is not misled about the nature of his own person,can see that there are many kinds of people,and can find those whose values and beliefs correspond most closely to his own.

...one mechanism which might underly people's need for an ambient culture like their own:Maslow has pointed out that the process of self actualisation can only start after other

实现（其中包括发现自己的个性等——译注）只有在其他各种需要诸如食物、爱情和安全已经得到满足之后才能开始（*Motivation and Personality*，pp.84～89）。现在，形形色色的人物在市内的某一地区越是混杂在一起，就越无法预测在你住宅附近出现的陌生人，你就越会感到害怕和不安全。在洛杉矶和纽约，安全问题竟然严重到如此地步：许多人家经常闭户锁着门窗；母亲派 15 岁的女儿去拐角处的邮箱投信就会感到提心吊胆。人们害怕周围都是素不相识的人；素不相识的人是危险的。但是，只要这种恐惧心理成为一个悬而未决的问题，他们就无法再过安安稳稳的太平日子。只有克服了这种恐惧心理，自我实现才能发生。而且，只有当人们在熟悉的土地上，在自己的同类中融洽相处，并且相互信任、彼此了解对方的待人接物的习惯和风土人情时，自我实现才会依次发生。

……无论如何，如果我们为了满足第一种假设的要求而鼓励各种不同的亚文化区相继出现，那么我们肯定鼓励这些亚文化区成为部落的或封闭式的。这与大城市如此吸引人的品质背道而驰。因此，人们很容易从一个亚文化区搬迁到另一个亚文化区，并选择最适合他们口味的亚文化区定居下来；他们在一生中的任何时刻想要迁居，都一定能如愿以偿。事实上，如果有必要的话法律必须保证每个人都能自由地迁往任何一个亚文化区。

Ⅳ.

看来非常明显，大都市应当包含大量的、相互沟通的亚文化区。但是，为什么这些亚文化区在空间上应当是分散的，对此一些人抱有偏见。他们很容易地争辩说，这些亚文化区可以并且应该在同一空间内共处，因为创造出各种文化的重要联系就是人们之间的联系。

我认为，这种观点如果提出来了，无疑是彻头彻尾错误的。我现在要提供论据来表明，亚文化区是生态学领域

needs,like the need for food and love,and security,have already been satisfied.[*Motivation and Personality*,pp.84-89.]Now the greater the mixture of kinds of persons in a local urban area,and the more unpredictable the strangers near your house,the more afraid and insecure you will become.In Los Angeles and New York this has reached the stage where people are constantly locking doors and windows,and where a mother does not feel safe sending her fifteen year old daughter to the comer mailbox.People are afraid when they are surrounded by the unfamiliar;the unfamiliar is dangerous.But so long as this fear is an unsolved problem,it will override the rest of their lives. Self-actualisation will only be able to happen when this fear is overcome;and that in turn,can only happen,when people are in familiar territory,among people of their own kind,whose habits and ways they know,and whom they trust.

...However,if we encourage the appearance of distinct subcultures,in order to satisfy the demands of the first assumption,*then we certainly do not want to encourage these subcultures to be tribal or closed.* That would fly in the face of the very quality which makes the metropolis so attractive.It must be possible,therefore,for people to move easily from one subculture to another,and for them to choose whichever one is most to their taste;and they must be able to do all of this at any moment in their lives.Indeed,if it ever becomes necessary,the law must guarantee each person freedom of access to every subculture....

Ⅳ.

It seems clear,then,that the metropolis should contain a large number of mutually accessible subcultures.But why should those subcultures be separated in space.Someone with an aspatial bias could easily argue that these subcultures

的事；如果各具特色的亚文化区，就总体而言在空间上是分散的，它们才能保持其特色而存在下去。

第一，无疑，来自不同亚文化区的人们实际上要求在他们各自的环境中有与众不同的东西。亨德里克斯早已阐明了这一观点。年龄不同、爱好各异、对家庭注重程度不一、民族背景相差悬殊的人们，需要不同的住宅，他们需要住宅周围有不同的户外环境，而最重要的是，他们需要门类齐全的社区服务行业。如果这些服务行业对顾客做到胸中有数，沿着有鲜明特色的亚文化区方向发展下去，就会变成高度专门化的行业了。如果同一个亚文化区的顾客高度集中地生活在一起，服务行业才能对他们做到了如指掌。想骑马飞奔的人都需要林间或林边宽阔的跑马道；想购买德国食品的德国人都会聚集靠拢过来，就像他们聚集在纽约的德人街一样；上了年纪的老人在小型疗养所附近，为了与少量的来往行人和车辆造成的交通污染作斗争，或许需要可供休憩的公园；单身汉或许需要光顾快餐小吃店；每天早晨想去做正统派教友弥撒的美国人都会拥簇在美国教堂的四周；街上的行人都会围拢在百货商店和集会场所；带领着幼儿的人们可能会集合在地方托儿所附近或露天游戏场上。

由此可见，不同的亚文化区都需要有各自的活动，需要有各自的环境。但是，为了允许集中必要的活动，亚文化区不仅在空间上需要集中，而且必须这样来集中：一个亚文化区的特色不会冲淡另一亚文化区的特色。的确，根据这一观点来看，亚文化区不仅在内部空间上需要集中，而且在外部空间上需要相互分散……

我们的引证到此为止。原文的其余部分为亚文化区必须在空间上分散这一观点提供了经验性的论证。而在本书中，我们把这一观点列为另一模式的一部分。我们

could,and should,coexist in the same space,since the essential links which create cultures are links between people.

I believe this view,if put forward,would be entirely wrong. I shall now present arguments to show that the articulation of subcultures is an ecological matter;that distinct subcultures will only survive,as distinct subcultures,if they are physically separated in space.

First,there is no doubt that people from different subcultures actually require different things of their environment. Hendricks has made this point clearly.People of different age groups,different interests,different emphasis on the family,different national background,need different kinds of houses,they need different sorts of outdoor environment round about their houses,and above all,they need different kinds of community services.These services can only become highly specialised,in the direction of a particular subculture,if they are sure of customers.They can only be sure of customers if customers of the same subculture live in strong concentrations. People who want to ride horses all need open riding;Germans who want to be able to buy German food may congregate together,as they do around German town,New York;old people may need parks to sit in,less traffic to contend with,nearby nursing services;bachelors may need quick snack food places;Armenians who want to go to the orthodox mass every morning will cluster around an Armenian church;street people collect around their stores and meeting places;people with many small children will be able to collect around local nurseries and open play space.

This makes it clear that different subcultures need their own activities,their own environments.But subcultures not only

以切身的经验，详尽地阐述了它的论据，见**亚文化区边界**（13）。

因此：

为充实城市的文化区和亚文化区而勤勤恳恳工作，尽可能把城市划分成数量众多的、小型的、彼此截然不同的亚文化镶嵌区。每一个亚文化区都有它自己的空间范围，并且每个亚文化区都有权创造它自己的不拘一格的生活方式。务必明白，亚文化区是小规模的，而且要小得足以使每个人都能到与之相邻的生活方式丰富多彩的亚文化区去。

千百个不同的亚文化区

亚文化区边界

&⊂⊃

我们认为，最小的亚文化区直径不要大于 150ft（1ft=0.3m）；最大的亚文化区直径为 0.25mi——**7000 人的社区**（12），**易识别的邻里**（14），**住宅团组**（37）。为了保证每个亚文化区的生活方式能够自由地发展，不受相邻的亚文化区的生活方式的抑制，在毗连的亚文化区之间的无居民地带，建立实质性的边界是极端重要的——**亚文化区边界**（13）……

need to be concentrated in space to allow for the concentration of the necessary activities.They also need to be concentrated so that one subculture does not dilute the next:indeed,from this point of view they not only need to be internally concentrated—but also physically separated from one another....

We cut the quote short here.The rest of the original paper presents ernpirical evidence for the need to separate subcultures spatially,and—in this book—we consider that as part of another pattern.The argument is given,with empirical details,in SUBCULTURE BOUNDARY(13).

Therefore:

Do everything possible to enrich the cultures and subcultures of the city,by breaking the city,as far as possible, into a vast mosaic of small and different subcultures, each with its own spatial territory,and each with the power to create its own distinct life style.Make sure that the subcultures are small enough,so that each person has access to the full variety of life styles in the subcultures near his own.

ഐരുബ

We imagine that the smallest subcultures will be no bigger than 150 feet across;the largest perhaps as much as a quarter of a mile—COMMUNITY OF 7000(12),IDENTIFIABLE NEIGHBORHOOD(14),HOUSE CLUSTER(37).To ensure that the life styles of each subculture can develop freely,uninhibited by those which are adjacent,it is essential to create substantial boundaries of nonresidential land between adjacent subcultures—SUBCULTURE BOUNDARY(13)...

模式9 分散的工作点**

9   SCATTERED WORK**

...this pattern helps the gradual evolution of MOSAIC OF SUBCULTURES(8),by placing families and work together,and so intensifying the emergence of highly differentiated subcultures, each with its individual character.

❧❀❧

The artificial separation of houses and work creates intolerable rifts in people's inner lives.

In modern times almost all cities create zones for "work" and other zones for "living" and in most cases enforce the separation by law.Two reasons are given for the separation.First,the workplaces need to be near each other,for commercial reasons.Second,workplaces destroy the quiet and safety of residential neighborhoods.

But this separation creates enormous rifts in people's emotional lives.Children grow up in areas where there are no men,except on weekends;women are trapped in an atmosphere where they are expected to be pretty,unintelligent housekeepers;men are forced to accept a schism in which they spend the greater part of their waking lives "at work,and away from their families" and then the other part of their lives "with their families,away from work."

Throughout,this separation reinforces the idea that work is a toil,while only family life is "living" —a schizophrenic view which creates tremendous problems for all the members of a family.

……本模式对逐步发展**亚文化区的镶嵌**（8）是有裨益的。它把家庭和工作点安排在一起，以便加速各具个性的、彼此迥然不同的亚文化区蓬勃出现。

୫୦୯ଓ

住宅和工作之间的人为分离，造成人们精神生活中无法忍受的创伤。

在现代，几乎所有的城市都分为"工作区"和"生活区"，并在极大多数情况下，通过法律手段，加强这种分离现象。分离的理由有二：第一，由于商业上的原因，工作区需要相互靠近。第二,工作区破坏附近居民的宁静和安全。

工作区的集中和隔离……导致死气沉沉的邻里
Concentration and segregation of work...lead to dead neighborhoods

但是，这种分离造成人们精神生活的巨大创伤。孩子们在没有大人照料的环境中成长，只有周末是例外；妇女们陷入这样一种窘境：一方面她们想长得俊俏可爱，另一方面又不得不成为平庸的家庭主妇；男人们被迫两地往返奔波：当他们醒着的时候，把大部分宝贵时间花在"离家上班和工作上"，而另一部分时间花在"下班回家以及与家人团聚上"。

In order to overcome this schism and reestablish the connection between love and work,central to a sane society, there needs to be a redistribution of all workplaces throughout the areas where people live,in such a way that children are near both men and women during the day,women are able to see themselves both as loving mothers and wives and still capable of creative work,and men too are able to experience the hourly connection of their lives as workmen and their lives as loving husbands and fathers.

What are the requirements for a distribution of work that can overcome these problems?

1.Every home is within 20-30 minutes of many hundreds of workplaces.

2.Many workplaces are within walking distance of children and families.

3.Workers can go home casually for lunch,run errands,work half-time,and spend half the day at home.

4.Some workplaces are in homes;there are many opportunities for people to work from their homes or to take work home.

5.Neighborhoods are protected from the traffic and noise generated by "noxious" workplaces.

The only pattern of work which does justice to these requirements is a pattern of scattered work:a pattern in which work is strongly decentralized.To protect the neighborhoods from the noise and trafic that workplaces often generate,some noisy work places can be in the boundaries of neighborhoods,communities and subcultures—see SUBCULTURE BOUNDARY(13);others,not noisy or noxious, can be built right into homes and neighborhoods.In both

这种分离状态会从各个方面强化如下的看法：工作是苦役，而只有家庭生活才是"充满生气的"——这是一个神志不清、精神失常的人的看法，它会给一个家庭的全体成员带来许多棘手的问题。

为了克服这种分离状况，重建工作和爱情之间的联系——这对一个心理健全的社会来说是至关重要的——需要在所有的居民区重新分布工作点，重新分布后应当实现：孩子们在白天可以和父母形影不离地生活在一起；妇女会意识到她们既是贤妻良母，又是进行创造性劳动的能手；男人也会时时刻刻体验到他们既是尊妻爱子的丈夫，又是能工巧匠。

合理分布工作点能解决生活和工作分离所带来的种种问题，其具体要求到底是哪些呢？

1. 每一家庭离开数以百计的工作点只有 20 ～ 30min 的路程。

2. 许多工作点离家近，儿童和家庭其他成员都可以步行去。

3. 家庭和工作点距离不远。工人能偶尔回家吃午饭，出差办事，半日工作，半日在家消磨时间。

4. 一些工作点设在家里；人们有许多机会可以在家里工作或取活到家里干。

5. 防止邻里遭到交通的干扰和"有害的"工作点产生的噪声污染。

符合上述要求的唯一的工作模式就是分散的工作点模式：这是一个把工作高度分散进行的模式。为了使邻里免遭工作点的交通干扰和噪声污染，一些扰民的工作点可设置在邻里、社区和亚文化区的边界内——参阅**亚文化区边界**（13），不是扰民的或有害的工作点可以设在家里和邻里内。在上述两种情况中，极为重要的一点是：**每个家庭离开数以百计的工作点只有几分钟的路程**。随后，每个家庭

cases,the crucial fact is this:*every home is within a few minutes of dozens of workplaces*.Then each household would have the chance to create for itself an intimate ecology of home and work:all its members have the option of arranging a workplace for themselves close to each other and their friends.People can meet for lunch,children can drop in,workers can run home. And under the prompting of such connections the workplaces themselves will inevitably become nicer places,more like homes,where life is carried on,not banished for eight hours.

This pattern is natural in traditional societies,where workplaces are relatively small and households comparatively self-sufficient.But is it compatible with the facts of high technology and the concentration of workers in factories?How strong is the need for workplaces to be near each other?

The main argument behind the centralization of plants,and their gradual increase in size,is an economic one.It has been demonstrated over and again that there are economies of scale in production,advantages which accrue from producing a huge number of goods or services in one place.

However,large centralized organizations are not intrinsic to mass production.There are many excellent examples which demonstrate the fact that where work is substantially scattered, people can still produce goods and services of enormous complexity.One of the best historical examples is the Jura Federation of watchmakers,formed in the mountain villages of Switzerland in the early 1870's.These workers produced watches in their home workshops,each preserving his independence while coordinating his efforts with other craftsmen from the surrounding villages.(For an account of this federation,see,for example,George

就会有机会为自身创造一种家庭和工作之间紧密融合的生态环境气氛：家庭中的全体成员都有权为自己选择工作点，以使他们彼此之间乃至和亲朋好友之间亲密接触。大家在用午餐时，可以碰头见面；孩子们能来串门儿；工人能赶回家去。由于鼓励各种交往，工作点本身就会不可避免地变成更加美好的地方，更像家庭似的，生活在那里继续进行着，在 8 小时工作时间内生活没有被淡忘。

这一模式在传统的社会里是很自然的：工作点相对较小，家家户户自给自足有余。但是这种情况与今天高度发达的技术和工厂中工人的集中这样一些事实是否相容呢？工作点相互靠拢这样一种需要的迫切程度如何呢？

在工厂集中化及其规模的逐渐扩大的背后所进行的争论，主要是经济上的争论。事实一再表明争论的焦点在于生产的经济规模的大小以及在一个地方生产大量产品或开发劳务中所取得的效益的高低。

可是，庞大的集权机构并不是大批量生产所固有的组织形式。许多出色的例子表明，在工作实际上分散进行的地方，人们仍能生产出结构复杂的产品，提供大量的劳务。最有说服力的历史例证之一就是瑞士钟表工人的吉拉联合会，该会于 19 世纪 70 年代初在瑞士的山村里成立。工人会员就在他们自己的家庭工作间里生产钟表。每个工人，当他与周围村庄上的手艺工人协调生产时，都保持住各自的独立性。关于这一联合会的来龙去脉，请参阅乔治·伍德科克所著的《无政府主义》一书（*George Woodcock*, Anarchism : A History of Libertarian Ideas and Movements, *Cleveland* : *Meridian Books*, 1962, pp.168 ～ 169）。

在我们的时代，雷蒙德·弗农已经指出，在纽约大都市的经济中，小型的、分散的工作点能更快地适应日新月异地变化着的市场的需求和供应状况，小工商企业会合在

Woodcock,Anarchism:A History of Libertarian Ideas and Movements,Cleveland:Meridian Books,1962,pp.168-169.)

In our own time,Raymond Vernon has shown that small, scattered workplaces in the New York metropolitan economy, respond much faster to changing demands and supplies,and that the degree of creativity in agglomerations of small businesses is vastly greater than that of the more cumbersome and centralized industrial giants.(See Raymond Vernon,*Metropolis* 1985,Chapter 7:External Economics.)

To understand these facts,we must first realize that the city itself is a vast centralized workspace and that all the benefits of this centralization are potentially available to every work group that is a part of the city's vast work community.In effect,the urban region as a whole acts to produce economies of scale by bringing thousands of work groups within range of each other.If this kind of "central-ization" is properly developed,it can support an endless number of combinations between small,scattered workgroups;and it can lend great flexibility to the modes of production. "Once we understand that modern industry does not necessarily bring with it financial and physical concentration,the growth of smaller centers and a more widespread distribution of genuine benefits of technology will,I think,take place" (Lewis Mumford,*Sticks and Stones*,New York,1924,p.216).

Remember that even such projects as complicated and seemingly centralized as the building of a bridge or a moon rocket,can be organized this way.Contracting and subcontracting procedures make it possible to produce complicated industrial goods and services by combining the efforts of hundreds of small firms.The Apollo project drew

一起，它们的创造性程度大大超过集中而又尾大不掉的大工业企业（Raymond Vernon，Metropolis 1985，Chapter 7：External Economics）。

为了理解这些事实，我们首先必须承认，城市本身就是一个巨大而又集中的工作空间。这种集中的全部实惠，每个工作小组（即城市广大的工作社区的一个组成部分）都有潜在的可能性去获取。实际上，市区作为一个整体来行动，成千上万个工作小组会合在一起，在经济规模上就会收到小中见大的经济效益。如果这种"集中"适当加以发展，它就能在多得不可胜数的、小型的、分散的工作组之间的联合方面发挥积极的作用，并能给予各种生产方式以极大的灵活性。"我认为，一旦我们懂得现代工业不必在资金和空间上集中的道理，则较小中心的发展和技术的真正成果更加广泛的传播才能成为现实。"（Lewis Mumford，*Sticks and Stones*，New York，1924，p.216.）

请记住，即使是建造桥梁或登月火箭这样一些极其复杂的、似乎很集中的工程项目，也可以按分散的办法来组织生产。通过签订承包合同和转包合同，联合成百上千的小公司通力合作，生产复杂的工业产品和提供劳务是完全可能的。阿波罗计划就吸引了 30000 多个独立的公司来生产复杂的登月宇宙飞船。

此外，有证据表明，许多承办多边合同业务的代理机构物色半自治的小企业。它们本能地知道，工作组越小，就越能自己管理自己，产品和劳务的质量就越好。（*Small Sellers and Large Buyers in American Industry*,Business Research Center,College of Business Administration,Syracuse University,New York,1961.）

让我们现在强调一下：我们并不认为工作点的分散应优先于尖端技术。我们认为两者是可以相容的，并行不悖的：人们要求感兴趣的创造性工作和现代的精湛技术结合

together more than 30,000 independent firms to produce the complicated spaceships to the moon.

Furthermore,there is evidence that the agencies which set up such multiple contracts look for small,semi-autonomous firms.They know instinctively that the smaller,more self-governing the group,the better the product and the service(*Small Sellers and Large Buyers in American Industry*,Business Research Center,College of Business Administration,Syracuse University,New York,1961).

Let us emphasize:we are not suggesting that the decentralization of work should take precedence over a sophisticated technology.We believe that the two are compatible:it is possible to fuse the human requirements for interesting and creative work with the exquisite technology of modern times.It is possible to make television sets,xerox machines and IBM typewriters, automobiles,stereo sets and washing machines under human working conditions.We mention in particular the xerox and IBM typewriters because they have played a vital role for us,the authors of this book. We could not have made this book together,in the communal way we have done,without these machines:and we consider them a vital part of the new decentralized society we seek.

Therefore:

Use zoning laws,neighborhood planning,tax incentives, and any other means available to scatter workplaces throughout the city.Prohibit large concentrations of work, without family life around them.Prohibit large concentrations of family life,without workplaces around them.

是可能办到的。在富有人情味的工作条件下，生产电视机、静电复印机、IBM 打字机、小轿车、立体声录音机和洗衣机是切实可行的。在这里我们要特别提一笔静电复印机和IBM 打字机，因为它们对我们（即本书的作者）起到了十分重要的作用。如果我们没有利用上述机器，互通信息，我们是无法共同完成此书的。我们认为，这些机器是我们寻求的一种新型的、分散权力的社会所不可或缺的组成部分。

南斯拉夫辛曼的一个小工厂；工作组正在生产拣谷机，这是工人们自己决定生产并在市场上销售的一项产品

A small factory in Zemin,Yugoslavia;the work group is building a com picking machine,an item they themselves decided to produce and sell in the marketplace

因此：

利用分区的法律、邻里规划、税收刺激和其他一切可以利用的手段，在全市范围内分散工作点。禁止工作点的过分集中，避免和家庭生活脱节。禁止家庭生活点的过分集中，避免和工作点脱节。

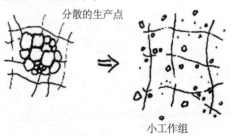

分散的生产点

小工作组

The scattered work itself can take a great variety of forms. It can occur in belts of industry,where it is essential for an industry to occupy an acre or more between subcultures— SUBCULTURE BOUNDARY(13),INDUSTRIAL RIBBON(42);it can occur in work communities,which are scattered among the neighborhoods—NEIGHBORHOOD BOUNDARY(15),WORK COMMUNITY(41);and it can occur in individu-al workshops,right among the houses—HOME WORKSHOP(157).The size of each workplace is limited only by the nature of human groups and the process of self-governance.It is discussed in detail in SELF-GOVERNING WORKSHOPS AND OFFICES(80)...

分散的工作点应采取灵活多样的形式。它可出现于工业带，对于工业来说，在亚文化区——**亚文化区边界**（13），**工业带**（42）——之间，占据一英亩或更多的土地是颇为重要的；它可出现在分散于邻里的工作社区内——**邻里边界**（15）、**工作社区**（41）；它可以出现于不折不扣地设在家里的个体工间——**家庭工作间**（157）。每个工作点的规模大小只受工作组的性质及其自我管理的进程的限制。这一点已在**自治工作间和办公室**（80）中做了详细的说明……

模式10　城市的魅力

　　……继**亚文化区的镶嵌**（8）之后，或许，城市最重要的结构特征就是城市生活最密集的那些中心的模式了。这些中心的千姿百态有助于形成亚文化区的镶嵌；如果每一个中心都处于若干指状带的天然交会点上，也同样有助于形成**指状城乡交错**（3）。本模式由路易斯·莱西昂纳罗首先写成，标题为《300000人的城市商业区》。

10   MAGIC OF THE CITY

...next to the MOSAIC OF SUBCULTURES(8),perhaps the most important structural feature of a city is the pattern of those centers where the city life is most intense.These centers can help to form the mosaic of subcultures by their variety;and they can also help to form CITY COUNTRY FINGERS(3),if each of the centers is at a natural meeting point of several fingers.This pattern was first written by Luis Racionero,under the name "Downtowns of 300,000."

8003

There are few people who do not enjoy the magic of a great city.But urban sprawl takes it away from everyone except the few who are lucky enough,or rich enough,to live close to the largest centers.

This is bound to happen in any urban region with a single high density core.Land near the core is expensive;few people can live near enough to it to give them genuine access to the city's life;most people live far out from the core.To all intents and purposes,they are in the suburbs and have no more than occasional access to the city's life.This problem can only be solved by decentralizing the core to form a multitude of smaller cores,each devoted to some special way of life,so that,even though decentralized,each one is still intense and still a center for the region as a whole.

The mechanism which creates a single isolated core is simple.Urban services tend to agglomerate.Restaurants,theater

　　几乎没有人不为大城市的迷人魅力所吸引。但是城市的扩展使得只有那些居住在最大中心附近的富有的幸运儿才能饱览它的美姿丰采，而其他人只好望洋兴叹了。

　　在任何一个市区内必然会产生一个高密度的中心。中心附近的土地价格昂贵；只有为数不多的人才能在其附近居住，才能真正过着城市生活；大多数人居住在远离中心的地方。他们所企求的就是居住在市郊，除偶尔进城外，就没有机会接触城市生活了。为了解决这个问题，只有分散这个中心以形成多个较小的中心，并且每个小中心都要致力于形成自己独特的生活方式，这样即使分散了，小中心依然是紧凑的、热闹非凡的，依然是整个地区的一个中心。

　　形成一个孤立的中心，其缘由是简单的。城市的服务行业趋向于集中。餐厅、戏院、商店、狂欢活动、咖啡馆、旅馆、夜总会、娱乐和特种服务均所趋向于密集。这些服务行业之所以密集，因为人人都想跻身于人最多的地方。一个核心在市内刚一形成，名目繁多的服务行业，尤其是那些最令人感到兴趣的、因而也是需要最大集中区的行业，纷纷在这一中心开设起来。这个核心不断发展。城市商业区迅速扩大。商业区内市场繁华，商品丰富多采、琳琅满目，令人陶醉，流连忘返。但是，逐渐地，随着大都市区的发展，从个人住宅到该中心的平均距离不断增大；而中心周围的土地价格猛涨，以致许多住宅被商店和办公室从那里排挤出去了——不用很久，直到没有一个人或几乎没有一个人能同这个得天独厚的中心再保持真正的接触为止。而这个中心却日日夜夜创造着令人向往的神秘的美。

s,shops,carnivals,cafes,hotels,night clubs,entertainment,special services,tend to cluster.They do so because each one wants to locate in that position where the most people are.As soon as one nucleus has formed in a city,each of the interesting services—especially those which are most interesting and therefore require the largest catch basin—locate themselves in this one nucleus.The one nucleus keeps growing.The downtown becomes enormous.It becomes rich,various,fascinating.But gradually,as the metropolitan area grows,the average distance from an individual house to this one center increases;and land values around the center rise so high that houses are driven out from there by shops and offices—until soon no one,or almost no one,is any longer genuinely in touch with the magic which is created day and night within this solitary center.

The problem is clear.On the one hand people will only expend so much effort to get goods and services and attend cultural events,even the very best ones.On the other hand,real variety and choice can only occur where there is concentrated,centralized activity;and when the concentration and centralization become too great,then people are no longer willing to take the time to go to it.

If we are to resolve the problem by decentralizing centers,we must ask what the minimum population is that can support a central business district with the magic of the city. Otis D.Duncan in "The Optimum Size of Cities" (Cities and Society,P.K.Hatt and A.J.Reiss,eds.,New York:The Free Press,1967,pp.759-772),shows that cities with more than 50,000 people have a big enough market to sustain 61 different kinds of retail shops and that cities with over 100,000 people can support sophisticated jewelry,fur,and fashion stores.He shows

问题是清清楚楚的。一方面，人们为了到各处购买商品，请人提供服务，参加文化活动，包括最佳的文化活动，就要耗费巨大的精力。另一方面，现实的千姿百态和各种选择的机会只能在集中的活动中心才能出现。当集中和中心化过度时，人们就不再愿意花时间去那里了。

如果我们想要用分散中心的办法来解决这个问题，那么我们必须问一问：维持一个具有城市魅力的中心商业区所需的最小人口数量是多少？奥蒂斯·D.邓肯在《城市的最佳规模》（*Cities and Society*，P.K.Hatt and A.J.Reiss，eds.，New York：The Free Press，1967，pp.759～772）一书中表明，超过50000人的城市要有一个足够大的市场来保持61种不同的零售商店，超过100000人的城市就能开设精致纤巧、玲珑剔透的珠宝饰品商店、毛皮商店和时装商店。他指出，100000人的城市可开办一所大学，建立一座博物馆、一所图书馆、一个动物园，组成一个交响乐队，创办一份日报和一个调频电台，人口为250000～500000的城市就能开设一所专门化的职业学校，如医专，建立一座歌剧院或全部电视网络。

布赖恩·K.贝利在研究大都市芝加哥的地区购物中心时发现，有70种不同的零售商店的中心为350000左右人口的居民基本区服务（*Geography of Market Centers and Retail Distribution*，New Jersey：Prentice-Hall，1967，p.47）。T.R.拉克希曼南和沃尔特·G.汉森在《零售的潜在模式》（*American Institute of Planners Journal*，May 1965，pp.134～143）一文中指明，在有各式各样的零售商店和职业性服务行业以及娱乐文化活动的设施齐全的中心，居住100000～200000人是适宜的。

因而，在为不超过300000人服务的集中区中心，发挥城市纷繁复杂的功能似乎是可能的。根据上述理由，建立尽可能多的中心是合乎需要的，所以，我们提议，每一市

that cities of 100,000 can support a university,a museum,a library,a zoo,a symphony orchestra,a daily newspaper,AM and FM radio,but that it takes a population of 250,000 to 500,000 to support a specialized professional school like a medical school,an opera,or all of the TV networks.

In a study of regional shopping centers in metropolitan Chicago,Brian K.Berry found that centers with 70 kinds of retail shops serve a population base of about 350,000 people(Geography of Market Centers and Retail Distribution,New Jersey:Prentice-Hall,1967,p.47).T.R.Lakshmanan and Walter G.Hansen,in "A Retail Potential Model" (American Institute of Planners Journal,May 1965,pp.134-143),showed that full-scale centers with a variety of retail and professional services,as well as recreational and cultural activities,are feasible for groups of 100,000 to 200,000 population.

It seems quite possible,then,to get very complex and rich urban functions at the heart of a catch basin which serves no more than 300,000 people.Since,for the reasons given earlier,it is desirable to have as many centers as possible,we propose that the city region should have one center for each 300,000 people,with the centers spaced out widely among the population,so that every person in the region is reasonably close to at least one of these major centers.

To make this more concrete,it is interesting to get some idea of the range of distances between these centers in a typical urban region.At a density of 5000 persons per square mile(the density of the less populated parts of Los Angeles) the area occupied by 300,000 will have a diameter of about nine miles;at a higher density of 80,000 persons per square mile(the density of central Paris)the area occupied by 300,000 people has a diameter of about two miles.Other patterns

区都有一个 300000 人的中心，还有许多小中心分散在全体居民之间，这样，每个市民就能合情合理地至少和这些较多的中心之一接近。

为了把这一点说得更加具体，对一个典型的市区各中心之间的距离范围有所了解是很有趣的。每平方英里的人口密度为 5000 人（洛杉矶人口较少地区的人口密度），300000 人所占据的土地面积的直径约为 9mi；每平方英里的较高的人口密度为 80000 人（巴黎市中心的人口密度），300000 人所占据的土地面积的直径约为 2mi。本语言中的其他模式表明，一座城市的人口密度要比洛杉矶的人口密度大得多，而比巴黎市中心的人口密度略低些——**不高于四层楼**（21）、**密度圈**（29）。因此，我们把这些粗略的估计作为人口密度的上限和下限。如果每一中心为 300000 人服务，各个中心至少相隔 2mi，并可能不会超过 9mi。

现在来论述最后一点。城市的魅力是人们大量劳动的结晶，劳动创造了应有尽有的专门化的行业。就在纽约这样的城市里，你能在餐馆品尝到蚁心巧克力的风味食品，在书市购买到 300 年前出版的诗集，在街头亲眼目睹加勒比人的钢鼓乐队演奏和美国乡村民谣歌手演唱的精彩场面。据比较，300000 人的城市，如果只拥有一个二流的歌剧院，几个大型百货商店和 6 个中等水平的饭馆，算是个乡下城市。如果新的城市商业中心，每一中心只为 300000 人服务，虽力图成为具有魅力的城市，结果以居民众多的二流乡下城市而告终，那将是荒谬可笑的。

只有每一中心不仅为集中区的 300000 人服务，还能提供别的中心所没有的特种优质服务，这个问题才能迎刃而解。这样一来，每个中心——即使规模很小——就能为几百万人服务，因而就能在这样大的城市别开生面，独树一帜了。

in this language suggest a city much more dense than Los Angeles,yet somewhat less dense than central Paris —FOUR-STORY LIMIT(21),DENSITY RINGS(29).We therefore take these crude estimates as upper and lower bounds.If each center serves 300,000 people,they will be at least two miles apart and probably no more than nine miles apart.

One final point must be discussed.The magic of a great city comes from the enormous specialization of human effort there.Only a city such as New York can support a restaurant where you can eat chocolate-covered ants,or buy three-hundred-year-old books of poems,or find a Caribbean steel band playing with American folk singers.By comparison,a city of 300,000 with a secondrate opera,a couple of large department stores,and half a dozen good restaurants is a hick town.It would be absurd if the new downtowns,each serving 300,000 people,in an effort to capture the magic of the city,ended up as a multitude of secondclass hick towns.

This problem can only be solved if each of the cores not only serves a catch basin of 300,000 people but also offers some kind of special quality which none of the other centers have,so that each core,though small,serves several million people and can therefore generate all the excitement and uniqueness which become possible in such a vast city.

Thus,as it is in Tokyo or London,the pattern must be implemented in such a way that one core has the best hotels,another the best antique shops,another the music,still another has the fish and sailing boats.Then we can be sure that every person is within reach of at least one downtown and also that all the downtowns are worth reaching for and really have the magic of a great metropolis.

由此可见，正如在东京和伦敦那样，本模式必须以如下的方式来加以实现；第一个中心有一流的高级宾馆，第二个中心有古色古香的古玩商店，第三个中心有美妙动听的音乐茶座，第四个中心有可以垂钓和划船的风景点。到那时，我们就能确有把握地说，每一市民都至少在一个城市商业区之内了，而其他所有的城市商业区也都统统值得去光顾一番，因为它们真正具有了大城市的神奇魅力。

因此：

在大城市区内每个人都能到达的地方设置具有城市魅力的商业区。利用共同的地区政策严格限制商业区的发展，使任何一个商业区都不能发展到为 300000 以上人口服务的规模。具有这一人口数量的居民区内的商业区，相互隔开的距离为 2 ~ 9mi。

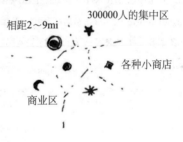

相距2~9mi　　　300000人的集中区

各种小商店

商业区

∾∾ﻌ

把每个商业区作为行人区和地方交通区来对待——**地方交通区**（11）、**散步场所**（31），并与郊区具有良好的运输线路的连接——**公共交通网**（16）；在每一商业区内鼓励丰富多彩的集中的夜生活——**夜生活**（33），并且至少划出商业区的一部分作为最狂热的街头活动场地——**狂欢节**（58）、**街头舞会**（63）……

Therefore:

Put the magic of the city within reach of everyone in a metropolitan area.Do this by means of collective regional policies which restrict the growth of downtown areas so strongly that no one downtown can grow to serve more than 300,000 people.With this population base,the downtowns will be between two and nine miles apart.

≈∽≈

Treat each downtown as a pedestrian and local transport area-LOCAL TRANSPORT AREAS(11), PROMENADE (31),with good transit connections from the outlying areas—WEB OF PUBLIC TRANSPORTATION (16);encourage a rich concentration of night life within each downtown—NIGHT LIFE(33),and set aside at least some part of it for the wildest kind of street life—CARNIVAL(58),DANCING IN THE STREET(63)...

模式11 地方交通区域**

　　……继**亚文化区的镶嵌**（8）之后，需要一种较大的蜂窝状结构：地方交通区域。这些区的直径为1～2mi，通过在市内建立天然边界，不仅有助于形成亚文化区，而且也有助于产生单独的城市指状带，见**指状城乡交错**（3），也有助于划定每一城市商业区为特殊的自给自足的地方交通区域——**城市的魅力**（10）。

11 LOCAL TRANSPORT AREAS**

...superimposed over the MOSAIC OF SUBCULTURES (8),there is a need for a still larger cellular structure:the local transport areas.These areas,1-2 miles across,not only help to form subcultures,by creating natural boundaries in the city,but they can also help to generate the individual city fingers in the CITY COUNTRY FINGERS(3),and they can help to circumscribe each downtown area too,as a special self-contained area of local transportation—MAGIC OF THE CITY(10).

ะงฉ

Cars give people wonderful freedom and increase their opportunities.But they also destroy the environment,to an extent so drastic that they kill all social life.

The value and power of the car have proved so great that it seems impossible to imagine a future without some form of private,high-speed vehicle.Who will willingly give up the degree of freedom provided by cars?At the same time,it is undeniably true that cars turn towns to mincemeat.Somehow local areas must be saved from the pressure of cars or their future equivalents.

It is possible to solve the problem as soon as we make a distinction between short trips and long trips.Cars are not very good for short trips inside a town,and it is on these trips that they do their greatest damage.But they are good for fairly long trips,where they cause less damage.The problem will

汽车给人们提供意料不到的行动自由并增加许多良好的机会。但是，汽车也会毁坏环境，从某种程度上说，可能严重到扼杀全部社会生活的地步。

汽车的价值和力量已经被证明是非常巨大的。如果没有某种私人的高速交通工具，未来似乎是很难想象的。谁会心甘情愿地放弃汽车所提供的自由度呢？同时，无可否认，汽车也会使城镇变得支离破碎。无论如何，局部地区必须从汽车的压力下摆脱出来，或用未来的交通工具取代现在的汽车。

我们一旦作出明确区分短途旅行和长途旅行之间的区别，这一问题就可迎刃而解了。在城镇内，汽车不适用于短途旅行。正是在短途旅行中汽车造成了最大的危害。但是，汽车适用于较长距离的旅行，而且造成的危害较小。如果城镇被划分成许多个直径约为 1mi 的区域，并设想，汽车可用于离开这些区域的旅行，而在这些区域内的各种旅行则采用其他的较缓慢的交通方式——步行、骑自行车、骑马、乘出租车，那么，这个问题就将顺利解决。它所需要的一切，从外部空间来说，就是一个街道模式：在这些区域内，劝阻使用私人汽车作旅行用，鼓励步行、骑自行车、骑马、乘出租车——但是，允许在离开这些区域的旅行中使用汽车。

现在让我们来列举一份由汽车引起的社会问题的统计表吧：

空气污染

噪声

危险性

健康不良

交通堵塞

停车问题

刺眼物

be solved if towns are divided up into areas about one mile across,with the idea that cars may be used for trips which leave these areas,but that other,slower forms of transportation will be used for all trips inside these areas—foot,bike,horse,taxi,All it needs,physically,is a street pattern that discourages people from using private cars for trips within these areas,and encourages the use of walking,bikes,horses,and taxis instead—but allows the use of cars for trips which leave the area.

Let us start with a list of the obvious social problems created by the car:

Air pollution

Noise

Danger

Ill health

Congestion

Parking problem

Eyesore

The first two are very serious,but are not inherent in the car;they could both be solved,for instance,by an electric car.They are,in that sense,temporary problems.Danger will be a persistent feature of the car so long as we go on using high-speed vehicles for local trips.The widespread lack of exercise and consequent ill health created by the use of motor-driven vehicles will persist unless offset by an amount of daily exercise at least equal to a 20 minute walk per day.And finally,the problems of congestion and loss of speed,difficulty and cost of parking,and eyesore are all direct results of the fact that the car is a very large vehicle which consumes a great deal of space.

The fact that cars are large is,in the end,the most serious

头两个问题是非常严重的，但并非汽车所固有的，例如，利用电动汽车，这两个问题就都可以得到解决。就这种意义来讲，这两个问题是暂时性的。只要我们继续把高速车辆用于地方旅行，危险性将是汽车的一个持久特征。广泛的缺乏体育锻炼和随之而来的健康不良的现象将长期存在下去，除非人们每天进行补偿性体育锻炼，其运动量至少相当于每天 20min 的步行。最后，交通堵塞问题、失速问题、停车困难和停车费昂贵问题以及刺眼物问题都是使用汽车的直接后果，因为汽车是体积很大的交通工具，占据大量的空间。

汽车体积大这一事实，归根结底，是以使用汽车为主的交通系统的最严重的问题，因为它是汽车本质上所固有的。现在让我们十分形象地来说明这一问题吧。一个人，当他站着静止不动的时候，占据约 $5ft^2$（$1ft^2=0.09m^2$）的空间，当他散步时，或许占据 $10ft^2$ 的空间。一辆汽车，当它停放着不动的时候，占据大约 $350ft^2$ 的空间（如果我们包括通道在内的话），当汽车前后相隔 3 辆车的距离，以每小时 30mi 的速度向前行驶，它占据的空间约为 $1000ft^2$。正如我们所知，在大部分时间内，汽车总是一个空间占有者。这就意味着，当人们使用汽车时，每个人占据的空间几乎为他步行时所占空间的 100 倍。

如果每个驾驶汽车的人所占据的空间面积为他步行时的 100 倍，那么，这意味着人们相隔的距离就会增大 10 倍。换句话说，使用汽车的总效应是使人们之间的距离扩大，并保持他们的分离状态。

汽车的这一特征对社会结构的影响是显而易见的。人们相互拉开距离。人口密度和相应的人际交往频率实际在减少。各种交往逐渐变得七零八落，并且专门化了，因为交往接触已被交往的性质局部化了，局限于明确限定的室内地点——家庭、工作场所，或许在几个孤独的朋友家里。

*aspect of a transportation system based on the use of cars,since it is inherent in the very nature of cars.*Let us state this problem in its most pungent form.A man occupies about 5 square feet of space when he is standing still,and perhaps 10 square feet when he is walking.A car occupies about 350 square feet when it is standing still(if we include access),and at 30 miles an hour,when cars are 3 car lengths apart,it occupies about 1000 square feet.As we know,most of the time cars have a single occupant.This means that when people use cars,each person occupies almost 100 times as much space as he does when he is a pedestrian.

If each person driving occupies an area 100 times as large as he does when he is on his feet,this means that people are 10 times as far apart.*In other words,the use of cars has the overall effect of spreading people out,and keeping them apart.*

The effect of this particular feature of cars on the social fabric is clear.People are drawn away from each other;densities and corresponding frequencies of interaction decrease substantially.Contacts become fragmented and specialized,since they are localized by the nature of the interaction into well-defined indoor places the home,the workplace,and maybe the homes of a few isolated fri ends.

It is quite possible that the collective cohesion people need to form a viable society just cannot develop when the vehicles which people use force them to be 10 times farther apart,on the average,than they have to be.This states the possible social cost of cars in its strongest form.*It may be that cars cause the breakdown of society,simply because of their geometry.*

At the same time that cars cause all these difficulties,they also have certain unprecedented virtues,which have in fact led to their enormous success.These virtues are:

当人们使用的车辆迫使他们自己原来间隔的距离平均扩大 10 倍时，他们为形成一个能生存下去的社会所需的内聚力就很可能几乎不再发展了。这一点很有力地说明了汽车对社会可能造成的巨大损害。汽车仅仅由于它的几何形状而引起社会瓦解是完全可能的。

汽车在惹起种种麻烦困难的同时，也具有一些前所未有的优点，事实上，这导致汽车获得巨大的成功。这些优点如下：

灵活性

私密性

无须换车的户至户旅行

即时性

这些优点本质上在二度空间的大都会地区内特别重要。公共交通在一些主干道上能够提供非常快速、频繁的户至户的旅行服务。但在更广的范围内，现在城市市区的二度空间性质和公共交通本身都还不能有效地和汽车竞争。即使在像伦敦和巴黎那样拥有世界上最完善的市内公共交通的城市里，火车和公共汽车每年的载容量也越来越少，因为人们转向驾驶汽车。他们宁愿忍受汽车的一切耽误、堵塞和昂贵的停车费，因为，很明显，汽车的方便性和私密性更有价值。

我们对此情况进行理论分析后发现，能满足一切需要的唯一交通系统就是私人车辆系统。这种系统能够利用一些高速公路作穿越城市的长途旅行，并且在地方交通区域内离开公共线路时私人车辆可以使用自己的动力。最接近这一理论模式的系统是各式各样的私人快速交通建议；其中一例就是威斯汀豪斯·斯塔卡系统：该系统提供小型双座汽车，这种车辆既可以在地方的街道上行驶，又可以开往高速公共车道作长途旅行。

可是，斯塔卡系统有一些缺点。相对来说，这种系统

Flexibility

Privacy

Door-to-door trips,without transfer

Immediacy

These virtues are particularly important in a metropolitan region which is essentially two-dimensional.Public transportation can provide very fast,frequent,door-to-door service,along certain arteries.But in the widely spread out,two-dimensional character of a modern urban region,public transportation by itself cannot compete successfully with cars. Even in cities like London and Paris,with the finest urban public transportation in the world,the trains and buses have fewer riders every year because people are switching to cars. They are willing to put up with all the delays,congestion,and parking costs,because apparently the convenience and privacy of the car are more valuable.

Under *theoretical* analysis of this situation,the only kind of transportation system which meets all the needs is a system of individual vehicles,which can use certain high-speed lines for long cross-city trips and which can use their own power when they leave the public lines in local areas. The systems which come closest to this theoretical model are the various Private Rapid Transit proposals;one example is the Westinghouse Starrcar—a system in which tiny two-man vehicles drive on streets locally and onto high-speed public rails for long trips.

However,the Starrcar-type systems have a number of disadvantages.They make relatively little contribution to the problem of space.The small cars,though smaller than a conventional car,still take up vastly more space than a person.

对解决空间问题帮助不大。小型汽车虽比一般的汽车小，但仍占据比一个人大得多的空间。因为这种私人汽车不能胜任长途越野旅行，所以，这些车辆一定被视为"二等车辆"——而且价格相当昂贵。这些车辆对解决健康问题也是无济于事的，因为当它们行驶时，人们仍然是坐着不动。相对来说，这种系统也是反社会的，因为当人们乘车旅行时，他们仍然被困在汽车的"透明罩"里。如果每人都有一辆斯塔卡轿车，那是最理想不过的了。但是交通多样化就不会再有存在下去的余地了。而交通多样化实在是人们朝思暮想的，他们希望使用自行车、马、老爷汽车、旧式汽车和家庭公共汽车等。

现在我们提出一种系统，它不仅具有斯塔卡系统的优点，而且更为现实，更易于付诸实施。我们认为，这种系统能较好地符合人们的需要。这一系统的实质在于下列两个主张：

1. 对于地方旅行来说，人们使用各种低速廉价的交通工具（自行车、三轮车、低座小摩托车、高尔夫轻便车、婴儿坐的手推车、马，等等），它们占据的空间比汽车占据的要小，并可使来往行人与他们周围的环境以及相互之间保持更加密切的接触。

地方短途旅行的多种方式
Many ways of getting around on local trips

2. 人们仍占有和使用轿车和卡车——但主要用于长途旅行。我们假定，这些汽车可以被制造成无噪声、不污染

Since the private cars will not be capable of long cross-country trips,they must be treated as a "second vehicle" — and are rather expensive.They make no contribution to the health problem,since people are still sitting motionless while they travel.The system is relatively antisocial,since people are still encapsulated in "bubbles" while they travel.It is highly idealistic,since it works if everyone has a Starrcar,but makes no allowance for the great variety of movement which people actually desire,i.e.,bikes,horses,jalopies,old classic cars,family buses.

We propose a system which has the advantages of the Starrcar system but which is more realistic,easier to implement,and,we believe,better adapted to people's needs.The essence of the system lies in the following two propositions:

1.For local trips,people use a variety of low-speed,low-cost vehicles(bicycles,tricycles,scooters,golf carts,bicycle buggies,horses,etc.),which take up less room than cars and which all leave their passengers in closer touch with their environment and with one another.

2.People still own,and use,cars and trucks—but mainly for long trips.We assume that these cars can be made to be quiet,non-polluting,and simple to repair,and that people simply consider them best suited for long distance travel.It will still be possible for people to use a car or a truck for a local trip,either in a case of emergency,or for some special convenience. However,the town is constructed in such a way that it is actually expensive and inconvenient to use cars for local trips-so that people only do it when they are willing to pay for the very great social costs of doing so.

环境且易于修理的，并且还假定，人们单纯认为这些汽车最适合于长途旅行。对于人们来说，使用轿车或卡车，无论是在紧急情况下还是为了某种特殊的方便，在地方上进行短途旅行将仍然是可能的。所以，无论如何，城镇建设要以如下的方式来进行：使用汽车进行短途旅行不仅费用昂贵，而且也是极不方便的，要不折不扣地做到这一点。做到了这一点，只有那些愿意支付极高的社会费用的人才能使用汽车作短途旅行了。

因此：

把城区划分为地方交通区域，每一区域的直径为 1 ～ 2mi，周围有一条环路环绕。在地方交通区内建造较小的地方内部道路和小路，供步行、骑自行车和骑马以及地方交通工具使用；建造主干道路，使轿车和卡车十分容易地来往于环路之上，但是要使它们在地方内部交通区域内旅行时放慢车速，并且行驶不便。

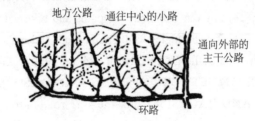

使主干公路用于长途交通，而不用于内部地方交通，将它们布置成为平行的单向公路，并使这些平行路离开地方交通区域的中心，使它们非常有利于与环路衔接，但不便于供地方短途旅行用——**平行路**（23）。布置大量的人行便道、自行车道和绿茵街道，与主干路成直角相交，并使这些地方交通区域内的小路直接穿过中心——**绿茵街道**（51）、**小路网络和汽车**（52）、**自行车道和车架**（56）；沿

Therefore:

Break the urban area down into local transport areas, each one between 1 and 2 miles across,surrounded by a ring road. Within the local transport area,build minor local roads and paths for internal movements on foot,by bike,on horseback,and in local vehicles;build major roads which make it easy for cars and trucks to get to and from the ring roads, but place them to make internal local trips slow and inconvenient.

శుమో

To keep main roads for long distance traffic,but not for internal local traffic,lay them out as parallel one way roads,and keep these parallel roads away from the center of the area,so that they are very good for getting to the ring roads,but inconvenient for short local trips—PARALLEL ROADS(23).Lay out abundant footpaths and bike paths and green streets,at right angles to the main roads,and make these paths for local traffic go directly through the center—GREEN STREETS(51),NETWORK OF PATHS AND CARS(52),BIKE PATHS AND RACKS(56);sink the ring roads around the outside of each area,or shield the noise they make some other way—RING ROADS(17);keep parking to a minimum within the area,and keep all major parking garages near the ring roads—NINE PER CENT PARKING(22),SHIELDED PARKING(97);and build a major interchange within the center of the area—INTERCHANGE(34)...

build up these larger city patterns from the grass roots,through action essentially controlled by two levels of self-governing communities,which exist as physically identifiable places;

每一地方交通区域的外围使环路低于地面或用某种其他的方法来屏蔽噪声——**环路**（17）；使区域内的停车场缩小到最小范围，并使所有的主要停车库设在环路附近——**停车场不超过用地的 9%**（22）、**有遮挡的停车场**（97）；并在地方交通区域的中心建造一个换乘站——**换乘站**（34）……

通过自治社区两级实际上控制的活动，把下面这些较大的模式从农业区建立起来，这些模式在外部空间上作为容易识别的地方而存在下去：

12.7000 人的社区

13. 亚文化区边界

14. 易识别的邻里

15. 邻里边界

12.COMMUNITY OF 7000

13.SUBCULTURE BOUNDARY

14.IDENTIFIABLE NEIGHBORHOOD

15.NEIGHBORHOOD BOUNDARY

模式12　7000人的社区*

……**亚文化区的镶嵌**（8）是由许多大大小小的自治社区和邻里组成的。7000人的社区有助于说明大社区的结构。

❧❧❧

在任何一个超过 5000 ~ 10000 人的社区里，个人的呼声是不会有任何效果的。

当地方政府的各个机构是自治的、自己管理自己的、自己编制预算的社区，并且这些社区很小，从而有可能使街道上的居民和他们的地方官员以及选民代表进行直接联系时，人民才能对地方政府施加真正的影响。

无论是公元前 4 世纪和公元前 3 世纪的雅典民主模式，还是美国杰斐逊的民主方案，抑或是孔夫子在《论语》中谈道的政府行动方针，这一切统统都是过时的陈旧观念而已。

12   COMMUNITY OF 7000*

...the MOSAIC OF SUBCULTURES(8)is made up of a great number of large and small self-governing communities and neighborhoods.Community of 7000 helps define the structure of the large communities.

&∞03

Individuals have no effective voice in any community of more than 5000-10,000 persons.

People can only have a genuine effect on local government when the units of local government are autonomous,self-governing,self-budgeting communities,which are small enough to create the possibility of an immediate link between the man in the street and his local officials and elected representatives.

This is an old idea.It was the model for Athenian democracy in the third and fourth centuries B.C.;it was Jefferson's plan for American democracy;it was the tack Confucius took in his book on government,*The Great Digest.*

For these people,the practice of exercising power over local matters was itself an experience of intrinsic satisfaction. Sophocles wrote that life would be unbearable were it not for the freedom to initiate action in a small community.And it was considered that this experience was not only good in itself,but was the only way of governing that would not lead to corruption.Jefferson wanted to spread out the power not because "the people" were so bright and clever,but precisely because they were prone to error,and it was therefore dangerous

这些人行使权力、管理地方事务仅仅是使自己的内心感到满足而已。沙孚克理斯（公元前 495 ？—前 406 ？，希腊悲剧作家——译者注）曾写道，若不是在小社区内为了自由而发起行动，生活就会变得无法忍受了。当时大家公认，这不仅是条好经验，而且也是不导致腐败的唯一有效的管理方法。杰斐逊要求分散权力，不是因为"人民"是何等的聪明伶俐，精明能干，而是因为他们易犯错误，因此，把权力授予不可避免地会犯大错误的少数人是十分危险的。"把国家划分成选区"就是他竞选总统的口号，这样，错误就易于控制，人民也将得到行使权力的实践并改进管理。

今天，人民和统治他们的权力中心之间的距离是很大的——既是心理上的，又是地理上的。一位杰斐逊崇拜者米尔顿·科特勒早已描述了这种感受：

城市施政的过程市民是完全看不见的，他也很少了解行政人员的构成情况，但却感到因苛捐杂税的沉重负担而引起的强烈痛苦。由于公共服务的日益缺乏，他以更加急不可耐的心情表达他的愿望和需要。可是，他需求的呼声在政府看来，只不过是一阵无关痛痒的耳旁风而已，根本不会予以理睬。市民和政府之间的这种脱节是市政府主要的政治问题，因为这具体表现出市政混乱的动向……（Milton Kotler, Neighborhood Foundations, Memorandum#24 ; "Neighborhood corporations and the reorganization of city government," unpub.ms.,August 1967.）

就目前的状况来说，外界环境是以两种方式促使市民与其政府分离，并保持这种分离。第一，政治社区的规模太大，结果造成它的成员和领导者之间的分离，简单地说，原因就在于它的成员的数量。第二,政府是老百姓见不到的，它设在大多数市民日常生活活动范围以外的地方。除非改变这两个条件，否则政治上的疏远大概是无法克服的。

to vest power in the hands of a few who would inevitably make big mistakes. "Break the country into wards" was his campaign slogan,so that the mistakes will be manageable and people will get practice and improve.

Today the distance between people and the centers of power that govern them is vast—both psychologically and geographically.Milton Kotler,a Jeffersonian,has described the experience:

The process of city administration is invisible to the citizen who sees little evidence of its human components but feels the sharp pain of taxation.With increasingly poor public service,his desires and needs are more insistently expressed.Yet his expressions of need seem to issue into thin air,for government does not appear attentive to his demands.This disjunction between citizen and government is the major political problem of city government,because it embodies the dynamics of civil disorder....(Milton Kotler,Neighborhood Foundations,Memoran dum#24; "Neighborhood corporations and the reorganization of city government," unpub.ms.,August 1967.)

There are two ways in which the physical environment,as it is now ordered,promotes and sustains the separation between citizens and their government.First,the size of the political community is so large that its members are separated from its leaders simply by their number.Second,government is invisible,physically located out of the realm of most citizens' daily lives.Unless these two con-ditions are altered,political alienation is not likely to be overcome.

1.*The size of the political community*.It is obvious that the larger the community the greater the distance between the average citizen and the heads of government.Paul Goodman

1. 政治社区的规模。很明显，社区越大，一般市民与政府的领导人之间的距离就越大。保罗·古德曼已经提出以像雅典那样的一些全盛期的城市为基础的拇指定则：地方机构的最高成员，与任何一个市民之间不应间隔着两个以上的朋友。假定每个人在他的地方社区认识 12 个人。根据这一见解和拇指定则，我们就会得出这一结论：一个政治社区的最佳规模应当是 1^{23}，即 1728 户或 5500 人。这一数据与旧芝加哥学派估计的 5000 人相符合。而且，这一数据与俄亥俄州哥伦布市经济委员会的规模同属一个数量级，也即科特勒所描绘的俄亥俄州哥伦布市 6000 ～ 7000 人的邻里有限公司的规模。(*Committee on Government Operations*,U.S.Senate,89th Congress,Second Session,Part 9,December 1966)

《生态学家》的编辑们对于地方政府各种机构的适度规模也具有同样的看法 (See their *Blueprint for Survival*, Penguin Books,1972，pp.50 ～ 55)。特伦斯·李在《作为社会空间模式的城市邻里》(Ekistics 177，August 1970) 一文中为空间社区的重要性提供了证据。李提出，社区的自然规模为 75acre (1acre=4046.86m²)。按每英亩 25 人计算，这样一个社区内，将住下大约 2000 人；按每英亩 60 人计算，将住下大约 4500 人。

2. 很醒目的地方政府所在地。即使地方政府机构在职能上分散了权力，它们在空间上往往仍然是集中的，它们隐藏在市、县的大楼内，脱离现实生活。这种地方是令人望而生畏、感情疏远的。而现在所需要的是造成这样一种气氛：每个人在地方政府所在地都感到像在家里一样亲切，他既可以畅所欲言，发表自己的看法，又可以吐吐苦水，发泄满腹的牢骚。每个人都必须感觉到，地方政府是一个论坛，确实是他自己的政府，他能把地方政府中负责这种那种工作的人员叫来，和他促膝谈心，并且经常能亲自见到他。

has proposed a rule of thumb,based on cities like Athens in their prime,that no citizen be more than two friends away from the highest member of the local unit.Assume that everyone knows about 12 people in his local community. Using this notion and Goodman's rule we can see that an optimum size for a political community would be about 123 or 1728 house holds or 5500 persons.This figure corresponds to an old Chicago school estimate of 5000.And it is the same order of magnitude as the size of ECCO,the neighborhood corporation in Columbus,Ohio,of 6000 to 7000,described by Kotler(*Committee on Government Operations*,U.S.Senate,89th Congress,Second Ses-sion,Part 9,December 1966).

The editors of *The Ecologist* have a similar intuition about the proper size for units of local government.(See their *Blueprint for Survival*,Penguin Books,1972,pp.50-55.)And Terence Lee,in his study, "Urban neighborhood as a socio-spatial schema," Ek-istics 177,August 1970,gives evidence for the importance of the spatial community.Lee gives 75 acres as a natural size for a community.At 25 persons per acre,such a community would accommodate some 2000 persons;at 60 persons per acre,some 4500.

2.*The visible location of local government.*Even when local branches of government are decentralized in function,they are often still centralized in space,hidden in vast municipal citycounty build-ings out of the realm of everyday life. These places are intimidating and alienating.What is needed is for every person to feel at home in the place of his local government with his ideas and complaints.A person must feel that it is a forum,that it is his directly,that he can call and talk to the person in charge of such and such,and see him personally

几千人的社区大会
Community meeting of several thousand

为此目的，作为论坛的地方政府必须位于人人看得见的、人人都会去的醒目的地方。例如，地方政府可设在5000～7000人社区的最繁华的市场。我们要在**地方市镇厅**（44）中更加充分地探讨这种可能性，但是，我们在这里要强调这一点，因为提供一个政治"心脏"，即具有吸引力的政治中心——是一个政治社区的重要的一部分。

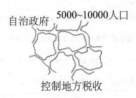

自治政府   5000~10000人口

控制地方税收

因此：

分散市政府的权力，让地方来管理 5000～10000 人的社区。尽可能利用自然地理的和历史的边界划分社区范围。授予每个社区以倡议、决定和实施各项重大事项的权力。这些事项包括土地利用、住宅建筑、维护保养、街道、公园、警察、学校教育、福利和邻里服务等，它们均和社区的切身利益息息相关。

⊰⊱

利用重要的区界——**亚文化区边界**（13），把各社区彼此分开；然后再把每一社区分成 10～20 个独立的邻里，每个邻里都有一名代表参加社区议会——**易识别的邻里**（14）；提供一个中心场所，让人们有机会在那里聚会——**偏心式核心区**（28）、**散步场所**（31），并在这个中心场所提供一个地方市镇厅，作为社区政治活动的中心点——**地方市镇厅**（44）……

within a day or two.

For this purpose,local forums must be situated in highly visible and accessible places.They could,for instance,be located in the most active marketplace of each community of 5000 to 7000.We discuss this possibility more fully under LOCAL TOWN HALL(44),but we emphasize it here,since the provision of a political "heart," a political center of gravity,is an essential part of a political community.

Therefore:

Decentralize city governments in a way that gives local control to communities of 5,000 to 10,000 persons.As nearly as possible,use natural geographic and historical boundaries to mark these communities.Give each community the power to initiate,decide,and execute the affairs that concern it closely:land use,housing,maintenance,streets,parks,police,sc hooling,welfare,neighborhood services.

৪৩৫৪

Separate the communities from one another by means of substantial areas—SUBCULTURE BOUNDARY (13); subdivide each community into 10 or 20 independent neighborhoods,each with a representative on the community council—IDENTIFIABLE NEIGHBORHOOD(14);provide a central place where people have a chance to come together— ECCENTRIC NUCLEUS(28),PROMENADE(31);and in this central place provide a local town hall,as a focal point for the community's political activity—LOCAL TOWN HALL(44)...

模式13 亚文化区边界*

 ……**亚文化区的镶嵌**（8）及其独立的亚文化区，无论它们是**7000人的社区**（12），还是**易识别的邻里**（14），都需要用边界来完善。事实上，根据本模式建立单纯的边界区，就会给边界以内的亚文化区注入活力，使它们各自都有机会发展自己的个性。

⚜

 亚文化区的镶嵌需要成百上千种不同的文化相互共存，它们以独特的方式充分表现出各自鲜明的特色。但是，亚文化区各有各的生态环境。它们只有在空间上以边界与外部分隔开，才能不受邻区的妨碍而充分显示出自己的鲜明特色。

 在**亚文化区的镶嵌**（8）中我们已经论证了，在一个城市内的许多不同的亚文化区不是民族聚居区的种族主义模

13   SUBCULTURE BOUNDARY*

...the MOSAIC OF SUBCULTURES(8)and its individual subcultures,whether they are COMMUNITIES OF 7000(12) or IDENTIFIABLE NEIGHBORHOODS(14),need to be completed by boundaries.In fact,the mere creation of the boundary areas,according to this pattern,will begin to give life to the subcultures between the boundaries,by giving them a chance to be themselves.

❧⳩⳩❧

The mosaic of subcultures requires that hundreds of different cultures live,in their own way,at full intensity,next door to one another.But subcultures have their own ecology.They can only live at full intensity,unhampered by their neighbors,if they are physically separated by physical boundaries.

In MOSAIC OF SUBCULTURES(8)we have argued that a great variety of subcultures in a city is not a racist pattern which forms ghettos,but a pattern of opportunity which allows a city to contain a multitude of different ways of life with the greatest possible intensity.

But this mosaic will only come into being if the various subcultures are insulated from one another,at least enough so that no one of them can oppress,or subdue,the life style of its neighbors,nor,in return,feel oppressed or subdued.As we shall see,this requires that adjacent subcultures are separated by swaths of open land,workplaces,public buildings,water,parks,or other natural boundaries.

式，而是为城市能兼容并蓄许多不同的、各具鲜明特色的生活方式提供良机的模式。

但是，这种镶嵌只有在不同的亚文化区相互分离时才能实现，而分离的程度至少达到：谁也不能压制或征服邻区的生活方式，谁也不会感到被压制或要被征服。正如我们将会看到的那样，这就要求相邻的亚文化区要由一条条长而宽的开阔地带、工作点、公共建筑物、水域、公园或其他天然边界分隔开来。

这一论据的要点来自如下的事实。在城市内，无论何地存在着相似的住宅区，其居民一定会对毗邻的各区施加巨大的压力，使它们适应自己区的价值观念和风格。例如，1976年住在旧金山"嬉皮士"黑特·阿什伯里区附近的"一本正经的"人们都曾经担心过，黑特·阿什伯会把他们的地价压低，所以他们对市镇厅施加压力，以便"肃清"黑特·阿什伯的影响，也就是说，要使黑特区更像他们自己的区。无论何时，一个亚文化区同另一个相邻的亚文化区在风格上截然不同，上述情况似乎就是不可避免总要发生的。人们一定会害怕邻区"潜入"他们自己的区，扰乱他们的地价，坑害他们的子女，把一些"好的"住户吓跑，如此种种，不一而足。他们千方百计力图使邻区就范，变得像他们自己的区一模一样。

卡尔·沃瑟曼、杰里·曼德尔和特德·迪恩斯特弗雷已经指出，即使在极为相似的亚文化区内也同样有这种现象。（*Planning and the Purchase Decision:Why People Buy in Planned Communities*,University of California,Berkeley,July 1965）他们在研究生活在森林、农田等地域辽阔的开发区内的居民时发现，在不同的社会集团之间，只要用广阔的田野、未开垦的土地、高速公路或水域隔开，由于毗邻而造成的紧张气氛就烟消云散了。总而言之，在相邻的亚文化区之间的空间障碍，如果大到一定程度，那就不用对邻区采取强硬措施了。

The argument hinges on the following fact.Wherever there is an area of homogeneous housing in a city, its inhabitants will exert strong pressure on the areas adjacent to it to make them conform to their values and style.For example,the"straight"people who lived near the "hippie"Haight Ashbury district in San Francisco in 1967 were afraid that the Haight would send their land values down,so they put pressure on City Hall to get the Haight "cleaned up" —that is,to make the Haight more like their own area.This seems to happen whenever one subculture is very different in style from another one next to it.People will be afraid that the neighboring area is going to "encroach" on their own area,upset their land values,undermine their children,send the "nice" people away,and so forth,and they will do everything they can to make the next door area like their own.

Carl Werthman,Jerry Mandel,and Ted Dienstfrey(*Planning and the Purchase Decision:Why People Buy in Planned Communities*,University of California,Berkeley,July 1965) have noticed the same phenomenon even among very similar subcultures.In a study of people living in tract developments, they found that the tension created by adjacencies between dissimilar social groups disappeared when there was enough open land,unused land,freeway,or water between them.In short,a physical barrier between the adjacent subcultures,if big enough,took the heat off.

Obviously,a rich mix of subcultures will not be possible if each subculture is being inhibited by pressure from its neighbors. *The subcultures must therefore be separated by land,which is not residential land,and by as much of it as possible.*

很明显，如果每个亚文化区都一直处于其邻区的高压之下，亚文化区之间丰富多彩的交流将成为不可能。因此，各亚文化区必须被一片尽可能大的无人居住的土地分隔开来。

现在我们亲自考察所得的结果也为这一最新的论述提供了证据。如果我们仔细察看一个大城市，并精确地找出截然不同的亚文化区及其特点，那么我们就会发现，它们总是位于边界附近，几乎从不靠近其他社区。例如，有两个极端不同的区——德律格拉夫山和唐人街。前者的两侧有一排排的船坞环绕着，后者的两侧与市内的银行区接界。在较大的海湾区，也有类似的情况。里奇蒙德岬和索萨利多是较大的海湾区中两个明显不同的社区，几乎完全是孤零零的。索萨利多背山面海；里奇蒙德岬三面临海，一面是工业地带。两个社区在某种程度上相互切断，各自都按自己的特点自由发展。

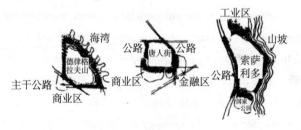

亚文化区边界
Subculture boundaries

我们的论据从生态学得到了进一步的支持。在自然界，一个生物物种逐渐演变成若干亚种，很大程度上是由于地理物种形成的结果，它遗传上的种种变化是在空间隔绝状态的时期发生的（see,for example,Ernst Mayr,*Animal Species and Evolution*,Cambridge,1963,Chapter 18:"The Ecology of Speciation",pp.556～585）。在一系列的生态研究中，人们早已观察到，同一物种的一部分成员和另一部分成员，彼此间被空间边界，诸如峻岭幽谷、大江巨川、荒漠莽原、

There is another kind of empirical observation which supports this last statement.If we look around a metropolitan area,and pinpoint the strongly differentiated subcultures,those with character,we shall always find that they are near boundaries and hardly ever close to other communities.For example,in San Francisco the two most distinctive areas are Telegraph Hill and Chinatown.Telegraph Hill is surrounded on two sides by the docks.Chinatown is bounded on two sides by the city's banking area.The same is true in the larger Bay Area.Point Richmond and Sausalito,two of the most distinctive communities in the greater Bay Area,are both almost completely isolated.Sausalito is surrounded by hills and water;Point Richmond by water and industrial land.Communities which are cut off to some extent are free to develop their own character.

Further support for our argument comes from ecology.In nature,the differentiation of a species into subspecies is largely due to the process of geographic speciation,the genetic changes which take place during a period of spatial isolation(see,for example,Ernst Mayr,Animal Species and Evolution,Cambridge, 1963,Chapter 18:"The Ecology of Speciation," pp.556-585). It has been observed in a multitude of ecological studies that members of the same species develop distinguishable traits when separated from other members of the species by physical boundaries like a mountain ridge,a valley,a river,a dry strip of land,a cliff,or a significant change in climate or vegetation. In just the same way,differentiation between subcultures in a city will be able to take place most easily when the flow of those elements which account for cultural variety—values, style,information,and so on—is at least partially restricted between neighboring subcultures.

悬崖峭壁所隔断，或由于气候或植被的重大变化的影响，会演变出具有明显区别的品质特性。同理，如果说明文化多样性的这样一些因素——价值观念、风格、信息等——的交流，在相邻的亚文化区内至少部分地受到限制，那么，市内各亚文化区之间的差别就能够极其容易地呈现出来。

因此：

用一条宽至少为 200ft 的狭长地带将相邻的亚文化区分开。让这条边界成为天然的边界——荒野、农田、天然水域或人工水域、铁道、主干公路、公园、学校和一些住宅建筑等。沿着两个亚文化区之间的边缘地带，建造集会场所，每个社区都能共享其功能也能相互接触。

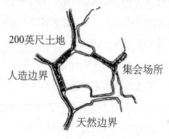

天然边界可能是这样一些地方——**乡村**（7）、**珍贵的地方**（24）、**通往水域**（25）、**僻静区**（59）、**近宅绿地**（60）、**水池和小溪**（64）和**池塘**（71）。人为边界可能包括**环路**（17）、**平行路**（23）、**工作社区**（41）、**工业带**（42）、**青少年协会**（84）和**有遮挡的停车场**（97）。亚文化区边界的内部组织应当遵循两条主要原则：它应当把土地的各种用途加以集中，以便环绕**活动中心**（30）和**工作社区**（41）形成功能性的团组。而且，边界应对两个相邻的社区来说都是畅通无阻的，因而边界也是它们的集合场所——**偏心式核心区**（28）……

Therefore:

Separate neighboring subcultures with a swath of land at least 200 feet wide.Let this boundary be natural— wilderness,farmland,water—or man-made—railroads,major roads,parks,schools,some housing.Along the seam between two subcultures,build meeting places,shared functions,touching each community.

෫ඏ෭

Natural boundaries can be things like THE COUNTRYSIDE(7),SACRED SITES(24),ACCESS TO WATER(25),QUIET BACKS(59),ACCESSIBLE GREEN(60),POOLS AND STREAMS(64),STILL WATER(71).Artificial boundaries can include RING ROADS(17),PARALLEL ROADS(23),WORK COMMUNITIES(41),INDUSTRIAL RIBBONS (42),TEENAGE SOCIETY(84),SHIELDED PARKING(97). The interior organization of the subculture boundary should follow two broad principles.It should concentrate the various land uses to form functional clusters around activity— ACTIVITY NODES(30),WORK COMMUNITY(41).And the boundary should be accessible to both the neighboring communities,so that it is a meeting ground for them— ECCENTRIC NUCLEUS(28)...

模式14 易识别的邻里**

……亚文化区的镶嵌（8）、7000人的社区（12）是由邻里构成的。本模式阐述邻里。它说明那些创造出活力和特性的小群体会使较大的模式——7000人的社区（12）和亚文化区的镶嵌（8）充满蓬勃的生机。

⊰⊱

人们需要属于他们自己的、容易识别的空间单元。

今日发展中的模式正在毁灭邻里
Today's pattern of development destroys neighborhoods

14   IDENTIFIABLE NEIGHBORHOOD**

...the MOSAIC OF SUBCULTURES(8)and the COMMUNITY OF 7000 (12)are made up of neighborhoods. This pattern defines the neighborhoods.It defines those small human groups which create the energy and character which can bring the larger COMMUNITY OF 7000 (12)and the MOSAIC OF SUBCULTURES(8)to life.

❧❧❧

People need an identifiable spatial unit to belong to.

They want to be able to identify the part of the city where they live as distinct from all others.Available evidence suggests,first,that the neighborhoods which people identify with have extremely small populations;second,that they are small in area;and third,that a major road through a neighborhood destroys it.

1.What is the right population for a neighborhood?

The neighborhood inhabitants should be able to look after their own interests by organizing themselves to bring pressure on city hall or local governments.This means the families in a neighborhood must be able to reach agreement on basic decisions about public services,community land,and so forth.Anthropological evidence suggests that a human group cannot coordinate itself to reach such decisions if its population is above 1500,and many people set the figure as low

人们都希望市内他们居住的那一部分地区与另外一些地区能明显区别开来。可得到的证据表明，人们识别出的邻里具有三个特点：第一，人口极少；第二，地域也很小；第三，主干路不能穿越邻里，否则，邻里会遭到破坏。

1. 一个邻里的合理人口是多少呢？邻里居民应当能够把自己组织起来，对市政府或地方政府施加压力以谋求自己的正当利益。这意味着邻里中的各个家庭对一些基本的决定，如关于公共服务、社区土地等，必须取得一致的意见。人类学提供的证据表明，如果一个群体的人口数量超过1500人，还有许多人提出的数字更小——为500人，则其本身就会众说纷纭，无法协调一致作出决定了。(See,for example,Anthony Wallace,*Housing and Social Structure*,Philadelphia Housing Authority,1952,available from University Microfilms,Inc.,Ann Arbor,Michigan,pp.21～24)根据组织地方一级的社区集会的经验表明，500人是比较现实的数字。

著名的邻里：奥格斯堡的福格莱
A famous neighborhood:the Fuggerei in Augsburg

2. 至于谈及空间的直径范围，在费城，从被问及的人提供的情况看，他们对住宅周围比较了解的地段范围很小，很少有超出2～3个街区的。(Mary W.Herman,"Comparative Studies of Identification Areas in Philadelphia," City of Philadelphia Community Renewal Program,Technical Report No.9,April 1964.) 威斯康星州的密尔沃基市的1/4的居民认为，邻里应当是不大于一个街区（300ft）的一个区，一

as 500.(See,for example,Anthony Wallace,*Housing and Social Structure*,Philadelphia Housing Authority,1952,available from University Microfilms,Inc.,Ann Arbor,Michigan,pp.21-24)The experience of organizing community meetings at the local level suggests that 500 is the more realistic figure.

2.As far as the physical diameter is concerned,in Philadelphia,people who were asked which area they really knew usually limited themsdves to a small area,seldom exceeding the two to three blocks around their own house. (Mary W.Herman, "Comparative Studies of Identification Areas in Philadelphia," City of Philadelphia Community Renewal Program,Technical Report No.9,April 1964.)One-quarter of the inhabitants of an area in Milwaukee considered a neighborhood to be an area no larger than a block(300 feet). One-half considered it to be no more than seven blocks. (Svend Riemer, "Villagers in Metropolis," British Journal of Sociology,2,No.1,March 1951,pp.31-43.)

3.The first two features,by themselves,are not enough.A neighborhood can only have a strong identity if it is protected from heavy traffic.Donald Appleyard and Mark Lintell have found that the heavier the traffic in an area,the less people think of it as home territory.Not only do residents view the streets with heavy traffic as less personal,but they feel the same about the houses along the street.("Environmental Quality of City Streets," by Donald Appleyard and Mark Lintell,Center for Planning and Development Research,University of California,Berkeley,1971.)

半居民则认为，邻里应当是不大于 7 个街区的一个区。
（Svend Riemer,"Villagers in Metropolis," *British Journal of Sociology*,2,No.I,March 1951,pp.31 ～ 43）

3. 一个邻里只具有前两个特点是不够的。只有使它免遭繁忙的交通之害，才会有鲜明的一致性。唐纳德·阿普尔亚德和马克·林特尔已经发现，在一个区内，交通越繁忙，把它看作安身立业之地的人就越少。居民不仅把交通繁忙的街道视为缺少个性，而且也感到沿街两侧的房屋同样缺少个性。（"Environmental Quality of City Strects," by Donald Appleyard and Mark Lintell,Center for Planning and Devclopment Research,University of California,Berkeley,1971）

小交通量邻里 每天 2000 辆次

高峰每小时 200 辆次 每小时双向 15 ～ 20mi

居民在谈论"与邻居相处和互访"时说：

"我感到像在家里一样。在这条街上有许多和蔼可亲的人。我不感到孤单寂寞。

人人都相互了解。"

这肯定是一条充满友爱的街道。

居民在谈论"家的领域感"时说：

"街道生活并不侵扰我的家……从街上带进来的只是一派欢乐。

我感到我的家扩展到整个街区。"

中交通量邻里 每天 6000 辆次

高峰每小时 550 辆次 每小时双向 25mi

居民在谈论"与邻居相处和互访"时说：

"你看得见邻居，但他们不是亲密无间的朋友。

除了人们见面互致问候外,你不会再感觉到有任何社区了。"

居民在谈论"家的领域感"时说：

"这是一块中间地带——不需要任何思索。"

大交通量邻里 每天 16000 辆次

neighborhood with light traffic 2000 vehicles/day

200 vehicles/peak hour 15-20 mph Two-way

Residents speaking on "neighboring and visiting"

I feel it's home.There are warm people on this street.I don't feel alone.

Everbody knows each other.

Definitely a friendly street.

Residents speaking on "home territory"

The street life doesn't intrude into the home...only happiness comes in from the street.

I feel my home extends to the whole block.

neighborhood with moderate traffic 6000 vehicles/day

550 vehicles/peak hour 25 mph Two-way

Residents speaking on "neighboring and visiting"

You see the neighbors but they aren't close friends.

Don't feel there is any community any note,but people say hello.

Residents speaking on "home territory"

It's a medium place—doesn't require any thought.

neighborhood with heavy traffic 16,000 vehicles/day

1900 vehicles/peak hour 35-40 mph One-way

Residents speaking on "neighboring and visiting"

It's not a friendly street—no one offers help.

People are afraid to go into the street because of the traffic.

Residents speaking on "home territory"

It is impersonal and public.

高峰每小时 1900 辆次 每小时单向 35 ～ 40mi

居民在谈论"与邻居相处和互访"时说:

"这不是一条充满友爱的街道——没有人会提供帮助。"

人们上街总是提心吊胆,惴惴不安,因为来往车辆太多。

居民在谈论"家的领域感"时说:

"这是没有个性的公共场所。

街上的噪声困扰着我的家庭。"

我们将怎样来说明主干路呢? 阿普尔亚德和林特尔在他们共同研究中发现, 每小时交通量超过 200 辆次, 邻里就开始蜕化。在这些街道上, 每小时交通量超过 550 辆次, 探亲访友的人就寥寥无几了, 在街上邂逅相遇而聊天的真是凤毛麟角。科林・布坎南的研究表明, 一旦"大多数居民(占 50 % 以上)……走路时都要小心翼翼地给车辆让路"主干路就会成为自由步行者的路障了。得出这一结论的根据是:"对于横穿马路的步行者来说, 平均耽误的时间为 2 秒……可作为介乎可接受的条件和不可接受的条件之间的两可情况的初步借鉴。"这一情况在交通量达到每小时 150 ～ 250 辆次时, 就会发生。(Colin D.Buchanan, *Traffic in Towns*, London:Her Majesty's Stationery Office,1963,p.204.) 因此, 任何街道的交通量每小时大于 200 辆次, 在任何时候, 这样的街道看起来大概就是"主干路"了, 并且开始损坏邻里的内部一致性。

最后要提一下如何贯彻执行的问题。几个月前, 我们对伯克利市的交通运输进行了一次实地考察, 这次考察涉及市内未来的所有主干路确定选址的问题。市民们曾被问及在哪些区域内他们想防止大量的交通车辆来往, 并且他们作出了回答。这种简单的要求已经引起四周广大的农业区纷纷成立政治组织:就在我们写本文的同时, 已有 30 多个小邻里统一起来, 连成一片, 其目的仅仅是他们在防止

Noise from the street intrudes into my home.

How shall we define a major road?The Appleyard-Lintell study found that with more than 200 cars per hour,the quality of the neighborhood begins to deteriorate.On the streets with 550 cars per hour people visit their neighbors less and never gather in the street to meet and talk.Research by Colin Buchanan indicates that major roads become a barrier to free pedestrian movement when "most people(more than 50%)...have to adapt their movement to give way to vehicles." This is based on "an average delay to all crossing pedestrians of 2 seconds...as a very rough guide to the borderline between acceptable and unacceptable conditions," which happens when the traffic reaches some 150 to 250 cars per hour.(Colin D.Buchanan,*Traffic in Towns*,London:Her Majesty's Stationery Office,1963,p.204.)Thus any street with greater than 200 cars per hour,at any time,will probably seem "major," and start to destroy the neighborhood identity.

A final note on implementation.Several months ago the City of Berkeley began a transportation survey with the idea of deciding the location of all future major arteries within the city.Citizens were asked to make statements about areas which they wanted to protect from heavy traffic.This simple request has caused widespread grass roots political organizing to take place:at the time of this writing more than 30 small neighborhoods have identified themselves,simply in order to make sure that they succeed in keeping heavy traffic out. In short,the issue of traffic is so fundamental to the fact of neighborhoods,that neighborhoods emerge,and crystallize,as soon as people are asked to decide where they want nearby traffic to be.Perhaps this is a universal way of implementing

大交通量方面能稳操胜券。简而言之，交通量问题是一个
亟待解决的重大问题，它关系到邻里的利益。一旦征求居
民们的意见，让他们自己决定靠近什么地方应当有交通干
线，邻里就会像雨后春笋般地出现，并日臻完善。或许这
是现存城市中实施本模式的一条普遍适用的途径。

因此：

**协助居民规定他们所居住的邻里的范围，其直径不得
超过 300yd，居民不得超过 400 ~ 500 人。在现有的城市内，
鼓励地方群体自己组织起来，以便形成这种邻里。至于税
收和土地控制，给邻里以一定程度的自治权。把主干路排
斥在这些邻里之外。**

最大人口数量为500人

最大直径为300yd

◈

首先用门道来标志邻里边界，门道是主要的道路通往
邻里的必经之地——**主门道**（53）；还要用邻里间非居住土
地的质朴无华的边界来标志——**邻里边界**（15）。把主干路
布置在这些边界内——**平行路**（23）；使邻里有一个惹人注
目的中心，也许是一块公共场地，或一块绿地——**近宅绿
地**（60）——或一个**小广场**（61）；把住宅和工间成团成组
地安排在邻里之内，每团组约 12 幢——**住宅团组**（37）、
工作社区（41）……

this pattern in existing cities.

Therefore:

Help people to define the neighborhoods they live in,not more than 300 yards across,with no more than 400 or 500 inhabitants.In existing cities,encourage local groups to organize the`nselves to form such neighborhoods.Give the neighborhoods some degree of autonomy as far as taxes and land controls are concerned.Keep major roads outside these neighborhoods.

❧❧

Mark the neighborhood,above all,by gateways wherever main paths enter it—MAIN GATEWAYS(53)—and by modest boundaries of non-residential land between the neighborhoods—NEIGHBORHOOD BOUNDARY(15). Keep major roads within these boundaries—PARALLEL ROADS(23);give the neighborhood a visible center,perhaps a common or a green—ACCESSIBLE GREEN(60)— or a SMALL PUBLIC SQUARE(61);and arrange houses and workshops within the neighborhood in clusters of about a dozen at a time—HOUSE CLUSTER(37),WORK COMMUNITY(41)...

模式15　邻里边界*

……空间边界是保护亚文化区不受彼此干扰并容许它们各自的生活方式标新立异、具有无与伦比的特色所不可缺少的。**亚文化区边界**（13）是**7000人的社区**（12）的可靠保证。但是，一种较小的二类边界是为创立较小的模式——**易识别的邻里**（14）所需要的。

〰️〰️

明显的边界对邻里是十分重要的。如果边界模糊不清，邻里将不能保持自己的容易识别的性质。

一个有机细胞的细胞膜，在大多数情况下，与细胞的内部一般大或较大。细胞膜不是将细胞分成内外两个部分的一个表面，而它本身就是一个自身有关联的统一体，它保持细胞功能上的完整性，并使细胞的内部和周围的液体

15 NEIGHBORHOOD BOUNDARY*

...the physical boundary needed to protect subcultures from one another,and to allow their ways of life to be unique and idiosyncratic,is guaranteed,for a COMMUNITY OF 7000(12),by the pattern SUBCULTURE BOUNDARY(13). But a second,smaller kind of boundary is needed to create the smaller IDENTIFIABLE NEIGHBORHOOD(14).

෨෬ଔ

The strength of the boundary is essential to a neighborhood.If the boundary is too weak the neighborhood will not be able to maintain its own identifiable character.

The cell wall of an organic cell is,in most cases,as large as,or larger,than the cell interior.It is not a surface which divides inside from outside,but a coherent entity in its own right,which preserves the functional integrity of the cell and also provides for a multitude of transactions between the cell interior and the ambient fluids.

We have already argued,in SUBCULTURE BOUNDARY (13),that a human group,with a specific life style,needs a boundary around it to protect its idiosyncrasies from encroachment and dilution by surrounding ways of life.This subculture boundary,then,functions just like a cell wall— it protects the subculture and creates space for its transactions with surrounding functions.

The argument applies as strongly to an individual neighborhood,which is a subculture in microcosm.

能够进行一系列的相互作用。

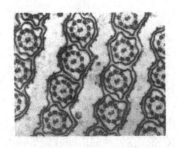

具有细胞膜的细胞：细胞膜适得其所
Cell with cell wall:The cell wall is a place in its own right

我们在**亚文化区边界**（13）中早已论证过，具有特殊生活方式的群体需要有它自己的边界，以免外界的生活方式冲击和削弱它自己的独特风格。由此可见，这种亚文化区边界所起的作用正如细胞膜一样：亚文化区边界保护亚文化区，并为亚文化区与周围的公共设施进行交流提供活动空间。

这一论据显然适用于有独特风格的邻里，它就是亚文化区的缩影。

可是，在亚文化区边界需要又宽又长的一片土地并有工商业活动的地方，邻里边界可能更加平淡无奇了。用商店和街道来划定一个有 500 或 500 以上人口的邻里的边界，确实是不可能的，而且社区的设施也简直无法把它围起来。当然，那里有若干邻里商店——**临街咖啡座**（88）、**街角杂货店**（89）——将有助于形成邻里的边界，但是，大体而言，邻里的边界将根据完全不同的形态学原理来形成。

纵观划定邻里边界的成功例子，我们从市民的体格和心理状态两个方面获知，邻里边界唯一最重要的特征是限制进入邻里的通路：因为成功地划定了边界的邻里有一定的、相对来说为数不多的小路和公路通入其内。

例如，下面是一幅伯克利市埃特纳邻里的地图。

However,where the subculture boundaries require wide swaths of land and commercial and industrial activity,the neighborhood boundaries can be much more modest.Indeed it is not possible for a neighborhood of 500 or more to bound itself with shops and streets and community facilities;there simply aren't enough to go around.Of course,the few neighborhood shops there are—the STREET CAFE (88),the CORNER GROCERY(89)—will help to form the edge of the neighborhood,but by and large the boundary of neighborhoods will have to come from a completely different morphological principle.

From observations of neighborhoods that succeed in being welldefined,both physically and in the minds of the townspeople,we have learned that the single most important feature of a neighborhood's boundary is restricted access into the neighborhood:neighborhoods that are successfully defined have definite and relatively few paths and roads leading into them.

For example,here is a map of the Etna Street neighborhood in Berkeley.

There are only seven roads into this neighborhood, compared with the fourteen which there would be in a typical part of the street grid.The other roads all dead end in T junctions immediately at the edge of the neighborhood. Thus,while the Etna Street neighborhood is not literally walled off from the community,access into it is subtly restricted. The result is that people do not come into the neighborhood by car unless they have business there;and when people are in the neighborhood,they recognize that they are in a distinct part of town.Of course,the neighborhood was

我们的邻里与一个典型的方格路网进行比较的情况
Our neighborhood, compared with a typical part of a grid system

　　如果将此邻里同一个具有14条道路的方格网街道的典型部分进行比较，那么，只有7条路进入该邻里。其他的路都有尽头，都在邻里边缘的丁字形交叉点上终止。因此，即使埃特纳街道邻里，从字面上讲，与社区没有用围墙隔开，但是，进入邻里的通路巧妙地受到了限制。结果是除非人们去那里办事，否则是不会有人开车去的。而且，当人们住在该邻里内，都异口同声地承认，他们是住在城市的一个独具特色的地方。当然，该邻里不是故意"创造"出来的。它过去是伯克利市的一个区，现在，由于在这街道系统中的这种偶然情况，它才变成了一个易识别的邻里。

　　这一原理的极端例子就是奥格斯堡的福格莱邻里，在**易识别的邻里**（14）中有一幅插图示出。福格莱邻里四周的边界都是由建筑物的背面和围墙围成，而且，通往该邻里的小路都很狭窄，都有门道标志。

　　的确，如果通往邻里的入口受到限制，这就意味着，根据定义，有可能进入的那几个点将具有特殊的重要性。它们实际上以这一种或另一种方式，更明显地说，将成为标志通往邻里的门道。我们将在**主门道**（53）中作更加充分的论述。然而事实是，每个成功的邻里是容易识别的，因为它有着标志它的边界的某种类型的门道。结果，这样的邻里边界在人们的脑海中就栩栩如生了，因为大家都认识门道。

　　为了避免门道这一概念有使人感到太封闭之嫌，我们就要刻不容缓地关注边界区，尤其是门道周围的那些地方，

not "created" deliberately.It was an area of Berkeley which has become an identifiable neighborhood because of this accident in the street system.

An extreme example of this principle is the Fuggerei in Augsburg,illustrated in IDENTIFIABLE NEIGHBORHOOD (14).The Fuggerei is entirely bounded by the backs of buildings and walls,and the paths into it are narrow,marked by gateways.

Indeed,if access is restricted,this means,by definition,that those few points where access is possible,will come to have special importance.In one way or another,subtly,or more obviously,they will be gateways,which mark the passage into the neighborhood.We discuss this more fully in MAIN GATEWAYS(53).But the fact is that every successful neighborhood is identifiable because it has some kind of gateways which mark its boundaries:the boundary comes alive in peoples'minds because they recognize the gateways.

In case the idea of gateways seems too closed,we remark at once that the boundary zone—and especially those parts of it around the gateways-must also form a kind of public meeting ground,where neighborhoods come together.If each neighborhood is a selfcontained entity,then the community of 7000 which the neighborhoods belong to will not control any of the land internal to the neighborhoods.But it will control all of the land between the neighborhoods—the boundary land—because this boundary land is just where functions common to all 7000 people must find space.In this sense the boundaries not only serve to protect individual neighborhoods,but simultaneously function to unite them in their larger processes.

也必须使之成为附近邻里的人都会来集聚的某种公共集会场所。如果每个社区都是自给自足的实体，那么，邻里所归属的 7000 人的社区，将不会控制邻里之内的任何土地，但是，它将控制邻里之间的全部土地，即边界地，因为这块边界地就是全社区 7000 人的公共设施正好要选址的空地。就此而言，邻里边界不仅为保护有独特个性的邻里而服务，而且同时起到联合许多邻里参加更大的社区发展活动的作用。

因此：

鼓励在每个邻里的周围形成一条边界，以便使与之毗邻的邻里分隔开来。用关闭街道和限制进入的通路的办法来形成这种边界，至少要砍掉街道正常数量的一半。在限制道路穿越边界的那些点上设置门道；并使边界区有足够的宽度，以便含有集会场所和若干邻里共用的公共设施。

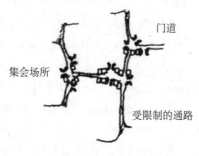

在邻里周围形成一条边界的最佳途径就是使建筑物的朝向向内；除了有一两条道路能通往设有门道的特定点——**主门道（53）**——之外，把穿越边界的全部道路统统砍掉。邻里边界内的公共土地可含有一个公园，几处道路交会点、若干小停车场和数个工作社区，包括可以形成天然边界的一切事物——**平行路（23）、工作社区（41）、**

TOWNS
城 镇
227

Therefore:

Encourage the formation of a boundary around each neighborhood,to separate it from the next door neighborhoods. Form this boundary by closing down streets and limiting access to the neighborhood—cut the normal number of streets at least in half.Place gateways at those points where the restricted access paths cross the boundary;and make the boundary zone wide enough to contain meeting places for the common functions shared by several neighborhoods.

ଈଓଔ

The easiest way of all to form a boundary around a neighborhood is by turning buildings inward, and by cutting off the paths which cross the boundary, except for one or two at special points which become gateways—MAIN GATEWAYS(53);the public land of the boundary may include a park,collector roads,small parking lots,and work communities— anything which forms a natural edge—PARALLEL ROADS(23),WORK COMMUNITY(41),QUIET BACKS(59),ACCESSIBLE GREEN(60),SHIELDED PARKING(97),SMALL PARKING LOTS (103). As for the meeting places in the boundary,they can be any of those neighborhood functions which invite gathering:a park,a shared garage,an outdoor room, a shopping street,a playground—SHOPPING STREET(32),POOLS AND STREAMS(64),PUBLIC OUTDOOR ROOM(69),GRAVE SITES(70),LOCAL SPORTS(72),ADVENTURE PLAYGROUND(73)...

僻静区（59）、近宅绿地（60）、有遮挡的停车场（97）和小停车场（103）。至于边界内的集会场所可以是邻里的如下公共设施中的任何一个，这些公共设施会吸引大家去聚会的：公园、公共停车库、户外亭榭、商业街和运动场——商业街（32）、小池和小溪（64）、户外亭榭（69）、墓地（70）、地方性运动场地（72）、冒险性的游戏场地（73）……

通过鼓励下列网络的发展，从而把各社区连接起来：

16.公共交通网

17.环路

18.学习网

19.商业网

20.小公共汽车

connect communities to one another by encouraging the growth of the following networks:

16.WEB OF PUBLIC TRANSPORTATION
17.RING ROADS
18.NETWORK OF LEARNING
19.WEB OF SHOPPING
20.MINI-BUSES

模式16　公共交通网*

……城市，如**指状城乡交错**（3）所规定的那样，呈带状向乡村扩展，并被分割成**地方交通区**（11）。为了连接各交通区，并保持人和物沿城市指状地带流动，就很有必要建立一个公共交通网。

⊱∘⊰

公共交通系统——飞机、直升飞机、汽垫船、火车、轮船、渡船、公共汽车、出租汽车、小火车、轻便马车、架空索道和自动人行道等的一个完整网络。只有各个部分协调地连接起来了，才能充分发挥作用。但是，通常它们并不会如此，因为掌管各种形式公共交通系统的不同机构，均无把它们相互连接起来的鼓励措施。

现在来扼要地谈一谈一般的公共交通问题。一个城市有大量的空间，相当均匀地分布在一张二维空间坐标的地图上。人们想确定旅行线路，典型的做法就是在这张地图上任意确定两点就行了。没有任何一种线性系统（像列车系统那样）能在市内密如蛛网的交通线两点之间提供直接连接。

因此，只有在各种不同的交通系统中存在着五花八门的连接，各种公共交通系统才有可能充分发挥效能。但是，除非这些连接是真正快速而又省时的，否则就是行不通的。等车的时间要短。从一种交通系统换乘另一种交通系统，步行距离要非常短。

这是再明显不过的了。而且，每个人想到公共交通，都会承认它的重要性。虽然道理人人明白，但真要实行起来却是困难重重。

16   WEB OF PUBLIC TRANSPORTATION*

...the city,as defined by CITY COUNTRY FINGERS
(3),spreads out in ribbon fashion,throughout the countryside,and
is broken into LOCAL TRANSPORT AREAS(11).To connect
the transport areas,and to maintain the flow of people and goods
along the fingers of the cities,it is now necessary to create a
web of public tranportation.

❧❧❧

**The system of public transportation—the entire web of
airplanes, helicopters,hovercraft,trains,boats,ferries,buses,
taxis, minitrains,carts,ski-lifts,moving sidewalks—can only work
if all the parts are well connected.But they usually aren't,because
the different agencies in charge of various forms of public
transportation have no incentives to connect to one another.**

Here,in brief,is the general public transportation problem.
A city contains a great number of places,distributed rather evenly
across a two-dimensional sheet.The trips people want to make are
typically between two points at random in this sheet.No one linear
system(like a train system),can give direct connections between
the vast possible number of point pairs in the city.

It is therefore only possible for systems of public
transportation to work,if there are rich connections between a
great variety of *different* systems.But these connections are not
workable,unless they are genuine fast,short,connections.The
waiting time for a connection must be short.And the walking
distance between the two connecting systems must be very short.

实际的困难有二，究其原因不外乎下面的这种情况：不同的机构将各种公共交通系统掌握在自己手中，它们不愿意进行合作，部分原因是它们实际上处于竞争之中，还有部分原因是合作会使它们更加难以生存。

在上下班的线路上情况尤为严重。火车、公共汽车、小公共汽车、快速运输、渡船，或许甚至还有飞机和直升飞机，都在这些指定线路上争夺同一旅客市场。一家独立的机构运营一种交通方式，再没有特殊鼓励措施来为比较不灵活的交通方式提供支线服务。许多运输公司甚至对快速运输、火车和渡轮都不愿意提供良好的支线连接，因为它们的这条上下班线路是盈利最多的。同样，在许多发展中国家的城市里，在各主要的上下班线路上，小公共汽车和巡回出租车都提供公共交通服务，招揽顾客，把许多本想乘公共汽车的旅客吸引过去了。这就造成了小型车辆为主要线路服务的局面，而开往市郊的公共汽车几乎都是空荡荡的，一般来说，公共汽车公司即使亏本也被要求为这些郊区服务。

由此可见，解决公共交通网的关键就在于是否可能解决各种不同系统的协调问题。这是问题的核心。我们现在提出一项解决办法。处理公共交通的传统方法是假设线路是第一位的，而把线路连接起来的换乘站是第二位的。而我们的建议恰恰与此相反：即换乘站是第一位因素，而和换乘站相连接的交通线路是第二位因素。

试设想如下的组织：每个换乘站都由使用它的社区来进行管理。社区给每个换乘站任命一位站长，拨给他预算，并规定服务守则。换乘站站长协调他站内的服务工作；他和不计其数的运输公司签订服务合同，这些运输公司本身为了打开工作局面而相互自由竞争。

在这个方案中，公共交通的责任性从交通线路转移到

This much is obvious;and everyone who has thought about public transportation recognizes its importance. However,obvious though it is,it is extremely hard to implement.

There are two practical difficulties,both of which stem from the fact that different kinds of public transportation are usually in the hands of different agencies who are reluctant to cooperate.They are reluctant to cooperate,partly because they are actually in competition,and partly just because cooperation makes life harder for them.

This is particularly true along commuting corridors. Trains,buses,minibuses,rapid transit,ferries,and maybe even planes and helicopters compete for the same passenger market along these corridors.When each mode is operated by an independent agency there is no particular incentive to provide feeder services to the more inflexible modes.Many services are even reluctant to provide good feeder connections to rapid transit,trains,and ferries,because their commuter lines are their most lucrative lines.Similarly,in many cities of the developing world,minibuses and *collectivos* provide public transportation along the main commuting corridors,pulling passengers away from buses.This leaves the mainlines served by small vehicles,while almost empty buses reach the peripheral lines,usually because the public bus company is required to serve these areas,even at a loss.

The solution to the web of public transportation,then,hinges on the possibility of solving the coordination problem of the different systems.This is the nut of the matter.We shall now propose a way of solving it.The traditional way of looking at public transportation assumes that lines are primary and that the interchanges needed to connect the lines to one another are

换乘站了。换乘站负责把线路彼此连接起来，使用换乘站的社区决定线路通过该站时要提供何种服务。从社区负责人到直接换乘站站长都要说服运输公司的负责人，让这些交通方式都经过换乘站。

连接各换乘站的运输公司将逐步建立起来。与我们的模式十分接近的一个实例就是闻名遐迩的瑞士铁路系统。这种系统表明这一模式同任何一种集中的代理机构相比，都能提供更高水平的服务。

瑞士铁路系统……是世界上最稠密的交通网。它票价贵且担风险，它始终如一地为最小的地方和最遥远的山谷服务。它这样做并不是为了赚钱，而是出于人民的意志。这是激烈的政治斗争的结果。在19世纪，"民主铁路运动"使瑞士的小型社区同已经拟定集中化计划的大城镇发生了冲突……如果我们将瑞士的铁路系统与法国的铁路系统（它有着令人羡慕的几何规则性）相比，后者完全集中于巴黎，所以，法国的繁荣和衰落，以及各地区的生死存亡都取决于同首都巴黎联系的性质如何，那么我们就会理解一个集权国家和一个联邦国家之间的差别了。铁路地图看似是最容易阅读的，一看便懂，其实不然，若我们现在把表明经济活动和人口流动的情况再添加到铁路地图上，就不那么容易读懂了。瑞士的工业分布遍及整个国土，就连偏远偏僻地区也不例外，这一点说明了这个国家的社会结构的强大有力和稳如磐石。它防止了19世纪那些令人恐惧的工业集中化、贫民窟和不定居的无产阶级的出现。

(Colin Ward, "The Organization of Anarchy," in *Patterns of Anarchy*, by Leonard I.Krimerman and Lewis Perry, New York, 1966)

因此：

视换乘站为第一位因素，交通线路为第二位因素。鼓励发展各种公共交通方式——飞机、直升飞机、渡船、轮船、

secondary.We propose the opposite:namely,that interchanges are primary and that the transport lines are secondary elements which connect the interchanges.

Imagine the following organization:each interchange is run by the community that uses it.The community appoints an interchange chief for every interchange,and gives him a budget,and a directive on service.The interchange chief coordinates the service at his interchange;he charters service from any number of transport companies—the companies,themselves,are in free competition with one another to create service.

In this scheme,responsibility for public transportation shifts from lines to interchanges.The interchanges are responsible for connecting themselves to each other,and the community which uses the interchange decides what kinds of service they want to have passing through it.It is then up to the interchange chief to persuade these transport modes to pass through it.

Slowly,a service connecting interchanges will build up.One example which closely follows our model,and shows that this model is capable of producing a higher level of service than any centralized agency can produce,is the famous Swiss Railway System.

The Swiss railway system...is the densest network in the world.At great cost and with great trouble,it has been made to serve the needs of the smallest localities and most remote valleys,not as a paying proposition but because such was the will of the people.It is the outcome of fierce political struggles.In the 19th century,the "democratic railway movement" brought the small Swiss communities into conflict with the big towns,which had plans for centralisation....And if we compare the Swiss system with the French which,with admirable geometrical regularity,is entirely centered on Paris so that the prosperities or the decline,the life or

火车、快速运输、公共汽车、小公共汽车、架空索道、自动扶梯、自动人行道和电梯——并进行规划，把它们的线路与各换乘站连接起来，同时希望许多不同类型的许多不同线路逐步地在各换乘站交会。

让地方社区来控制换乘站，以便社区只同为这些换乘站服务的运输公司签订合同，以求达到实现这一模式的目的。

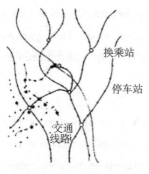

将交会于一个单一的换乘站的各种不同的交通线路及其停车场，都集中在 600 英尺的范围之内，以便人们能够步行去转车——**换乘站**（34）。十分重要的是，应当有良好的支线系统为主要车站服务，这样一来，人们就根本不会再被迫使用私人汽车了——**小公共汽车**（20）······

death of whole regions has depended on the quality of the link with the capital,we see the difference between a centralised state and a federal alliance.The railway map is the easiest to read at a glance,but let us now superimpose on it another showing economic activity and the movement of population.The distribution of industrial activity all over Switzerland,even in the outlying areas,accounts for the strength and stability of the social structure of the country and prevented those horrible 19th century concentrations of industry,with their slums and rootless proletariat.(Colin Ward, "The Organization of Anarchy," in Patterns of Anarchy,by Leonard I.Krimerman and Lewis Perry,New York,1966.)

Therefore:

Treat interchanges as primary and transportation lines as secondary.Create incentives so that all the different modes of public transportation—airplanes,helicopters,ferries,boats,trains,rapid transit,buses,minibuses,skilifts,escalators,travelators,elevatorsplan their lines to connect the interchanges,with the hope that gradually many different lines,of many different types,will meet at every interchange.

Give the local communities control over their interchanges so that they can implement the pattern by giving contracts only to those transportation companies which are willing to serve these interchanges.

❧❧❧

Keep all the various lines that converge on a single interchange,and their parking,within 600 feet,so that people can transfer on foot—INTERCHANGE(34).It is essential that the major stations be served by a good feeder system,so people are not forced to use private cars at all—MINIBUSES(20)...

模式17　环路

　　……本模式所详述的环路有助于规定并形成**地方交通区**（11）；如果环路为了起到连接作用而布置在**换乘站**（34）之间，则同样有助于形成**公共交通网**（16）。

❧❦

　　现代社会需要高速公路，这是无法回避的。而十分重要的是布置和修建高速公路不能毁坏社区或乡村。

　　由于地方的广泛抗议，修建快车道和超级公路的热潮于20世纪五六十年代渐趋缓和，但是我们还是不能完全回避高速公路。现在还未呈现出另一种切实可行的方案的前景——这种方案要能为现代城市在经济上和社会上赖以生存的轿车、卡车和公共汽车提供巨大的移动容量。

17 RING ROADS

...the ring roads which this pattern specifies,help to define and generate the LOCAL TRANSPORT AREAS (11);if they are placed to make connections between INTERCHANGES(34),they also help to form the WEB OF PUBLIC TRANSPORTATION(16).

❧❦

It is not possible to avoid the need for high speed roads in modern society;but it is essential to place them and build them in such a way that they do not destroy communities or countryside.

Even though the rush of freeways and superhighways built in the 1950's and I960's is slowing down,because of widespread local protest,we cannot avoid high speed roads altogether.There is,at present,no prospect for a viable alternative which can provide for the vast volume of movement of cars and trucks and buses which a modern city lives on economically and socially.

At the same time,however,high speed roads do enormous damage when they are badly placed.They slice communities in half;they cut off waterfronts;they cut off access to the countryside;and,above all,they create enormous noise.For hundreds of yards,even a mile or two,the noise of every superhighway roars in the background.

To resolve these obvious dilemmas that come with the location and construction of high speed roads,we must find

可是,高速公路布置不当,确实会同时造成严重的破坏。高速公路会把社区分割成两半,切断通往滨水区和乡村之路,而尤为严重的是会产生大量的噪声。每条超级公路都会发出震耳欲聋的巨响,在周围几百码甚至 1 ~ 2mi 范围内的地方都清晰可闻。

为了摆脱这种困境,我们必须寻找出布置和修筑高速公路的新途径,以免它们毁坏社区,人民的身心健康受到噪声干扰的伤害。为此,我们提出三项要求,我们认为,这三项要求可作为下列政策的核心内容。

1. 每一个具有一致性的社区作为一个地方交通区——**地方交通区**(11)——永远不再被高速公路所分裂,更确切地说,它至少和一条高速公路邻接。这就使快速汽车能从这样一个社区到另外一些社区,在整个区域畅通无阻地驰骋了。

2. 每一地方交通区要保证居民不横越高速公路,就能到达辽阔的乡村——参阅**指状城乡交错**(3)。大概,这意味着高速公路的选址应当做到:每一地方交通区至少一侧可直通广阔无垠的乡村。

3. 为了保护高速公路周围地区的生命安全,高速公路的隔声问题必须首先加以解决。这意味着高速公路必须修筑成低于地面的公路,这样才能屏蔽噪声,或用小土路、停车结构物、货栈等阻挡噪声,而这些结构物将不会受到噪声的破坏。

因此:

将高速公路(快车道或其他主要干线)这样来安排,以便:

1. 至少有一条高速公路位于与一地方交通区相切的地方。

2. 每一地方交通区至少有一侧不和高速公路相邻接,而能直通乡村。

ways of building and locating these roads,so that they do not destroy communities and shatter life with their noise.We can give three requirements that,we believe,go to the heart of this policy:

1.Every community that has coherence as an area of local transportation—LOCAL TRANSPORT AREAS(11)—is never split by a high speed road,but rather has at least one high speed road adjacent to it.This allows rapid auto travel from one such community out to other communities and to the region at large.

2.It must be possible for residents of each local transport area to reach the open countryside without crossing a high speed road—see CITY COUNTRY FINGERS(3).This means,very roughly,that high speed roads must always be placed in such positions that at least one side of every local transport area has direct access to open country.

3.Most important of all,high speed roads must be shielded acoustically to protect the life around them.This means that they must either be sunken,or shielded by earth berms,parking structures,or warehouses,which will not be damaged by the noise.

Therefore:

Place high speed roads(freeways and other major arteries)so that:

1.At least one high speed road lies tangent to each local transport area.

2.Each local transport area has at least one side not bounded by a high speed road,but directly open to the countryside.

3.The road is always sunken,or shielded along its length by berms,or earth,or industrial buildings,to protect the nearby neighborhoods from noise.

3. 为了保护相邻的邻里不受噪声危害，高速公路一般都低于地面，或在它的沿线用小土路、土堆或工业建筑来屏蔽噪声。

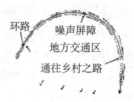

环路
噪声屏障
地方交通区
通往乡村之路

⌘⌘⌘

永远把高速公路布置在亚文化区之间的边界上——**亚文化区边界**（13），而且绝不通过滨水区——**通往水域**（25）。把工业建筑和大型停车库建造在高速公路两侧，无论何时，都要尽量利用它们作为特大的噪声屏障——**工业带**（42）、**有遮挡的停车场**（97）……

❧❀❦

Always place the high speed roads on boundaries between subcultures—SUBCULTURE BOUNDARY(13)and never along waterfronts—ACCESS TO WATER(25).Place industry and big parking garages next to the roads,and use them,whenever possible,as extra noise shields—INDUSTRIAL RIBBONS(42),SHIELDED PARKING(97)...

模式18　学习网*

　　……与公共交通网络可相提并论的就是学习网络，它不是在空间上而是在概念上与前者具有同等的重要性。全市层出不穷的、相互间有着千丝万缕联系的种种情况，事实上编成了一份城市"课程表"：城市要向它的青年传授生活方式。

✂✄✆

　　在一个强调教授知识的社会中，儿童、学生和成年人都会变成消极的人，他们不能进行独立思考或独立行动。而且有创造性的、积极进取的人只有在强调学习而不是强调教授的社会中才能茁壮成长。

18   NETWORK OF LEARNING*

...another network,not physical like transportation,but conceptual,and equal in importance,is the network of learning: the thousands of interconnected situations that occur all over the city,and which in fact comprise the city's "curriculum" :the way of life it teaches to its young.

⊰⊱

In a society which emphasizes teaching,children and students—and adults—become passive and unable to think or act for themselves.Creative,active individuals can only grow up in a society which emphasizes learning instead of teaching.

There is no need to add to the criticism of our public schools.The critique is extensive and can hardly be improved on.The processes of learning and teaching,too,have been exhaustivdy studied...The question now is what to do.(George Dennison,*Lives of Children*,New York:Vintage Books,1969,p.3.)

To date,the most penetrating analysis and proposal for an alternative framework.for education comes from Ivan Illich in his book,*De-Schooling Society*,and his article, "Education without Schools:How It Can Be Done," in the *New York Review of Books*,New York,15(12):25-31,special supplement,July 1971.

Illich describes a style of learning that is quite the opposite from schools.It is geared especially to the rich opportunities for

不必再批评我们的公立学校了。泛泛的批评几乎无济于事。对学习的过程和教学的过程都做了详尽无遗的研究……现在的问题是能做些什么。(George Dennison, *Lives of Children*, NewYork:Vintage Books,1969,p.3.)

迄今为止，只有伊凡·伊里奇在《无正规学校教育的社会》一书中以及在《无学校的教育：怎样才能办到这一点》一文中对教与学作了剔透入里的分析并提出了精辟的见解。(*New York Review of Books*,New York,15（12）：25～31,special supplement,July 1971)

伊里奇描写了一种学习方式，它和目前的学校教育呈鲜明的对照。这种学习方式尤其适用于丰富多彩的学习机会，而这些机会自然地存在于每一个大都会中：

由社会控制学校的另一种比较方案是人们自愿地参与社会生活，这种社会生活由各种网络交织而成，从而为学习社会的各个方面提供了良好的机会。事实上，现在存在着这些网络，但很少用来为教育目的服务。正规教育的危机，如果将来会有任何肯定的后果的话，则将不可避免地导致各种网络和教育过程的结合……

现在的学校是根据下列假设设计出来的：世间万物都有奥秘；人一生的品格有赖于对那奥秘认识的深浅；只有在秩序井然的世代相传中才能逐渐认识奥秘；而且只有教师才能恰如其分地揭示这些奥秘。受过学校正规教育的人，往往把世界想象成为一座具有分门别类奥秘的金字塔，只有拿到文凭的人才能到达这座金字塔。新的教育机构会瓦解这座金字塔。其目的是帮助初学者在通往知识宝库的路途中排忧解难：如果初学者不能登堂入室，也要让他透过控制室或议会的窗户窥豹一斑。而且这些新的机构应当是既无文凭又非名门出身的初学者都可以去的知识渠道——公共场所。在那里，贵族和元老们在他的眼皮底下统统消声匿迹了……

learning that are natural to every metropolitan area:

The alternative to social control through the schools is the voluntary participation in society through networks which provide access to all its resources for learning.In fact these networks now exist,but they are rarely used for educational purposes.The crisis of schooling,if it is to have any positive consequence,will inevitably lead to their incorporation into the educational process...

Schools are designed on the assumption that there is a secret to everything in life;that the quality of life depends on knowing that secret;that secrets can be known only in orderly successions;and that only teachers can properly reveal these secrets.An individual with a schooled mind conceives of the world as a pyramid of classified packages accessible only to those who carry the proper tags.*New educational institutions would break apart this pyramid.Their purpose must be to facilitate access for the learner:to allow him to look into the windows of the control room or the parliament,if he cannot get in the door.*Moreover,such new institutions should be channels to which the learner would have access without credentials or pedigree—public spaces in which peers and elders outside his immediate horizon now become available...

While network administrators would concentrate primarily on the building and maintenance of roads providing access to resources,the pedagogue would help the student to find the path which for him could lead fastest to his goal.If a student wants to learn spoken Cantonese from a Chinese neighbor,the pedagogue would be available to judge their proficiency,and to help them select the textbook and methods most suitable to their talents,character,and the time available for study.He can counsel

正如市政网络的行政人员为提供接近资源的通道而把注意力首先集中于道路的修筑和养护一样，教育家也会帮助学生寻找一条达到其目标的捷径。如果学生想从中国邻居那里学习广东话，教育家会亲临评判他们说的熟练程度，并协助他们挑选最适合于他们的天资、性格与时间的教科书，并指导他们的学习方法。教育家能向一心一意想成为飞机技师的人建议一个最适合当学徒的地方。教育家能向那些想寻找爱争论的元老去讨论非洲历史的人介绍书籍。教育家和市政网络的行政人员一样，视自己为职业教育家。任何个人使用教育证书，既可去学校，也可找这些教育家学习……

　　卡耐基（1835～1919，生于苏格兰，美国钢铁工业家及慈善家——译注）委员会除了在它的报告中作出的试验性结论外，又提出了一系列的重要文件，表明负责人们正在逐步意识到这样一个事实：以毕业证书为目标的正规学校教育不能继续被指望为现代社会的中心教育机构了。坦桑尼亚的朱利叶斯·奈厄已经宣布把教育同农村生活相结合的计划。在加拿大，赖特委员会在关于高中以后的教育的报告中说，没有一种已知的正规教育系统能为安大略的公民提供均等的机会。秘鲁总统已经接受其委员会的推荐书，该委员会建议，为了有利于提供终身免费教育的机会，取消免费的学校。事实上，该委员会报告总统是希望他坚持：这一方案在最初实施阶段不要操之过急，这样可以稳住学校的教师，并使他们脱离纯粹的教育家的道路。（Abridged from pp.76 and 99 in Deschooling Society by Ivan Illich.Vol.44 in World Perspectives Series,edited by Ruth Nanda Anshen,New York:Harper&Row,1971.）

　　总之，如此迅猛分散的教育制度本身与城市的结构日趋一致。各行各业的人们不断涌现，并传授他们所熟知和热爱的各种学问：专业人员和工作小组在他们的办

the would-be airplane mechanic on finding the best places for apprenticeship.He can recommend books to somebody who wants to find challenging peers to discuss African history.Like the network administrator,the pedagogical counselor conceives of himself as a professional educator.Access to either could be gained by individuals through the use of educational vouchers...

In addition to the tentative conclusions of the Carnegie Commission reports,the last year has brought forth a series of important documents which show that responsible people are becoming aware of the fact that schooling for certification cannot continue to be counted upon as the central educational device of a modern society.Julius Nyere of Tanzania has announced plans to integrate education with the life of the village.In Canada,the Wright Commission on post—secondary education has reported that no known system of formal education could provide equal opportunities for the citizens of Ontario.The president of Peru has accepted the recommendation of his commission on education,which proposes to abolish free schools in favor of free educational opportunities provided throughout life.In fact he is reported to have insisted that this program proceed slowly at first in order to keep teachers in school and out of the way of true educators.(Abridged from pp.76-99 in Deschooling Society by Ivan Illich.Vol.44 in World Perspectives Series,edited by Ruth Nanda Anshen,New York:Harper&Row,1971.)

In short,the educational system so radically decentralized becomes congruent with the urban structure itself.People of all walks of life come forth,and offer a class in the things they know and love:professionals and workgroups offer apprenticeships in their offices and workshops,old

公室和车间里提供业务培训，老年人传授他们的亲身经历、工作经验和感兴趣的东西，专家们当家庭教师讲授专业课程。生活和学习已融为一体。不难想象，最后，每3户或4户人家将至少有一人在家里授课或从事某种培训。

因此：

逐步实现学习过程的非集中化，以此来取代在一个固定的地方的因循守旧的义务教育，并通过与全市的许多地方和市民的广泛接触，来丰富学习的内容：车间、家庭教师或走街串巷的教师、热心帮助青年的行家、教小孩子的大孩子、博物馆、旅游的青年小组、学术讨论会、工厂、老年人等。设想所有这些情况构成学习过程的主要内容；考察所有这些情况，描述它们，并将它们排成城市的"课程表"而公诸于众；然后，让学生、儿童、他们的家庭和邻里把这些情况编排在一起，他们就有学可上了，当他们获得合格证书时，要付"学费"，这是为社区纳税而筹集的款项。建造新的教育设施，以便扩大和丰富这种学习网。

学习网点名称地址示意图

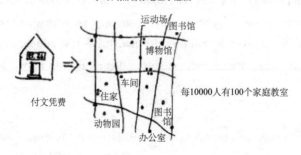

付文凭费　　运动场　图书馆　博物馆　车间　每10000人有100个家庭教室　住家　动物园　图书馆　办公室

首先，鼓励在老百姓家里组织讨论会和建立车间——**家庭工作间**（157）；确保每一城市有一条"小路"，供幼儿

people offer to teach whatever their life work and interest has been,specialists offer tutoring in their special subjects. Living and learning are the same.It is not hard to imagine that eventually every third or fourth household will have at least one person in it who is offering a class or training of some kind.

Therefore:

Instead of the lock-step of compulsory schooling in a fixed place,work in piecemeal ways to decentralize the process of learning and enrich it through contact with many places and people all over the city:workshops,teachers at home or walking through the city,professionals willing to take on the young as helpers,older children teaching younger children,museums,youth groups traveling,scholarly seminars,industrial workshops,old people,and so on. Conceive of all these situations as forming the backbone of the learning process;survey all these situations,describe them,and publish them as the city's "curriculum" ;then let students,children,their families and neighborhoods weave together for themselves the situations that comprise their "school" paying as they go with standard vouchers,raised by community tax.Build new educational facilities in a way which extends and enriches this network.

☜✿☞

Above all,encourage the formation of seminars and workshops in people's homes—HOME WORKSHOP (157);make sure that each city has a "path" where young children can safely wander on their own—CHILDREN IN THE CITY(57);build extra public "homes" for children,one to

安全漫步之用——**市区内的儿童**（57）；建造特大的公共的儿童之"家"，每个邻里至少一个——**儿童之家**（86）；在以工作和商业活动为主的城镇，创建大量的以工作为导向的小学——**店面小学**（85）；鼓励青少年成立自我组织的学习团体——**青少年协会**（84）；把大学办成全区域所有成年人分散的学习场所——**像市场一样开放的大学**（43）；并将专业人士和商人的实际工作作为网络中的基本节点——**师徒情谊**（83）……

every neighborhood at least—CHILDREN'S HOME(86);create a large number of work-oriented small schools in those parts of town dominated by work and commercial activity—SHOPFRONT SCHOOLS(85);encourage teenagers to work out a self-organized learning society of their own—TEENAGE SOCIETY(84);treat the university as scattered adult learning for all the adults in the region—UNIVERSITY AS A MARKETPLACE(43);and use the real work of professionals and tradesmen as the basic nodes in the network—MASTER AND APPRENTICES(83)...

模式19 商业网*

……本模式说明一种逐步发展的过程，它有助于商店和服务行业在需要它们的地方选定地址和合理布局，以便加强**亚文化区的镶嵌**（8）、**亚文化区边界**（13）和**分散的工作点**（9）所需的非集中化经济及**地方交通区**（11）。

∞०३

商店所处的位置要有利于其本身最佳地为人民的需要服务，并且也保证其自身的稳定性，这样的两全其美实属罕见。

城镇的大部分地区服务网点不足。能够提供各种服务的新商店往往开设在其他商店和主要中心附近，而不是坐落在需要它们的地方。在一个理想的城镇中，商店被视为社会需要的一部分，而不单单是商业链中的一种盈利手段，商店本应当比今天分布得更为广泛和均匀。

许多小商店是不稳固的，情况确实也是如此。人们开办的小商店中的三分之二通常冷冷清清，一年内没有做成一笔生意。显而易见，不稳定的商业不能为社区服务得很周到，再重复一遍，这些小商店经济上的不稳定性，在很大的程度上是与选址错误密切有关的。

为了保证商店是稳定的，而且能满足社区的需要，每个新商店必须开设在提供大致相同服务的各商店之间的空档内，并确保在顾客所必经的地方，这是为了生存所必需的。现在我们试图用准确无误的术语来阐明这一原则。

商店的稳定系统的特征是众所周知的。这种特征实质

19   WEB OF SHOPPING*

...this pattern defines a piecemeal process which can help to locate shops and services where they are needed,in such a way that they will strengthen the MOSAIC OF SUBCULTURES(8),SUBCULTURE BOUNDARIES(13),and the decentralized economy needed for SCATTERED WORK(9) and LOCAL TRANSPORT AREAS(11).

୫୬ଓଷ

Shops rarely place themselves in those positions which best serve the people's needs,and also guarantee their own stability.

Large parts of towns have insufficient services.New shops which could provide these services often locate near the other shops and major centers,instead of locating themselves where they are needed.In an ideal town,where the shops are seen as part of the society's necessities and not merely as a way of making profit for the shopping chains,the shops would be much more widely and more homogeneously distributed than they are today.

It is also true that many small shops are unstable.Two-thirds of the small shops that people open go out of business within a year.Obviously,the community is not well served by unstable businesses,and once again,their economic instability is largely linked to mistakes of location.

To guarantee that shops are stable,as well as distributed to meet community needs,each new shop must be placed where it

上是受下列思想支配的结果：每个商业单位都有一个顾客会集点，它为了生存需要顾客，因此，任何规模和任何类型的商店，如果分布均匀，而且每个商店都处于一个大得足以使它能维持生存的顾客会集点的中心，那么，它们都将是稳定的。

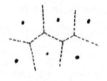

顾客会集点
Catch basins

商店和购物中心不总是自动地按照它们专有的顾客会集点来分布的，这种情况的原因是很容易用所谓"霍特林难题"来解释清楚的。请畅想一番夏日的海滨，海滨某处有一冰激凌销售者。现在假设你也是一个冰激凌销售者。你到达了海滨。你相对于第一个销售者,应在何处设点叫卖? 有两种可能的解决方案。

第一种情况：你和他平分秋色，把海滨一分为二，各占一半。在此情况下，你会尽量离开他，在海滩上有一半人更靠近你；而不是靠近他的地方停下来和他竞争。

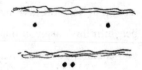

解决这个冰激凌问题的两种办法
Two approaches to the ice-cream problem

第二种情况：你选的点正好挨着他。总而言之，你决定和他竞争，决一雌雄，一比高低。你现在所处的位置是指挥整个海滩的游人而不是一半的游人来购买你的冰激凌了。

每当一个商店或一个购物中心开设的时候，都面临着

will fill a gap among the other shops offering a roughly similar service and also be assured that it will get the threshold of customers which it needs in order to survive.We shall now try to express this principle in precise terms.

The characteristics of a stable system of shops is rather well known.It relies,essentially,on the idea that each unit of shopping has a certain catch basin—the population which it needs in order to survive—and that units of any given type and size will therefore be stable if they are evenly distributed,each one at the center of a catch basin large enough to support it.

The reason that shops and shopping centers do not always,automatically,distribute themselves according to their appropriate catch basins is easily explained by the situation known as Hotelling's problem.Imagine a beach in summer time—and,somewhere along the beach,an ice-cream seller. Suppose now,that you are also an ice-cream seller.You arrive on the beach.Where should you place yourself in relation to the first ice-cream seller?There are two possible solutions.

In the first case,you essentially decide to split the beach with the other ice-cream seller.You take half the beach,and leave him half the beach.In this case,you place yourself as far away from him as you can,in a position where half the people on the beach are nearer to you than to him.

In the second case,you place yourself right next to him. You decide,in short,to try and compete with him—and place yourself in such a way as to command the whole beach,not half of it.

Every time a shop,or shopping center opens,it faces a similar choice.It can either locate in a new area where there

同样的选择。它既可以开设在没有竞争对手的新区，又可以凑巧开设在其他各行各业都希望能吸引对方顾客的剧烈竞争区。

很简单，麻烦就在于人们倾向于选择两者之中的第二种，因为，从表面上看来，这似乎是更加安稳的。而事实上第一种选择更好、更安稳。这对顾客更好，因为他们可以在靠近家门和工作地点的为他们服务的商店中购买物品；这对店主也更安稳，因为他们的商店，不管表面上如何，一旦站稳脚跟，则更加可能生存下去，他们不会有需要他们服务的顾客会集点中心那样的剧烈竞争。

现在让我们来考虑一下商业网的总体性质吧。在现代的城市中，同类商店倾向于聚集在购物中心。它们被迫聚集在一起，部分原因是分区法令禁止它们在所谓的居民区选址；店主错误的经营思想驱使他们把商店集中：对与别的商店竞争和共同拥有大致相等人数的前来购物的顾客这两者权衡利弊，店主们认为，前者对于他们则更加有利可图。在"形形色色的人"网中，我们一直建议，商店要均匀地扩散开去，少强调竞争而多强调服务。当然，仍然要有足够的竞争，以便确保经营不善的商店停业，因为每个商店，如果善于经营，提供优质服务，都能将顾客会集点的顾客吸引过来——但着重点是强调合作而不是竞争。

现存的商业网
The existing web

are no other competing businesses,or it can place itself exactly where all the other businesses are already in the hope of attracting their customers away from them.

The trouble is,very simply,that people tend to choose the second of these two alternatives,because it seems,on the surface,to be safer.In fact,however,the first of the two choices is both better and safer.It is better for the customers, who then have stores to serve them closer to their homes and work places than they do now;and it is safer for the shopkeepers themselves since—in spite of appearances— their stores are much more likely to survive when they stand,without competition,in the middle of a catch basin which needs their services.

Let us now consider the global nature of a web which has this character.In present cities,shops of similar types tend to be clustered in shopping centers.They are forced to cluster,in part because of zoning ordinances,which forbid them to locate in socalled residential areas;and they are encouraged to cluster by their mistaken notion that competition with other shops will serve them better than roughly equal sharing of the available customers.In the "peoples" web we are proposing,shops are far more evenly spread out,with less emphasis on competition and greater emphasis on service. Of course,there will still be competition,enough to make sure that very bad shops go out of business,because each shop will be capable of drawing customers from the nearby catch basins if it offers better service—but the accent is on cooperation instead of competition.

To generate this kind of homogeneous people's web,it is only necessary that each new shop follow the following three-

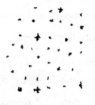

形形色色的人网
The peoples' web

为了形成这种均匀的人网，每个新商店选址时，只要遵循下列三步骤程序就行了：

1. 对你所感兴趣的服务项目的所有其他商店一一进行鉴定，并把它们逐一标示在地图上。

2. 鉴定并标出潜在顾客的位置。无论何地，只要有可能，都要标明任何一个划定区的潜在顾客的密度或总数。

3. 在有潜在顾客的那些区的现存商店网中，寻找最大的空档。

各服务行业间的空档
The gap in services

我们的两位同事已经试验了由这种程序所创建的商业网的效率和潜在稳定性。（"Computer Simulation of Market Location in an Urban Area," S.Angel and F.Loetterle,CES files,June 1967.）他们选择了一些市场进行研究。他们是从下列几方面着手研究的：一个划定的区、已知的人口密度和购买力，以及不同规模的市场的随机分布。他们接着创建了新的市场，并根据下面的规则砍掉了旧市场。这些规则是：（1）在所有的现存市场中，抹掉那些买卖不兴隆而无法维持现有规模的市场；（2）在各种可能的位置中为新

step procedure when it chooses a location:

1.Identify all other shops which offer the service you are interested in;locate them on the map.

2.Identify and map the location of potential consumers. Wherever possible,indicate the density or total number of potential consumers in any given area.

3.Look for the biggest gap in the existing web of shops in those areas where there are potential consumers.

Two colleagues of ours have tested the efficiency and potential stability of the webs created by this procedure. ("Computer Simulation of Market Location in an Urban Area," S.Angel and F.Loetterle,CES files,June 1967.)They chose to study markets.They began with a fixed area,a known population density and purchasing power,and a random distribution of markets of different sizes.They then created new markets and killed off old markets according to the following rules.(1)Among all of the existing markets,erase any that do not capture sufficient business to support their given size;(2) among all of the possible locations for a new market,find the one which would most strongly support a new market;(3)find that size for the new market that would be most economically feasible;(4)find that market among all those now existing that is the least economically feasible,and erase it from the web;(5) repeat steps(2)through(4)until no further improvement in the web can be made.

Under the impact of these rules,the random distribution of markets at the beginning leads gradually to a fluctuating, pulsating distribution of markets which remains economically stable throughout its changes.

Now of course,even if shops of the same kind are kept

市场寻找一个最有利的位置，以便使新市场获得最强有力的支持；（3）为新市场找到一个在经济上最切实可行的规模；（4）从现存的市场中找出在经济上最不合算的市场，并把它从商业网中抹掉；（5）通过规则（4）重复规则（2）直至在商业网中无法再进一步改进为止。

在这些规则的影响下，市场的随机分布在开始阶段逐渐地导致市场的波动，即脉动分布，这种脉动分布贯穿市场随机分布的始终，在经济上仍然是稳定的。

现在，当然，由于采用这一程序，同类的商店会分离开来，而不同类的商店将趋于聚集。理由十分简单，这是方便顾客的结果。如果我们遵循上述选址的规则——总是把新商店开设在同类商店网之间的最大空档内——那么，在空档内还有大量的空间可供选址之用：很自然，我们将设法把商店的位置选在空档内其他商店最大的团组附近，以求增加新商店的顾客流量，为顾客提供更加方便的服务。

贝利非常透彻地研究了不断出现的商店团组，结果证明：聚集的等级明显相似，即使它们之间的间距随人口密度的大小而有极大的不同。(See *Geography of Market Centers and Retail Distribution*,B.Berry,Englewood Cliffs,New Jersey:Prentice-Hall,Inc.,1967，pp.32～33) 在这一商店的聚集网中，这些因素与本语言中所描述的若干模式完全相符。

因此：

当你为任何一个个体商店选择店址时，要遵循上述的三步骤程序。

1. 对为你提供你所感兴趣的服务项目的所有其他商店一一进行鉴定，并把它们逐一标示在地图上。

2. 鉴定并标出潜在顾客的位置。无论何地，只要有可能，都要标明任何一个划定区的潜在顾客的密度或总数。

3. 在有潜在顾客的那些区的现存商店网中，寻找最大的空档。

apart by this procedure,shops of different kinds will tend to cluster.This follows,simply,from the convenience of the shopper.If we follow the rules of location given above—always locating a new shop in the biggest gap in the web of similar shops—then,within that gap there are still quite a large number of different possible places to locate:and naturally,we shall try to locate near the largest cluster of other shops within that gap,to increase the number of people coming past the shop,in short,to make it more convenient for shoppers.

The clusters which emerge have been thoroughly studied by Berry.It tums out that the levels of clustering are remarkably similar,even though their spacing varies greatly according to population density.(See Geography of Market Centers and Retail Distribution,B.Berry,Englewood Cliffs,New Jersey:Prentice-Hall,Inc.,1967,pp.32-33.)The elements in this web of clustering correspond closely to patterns defined in this language.

Therefore:

When you locate any individual shop,follow a threestep procedure:

1.Identify all other shops which offer the service you are interested in;locate them on the map.

2.Identify and map the location of potential consumers. Wherever possible,indicate the density or total number of potential consumers in any given area.

3.Look for the biggest gap in the existing web of shops in those areas where there are potential consumers.

4.Within the gap in the web of similar shops,locate your shop next to the largest cluster of other kinds of shops.

4. 在同类商店网的空档内，把你的商店开设在其他类型的商店最大的团组附近。

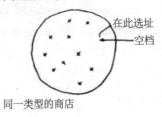

在此选址
空档

同一类型的商店

&oc&

我们估计，在这一规则的影响下，具有下列综合特征的商业网定将出现：

	人口	相隔距离/mi
城市的魅力（10）	300000	10*
散步场所（31）	50000	4*
商业街（32）	10000	1.8*
综合商场（46）	4000	1.1*
街角杂货店（89）	1000	0.5*

* 上述距离是以每平方英里的总人口密度为 5000 人求算出来的。取人口密度为 D 人 /mi^2，以 $\sqrt{D/5000}$ 除以距离……

We estimate,that under the impact of this rule,a web of shopping with the following overall characteristics will emerge:

	Population	Distance Apart (Miles)
MAGIC OF THE CITY(10)	300,000	10*
PROMENADES(31)	50,000	4*
SHOPPING STREETS(32)	10,000	1.8*
MARKETS OF MANY SHOPS(46)	4,000	1.1*
CORNER GROCERIES(89)	1,000	0.5*

*These distances are calculated for an overall population density of 5000 per square mile.For a population density of D persons/square mile,divide the distances by $\sqrt{D/5000}$ ...

模式20　小公共汽车*

……本模式有助于完善**地方交通区**（11）和**公共交通网**（16）。地方交通区主要依靠步行、自行车、马车和马。公共交通网依靠火车、飞机和公共汽车。这两种交通模式都需要一种更为灵活的公共交通来支援，以弥补自己的不足。

∞∞

公共交通必须有能力在大都会地区把旅客从一点运往任何一点。

在公路和铁路上往返奔驶的公共汽车和火车，因离大多数旅客的起点、终点及车站，都未充分发挥效用。出租汽车能将旅客从一处运送到另一处，但是车费太昂贵了。

为了解决这一问题，一种折中的交通工具是必不可少的。这种车辆一半像公共汽车，一半像出租车——这是一种小型的公共汽车，它能把旅客从任何一处送往另一处，并可在途中接客上车，而车费却比出租车便宜。

近来的研究和全面的试验表明，对使用电话雇车的小公共汽车系统要刮目相看了，它能起到这样的功能：它能在 15min 内把旅客从这一处送到另一处，一趟的车费不超过 50 美分（1974 年），收费不高是因为这一系统效率高，足以维持经营。这一系统，除了在乘坐的途中可让旅客上车和下车外，如同出租车一样工作。为了节省时间，小公共汽车会把你送到离你家最近的拐角处，而不是你家的正大门前。它收的车费只及平均出租车费的 1/4。

20 MINI-BUSES*

...this pattern helps complete the LOCAL TRANSPORT AREAS(11) and the WEB OF PUBLIC TRANSPORTATION (16).The local transport areas rely heavily on foot traffic,and on bikes and carts and horses.The web of public transportation relies on trains and planes and buses.Both of these patterns need a more flexible kind of public transportation to support them.

෴

Public transportation must be able to take people from any point to any other point within the metropolitan area.

Buses and trains,which run along lines,are too far from most origins and destinations to be useful.Taxis,which can go from point to point,are too expensive.

To solve the problem,it is necessary to have a kind of vehicle which is half way between the two-half like a bus,half like a taxi—a small bus which can pick up people at any point and take them to any other point,but which may also pick up other passengers on the way,to make the trip less costly than a taxi fare.

Recent research,and full-scale experiments,have shown that a system of mini-buses,on call by telephone,can function in this fashion,taking people from door to door in 15 minutes,for no more than 50 cents a ride(1974):and that the system is efficient enough to support itself.It works just like a taxi,except that it picks up and drops off other passengers while you are

这一系统在某种程度上要依赖于最新的、尖端的计算机程序的发展。当旅客打电话雇车时，计算机就会检查各种不同的小公共汽车在各条专线上运行的状况，并做出决定：哪一辆小公共汽车能最佳地接这位新旅客上车而跑的冤枉路最少。双向无线电通话使小公共汽车同在计算机开关屏前的调度员一直保持着联系。所有这一切和其他细节在关于拨号公共汽车的近期调查报告中都——加以说明了。（*Summary Report-The Dial-a-Ride Transportation System*，M.I.T.Urban Systems Laboratory,Report#USL-TR-70-10，March 1971.)

加拿大的小公共汽车
Canadian mini-bus

公共汽车的拨号系统实际上已经存在，这种系统应运而生是因为从经济角度看是切实可行的。目前，当常规的指定线路上的公共交通系统正在经受着一种恶性循环——服务水平越来越差、乘客越来越少和增加的公共津贴越来越不足——之际，全世界有三十多个正常运行的拨号公共汽车系统正在成功地经营着业务。例如，加拿大的萨克其万河附近的利宅那市，拨号公共汽车系统是该市交通系统中能自我维持下去的唯一的一部分。（*Regina Telebus Study:Operations Report*,and Financial Report,W.G.Atkinson et al.,June 1972）。在纽约州的巴达维亚，拨号公共汽车是唯一的公共交通工具，为16000人服务。每次车费为60美分。

我们在行将叙述完毕本模式之时，想提醒读者注意，在公共交通中赋予小公共汽车交通方式以重要性的两个极其重要的问题。

riding;it goes to the nearest corner to save time—not to your own front door;and it costs a quarter of an average taxi fare.

The system hinges,to a certain extent,on the development of sophisticated new computer programs.As calls come in,the computer examines the present movements of all the various mini buses,each with its particular load of passengers,and decides which bus can best afford to pick up the new passenger,with the least detour.Two-way radio contact keeps the mini-buses in communication with the dispatcher at the computer switchboard.All this,and other details,are discussed fully in a review of current dial-a-bus research:*Summary Report—The Dial-a-Ride Transportation System*,M.I.T.Urban Systems Laboratory,Report#USL-TR-70-10,March 1971.

Dial systems for buses are actually coming into existence now because they are economically feasible. While conventional fixed route public transport systems are experiencing a dangerous spiral of lower levels of service,fewer passengers,and increased public subsidies,over 30 working dial-a-bus systems are presently in successful operation throughout the world.For example,a dial-a-bus system in Regina,Saskatchewan,is the only part of the Regina Transit System which supports itself(*Regina Telebus Study:Operations Report,and Financial Report*,W. G.Atkinson et al.,June 1972).In Batavia,New York,dial-a-bus is the sole means of public transport,serving a population of 16,000 at fares of 40 to 60 cents per ride.

We finish this pattern by reminding the reader of two vital problems of public transportation,which underline the importance of the mini-bus approach.

第一，在城市里有大量的人不能驾驶轿车。我们认为，小公共汽车系统是满足所有这些人的需要的唯一现实的途径。

这些人数量之多大大超出人们的预料。事实上，他们是一个保持沉默的少数，他们是一些不怨天尤人的老年人和生理上有残疾的年轻人和穷人。在 1970 年，20% 以上的美国家庭没有轿车，在收入低于 3000 美元的所有家庭中，57.5% 没有轿车。以 65 岁或 65 岁以上的老人为主的家庭中，44.9% 没有轿车。10 ~ 18 岁的青少年中，有 80% 的人依赖他人——包括公共交通——来进行流动的。如果公共交通系统能向残废人提供户至户直达接送，残障人士中约有 57000000 人是公共交通的潜在乘客。（Sumner Myers，"Turning Transit Subsidies into 'Compensatory Transportation,'" City，Vol.6，No.3，Summer 1972，p.20）

第二，事实是明摆着的，公共交通网对于大型公共汽车、轮船和火车的特殊需求是各不相同的。如果没有小公共汽车系统助以一臂之力，它将无法实现有效运行。大的交通系统需要支线：即能到达各个车站的某种交通方式。如果人们不得不驾驶自己的轿车去乘火车，那么，一旦他们坐上轿车，就会一直乘坐下去，根本不会再去利用火车了。小公共汽车系统在较大的公共交通网中，在为提供支线服务方面，有举足轻重的意义。

因此：

建立一种小型的出租车式的公共汽车系统，每辆小公共汽车载客多至 6 人，由无线电控制，通过拨号打电话来雇车。这种拨号公共汽车系统能根据旅客的需要提供点至点直达接送服务，并由计算机辅助系统保证在各条专线上的小公共汽车跑的弯路距离最短，旅客等车的时间最少。在每条线路上每隔 600ft 建立一个小公共汽车站，并在站上安装一门电话，以便旅客使用电话来雇小公共汽车。

First,there are very large numbers of people in cities who cannot drive;we believe the mini-bus system is the only realistic way of meeting the needs of all these people.

Their numbers are much larger than one would think.They are,in effect,a silent minority comprising the uncomplaining old and physically handicapped,the young and the poor.In 1970,over 20 percent of U.S.households did not own a car. Fifty-seven and five-tenths percent of all households with incomes under $3000 did not own a car.For households headed by persons 65 years of age or older,44.9 percent did not own a car.Of the youths between 10 and 18 years of age,80 percent are dependent on others,including public transit,for their mobility.Among the physically disabled about 5.7 million are potential riders of public transportation if the system could take them door-to-door.(Sumner Myers, "Turning Transit Subsidies into 'Compensatory Transportation,'" City,Vol.6,No.3,Summ er 1972,p.20.)

Second,quite apart from these special needs,the fact is that a web of public transportation,with large buses,boats,and trains,will not work anyway,without a mini-bus system.The large systems need feeders:some way of getting to the stations. If people have to get in their cars to go to the train,then,once in the car,they stay in it and do not use the train at all.The mini-bus system is essential for the purpose of providing feeder service in the larger web of public transportation.

Therefore:

Establish a system of small taxi-like buses,carrying up to six people each,radio-controlled,on call by telephone, able to provide point-to-point service according to the passengers' needs,and supplemented by a computer system

载6名乘客的小公共汽车

电话无线电调度

每隔600ft设置一个小公共汽车站

⚜

　　主要在主干公路沿线设置公共汽车站，尽可能满足下列要求：再也不会有旅客必须步行 600 多米才能走到最近一个公共汽车站了——**平行路**（23）；在每一**换乘站**（34）内设一公共汽车站：并使每个公共汽车站成为只需候车几分钟的令人愉快的地方——**公共汽车站**（92）……

　　遵照下列基本原理，制定社区和邻里的政策，以便控制地方环境的性质。

21. 不高于四层楼

22. 停车场不超过用地的 9%

23. 平行路

24. 珍贵的地方

25. 通住水域

26. 生命的周期

27. 男人和女人

which guarantees minimum detours,and minimum waiting times.Make bus stops for the mini-buses every 600 feet in each direction,and equip these bus stops with a phone for dialing a bus.

⫯⫯

Place the bus stops mainly along major roads,as far as this can be consistent with the fact that no one ever has to walk more than 600 feet to the nearest one—PARALLEL ROADS(23);put one in every INTERCHANGE(34);and make each one a place where a few minutes' wait is pleasant—BUS STOP(92)...

establish community and neighborhood policy to control the character of the local environment according to the following fundamental principles:

21.FOUR-STORY LIMIT

22.NINE PER CENT PARKING

23.PARALLEL ROADS

24.SACRED SITES

25.ACCESS TO WATER

26.LIFE CYCLE

27.MEN AND WOMEN

模式21 不高于四层楼**

……在一市区内，建筑密度是呈波状起伏的。一般来说，趋近中心区，建筑密度较高，接近边缘区，建筑密度较低——**指状城乡交错**（3）、**乡村沿街建筑**（5）、**城市的魅力**（10）。但是，在全市范围内，甚至在建筑密度最大的一些地方，人们由于人情味方面的种种原因，强烈要求限制一切建筑的高度。

∞∞∞

有大量证据表明，高耸入云的建筑会使人发狂。

21 FOUR-STORY LIMIT**

...within an urban area,the density of building fluctuates. It will,in general,be rather higher toward the center and lower toward the edges—CITY COUNTRY FINGERS(3),LACE OF COUNTRY STREETS(5),MAGIC OF THE CITY(10). However,throughout the city,even at its densest points,there are strong human reasons to subject all buildings to height restrictions.

৪০০৪

There is abundant evidence to show that high buildings make people crazy.

High buildings have no genuine advantages,except in speculative gains for banks and land owners.They are not cheaper,they do not help create open space,they destroy the townscape,they destroy social life,they promote crime,they make life difficult for children,they are expensive to maintain,they wreck the open spaces near them,and they damage light and air and view.But quite apart from all of this,which shows that they aren't very sensible,empirical evidence shows that they can actually damage people's minds and feelings.

There are two separate bodies of evidence for this.One shows the effect of high-rise housing on the mental and social well being of families.The other shows the effect of large buildings,and high buildings,on the human relations in offices and workplaces.We present the first of these two bodies of

高层建筑，除了在投机中为银行主和土地占有者谋取暴利外，实在没有什么长处可谈。它们造价昂贵，它们无助于形成开阔的空间，它们毁坏城镇的秀丽风光，它们破坏社会生活，它们促使犯罪率猛增，它们使孩子们的生活蒙受重重困难，它们的维护修缮需要耗费巨资，它们损坏附近的开阔空间，它们阻挡光线、空气和景观。但是，撇开凡此种种不是容易察觉的弊端外，切身的感受表明，高层建筑实际上能毁坏人们的心灵和感觉。

"真理部——用纽斯皮克（*Newspeak* 是乔治·奥威尔 *Gevrge Orwell* 在 1984 年的小说中创造的一种官僚政客们的语言，其意义含混不清，模棱两可，指东说西，十分令人费解，以此来达到某种政治目的——译者注）说，就是米尼特鲁了。它是一座矗立着的触目惊心的高层建筑，在我们的视野中它和其他的一切东西都格格不入。它那庞大的结构像金字塔一般，表面白色的混凝土在闪闪发光。它一层又一层地拔地而起，高高耸入云霄，直刺青天 300 米。"（乔治·奥威尔，1984）

"*The Ministry of Truth—Minitrue, in Newspeak—was startlingly different from any other object in sight. It was an enormous pyramidal structure of glittering white concrete, soaring up terrace after terrace 300 metres in the air.*"(George Orwell,1984)

有着两类各不相关的例证可引用来说明这点。一类例证

evidence in the text which follows.The second,concerning offices and workplaces,we have placed in BUILDING COMPLEX(95),since it has implications not just for the height of buildings but also for their total volume.

We wish to stress,however,that the seemingly one-sided concern with housing in the paragraphs which follow,is only apparent.The underlying phenomenon—namely,mental disorder and social alienation created by the height of buildings—occurs equally in housing and in workplaces.

The strongest evidence comes from D.M. Fanning ("Families in Flats," *British Medical Journal*, November 18,1967,pp.382-386).Fanning shows a direct correlation between incidence of mental disorder and the height of people's apartments.The higher people live off the ground,the more likely are they to suffer mental illness. And it is not simply a case of people prone to mental illness choosing high-rise apartments.Fanning shows that the correlation is strongest for the people who spend the most time in their apartments.Among the families he studied,the correlation was strongest for women,who spend the most time in their apartments;it was less strong for children,who spend less time in the apartments;and it was weakest for men,who spend the least amount of time in their apartments. This strongly suggests that sheer time spent in the high-rise is itself what causes the effect.

A simple mechanism may explain this:high-rise living takes people away from the ground,and away from the casual,everyday society that occurs on the sidewalks and streets and on the gardens and porches.It leaves them alone in their apartments.The decision to go out for some public life

说明高层建筑对各个家庭精神和社会安宁的影响。另一类例证说明庞大的建筑和高层建筑对在办公室和车间工作的人员之间人情关系的影响。我们在本文中将提供第一类例证。第二类例证我们将在**建筑群体**（95）中提供，因为它不仅要涉及建筑的高度，而且还要涉及建筑的总体量。

可是，我们想要强调一下，本文下面各段中似乎只片面地谈到有关住户方面的问题，仅仅由于问题比较明显。基本的现象——由高层建筑所引起的精神错乱和社会疏远——在住宅和工作地点都是同样存在的。

最强有力的证据来自 D.M·范宁（D.M.Fanning）（"Families in Flats," British Medical Journal, November 18, 1967, pp.382～386），他指明精神错乱和居民公寓高度之间的直接关系。人们离地而居越高，患精神病的可能性就越大。这并不是选择了高层公寓居住的人中易犯精神病的唯一案例。范宁指出，大部分时间在自己的（高层）公寓中度过的人在精神上受到的影响最为强烈。在家庭中妇女在精神上受到的影响最深，因为她们的大部分时间是在自己的公寓中消磨掉的；其次是孩子，因为他们在自己公寓中度过的时间较少；男人在精神上受到的影响最浅，因为他们在公寓中度过的时间最少。这就令人信服地说明：在高层建筑中度过的绝对时间量本身就会对人的精神产生影响。

一个简单的机制即可说明：人身居高楼就离开了地面，离开了自然的日常社会生活，而这种生活情景在人行道上、街道上、花园里和走廊内却比比皆是。人待在公寓里会感到孤独寂寞。外出参加公共活动的决定，就成为拘谨和尴尬的事；要不是有某种特殊任务必须外出，就倾向于独自深居简出。这种人为的孤立状态会引起个人的精神崩溃。

范宁的发现为卡彭博士的临床实验所证实。D·卡彭在他的报告《精神保健和高层楼房》"Mental Health and the High Rise," Canadian Public Health Association,April 1971） becomes formal and awkward;and unless there is some specific task which brings people out in the world,the tendency is to stay home,alone.The forced isolation then causes individual breakdowns.

Fanning's findings are reinforced by Dr.D.Cappon's clinical experiences reported in "Mental Health and the High Rise," Canadian Public Health Association,April 1971:

There is every reason to believe that high-rise apartment dwelling has adverse effects on mental and social health.And there is sufficient clinical,anecdotal and intuitive observations to back this up.Herewith,in no particular order ranking,a host of factors:

In my experience as Mental Health Director in a child guidance clinic in York Township,Toronto,for 5 years,I saw numerous children who had been kinetically deprived...and kinetic deprivation is the worst of the perceptual,exploratory kinds,for a young child,leaving legacies of lethargy,or restlessness,antisocial acting out or withdrawal, depersonalization or psychopathy.

Young children in a high-rise are much more socially deprived of neighborhood peers and activities than their S.F.D.(Single Family Dwelling)counterparts,hence they are poorly socialized and at too close quarters to adults,who are tense and irritable as a consequence.

Adolescents in a high-rise suffer more from the "nothing-to-do" ennui than those of a S.F.D.,with enhanced social needs for "drop in centres" and a greater tendency to escapism...

Mothers are more anxious about their very young ones, when they can't see them in the street below,from a convenient

中说道：

有一切理由认为，在高层公寓居住这一点对精神保健和社会保健都具有不良的影响。我对病人所作的临床观察以及研究他们的趣闻轶事后得出的结论足以支持这一观点。在这里我不按严格的先后顺序罗列一些致病的因素：

在我的经历中，有 5 年我是在多伦多市的约克区的儿童指导诊疗所担任精神科主任，我看见许多儿童丧失了活动能力……在感知和探查过的疾病中，丧失正常活动能力是最坏的病情，会给幼儿留下嗜眠症或多动症、反社会行为、见人躲闪隐退、失去个性或精神变态的病根。

住在高层公寓的与住在单层独院型住宅的幼儿，从社会联系的角度看，前者和邻里的幼儿们接触和活动的机会比后者少得多。因此，这些幼儿接触社会少得可怜，离不开成人的住所，容易心情烦躁，精神紧张。

住在高层公寓的青少年较住在单层独院型住宅中的更加感到"无所事事"的厌倦和无聊，他们具有强烈的社会需要"顺便光顾一下中心区"，和一种更加明显的逃避现实的倾向……

当母亲们从厨房的窗户看不见自己在街上的孩子时，心里就焦急担心。

由于从高层通往底层的主要出口的途中存在着各种障碍，如电梯、走廊等，居住高层公寓的消极性就更多了；一般来说，要越过这一段垂直距离得耗去一定时间。在高层公寓内电视已经普及。这大概会给老年人带来最不良的后果，因为他们需要活动和运动疗法以保持健康，按运动量来说，他们和青少年需要的一样多。虽然他们在高楼内深居简出可免于车祸，但会缩短他们的寿命……

珍尼·莫维勒在一份研究丹麦情况的报告 (Borns Brug of Friarsaler，Disponering Af Friarsaler,Etageboligomrader, Denmark,Med Saerlig Henblik Pa Borns Legsmuligheder,S.

kitchen window.

There is higher passivity in the high-rise because of the barriers to active outlets on the ground;such barriers as elevators,corridors;and generally there is a time lapse and an effort in negotiating the vertical journey.TV watching is extended in the high-rise.This affects probably most adversely the old who need kinesia and activity,in proportion,as much as the very young do.Though immobility saves them from accidents,it also shortens their life in a high-rise...

A Danish study by Jeanne Morville adds more evidence(Borns Brug af Friarsaler,Disponering Af Friarsaler, Etageboligomrader Med Saerlig Henblik Pa Borns Legsmuligheder,S.B.I.,Denmark,1969):

Children from the high blocks start playing out of doors on their own at a later age than children from the low blocks:Only 2% of the children aged two to three years in the high point blocks play on their own out of doors,while 27% of the children in the low blocks do this.

Among the children aged five years in the high point blocks 29% do not as yet play on their own out of doors,while in the low blocks all the children aged five do so...The percentage of yotmg children playing out of doors on their own decreases with the height of their homes;90% of all the children from the three lower floors in the high point blocks play on their own out of doors,while only 59% of the children from the three upper floors do so...

Young children in the high blocks have fewer contacts with playmates than those in the low blocks:Among children aged one,two and three years,86% from the low blocks have daily contact with playmates;this applies to only 29% from the

B.I.,Denmark,1969）中补充了更多的证据：

高层公寓的儿童开始在户外独立玩耍要比低层公寓的儿童晚：2～3岁的儿童，前者只占2%，而后者却占27%。

年满5岁的儿童至今未独自在户外玩耍的，高层公寓的占29%，低层公寓的则一个也没有……幼儿独自在户外玩的百分比随他们自己的家所在公寓的高度而递减：在户外独自玩的儿童中，住在高层公寓三层以下的占90%，而三层以上的只占59%……

高层公寓的幼儿同小朋友的接触比低层公寓的少：一岁、两岁、三岁的幼儿中，每日和小朋友保持接触的，低层公寓的占86%，而高层公寓的只占29%。

尤其在最近，奥斯卡·纽曼在《可防护的空间》一文中提出了证据。纽曼对纽约的两座相邻的住宅建筑——一座是高层公寓楼房，另一座是比较小的无电梯的三层楼房——进行了比较。两座建筑内的人口总密度是一样的，居民的收入也大致相同。但是，纽曼发现，高层楼房内的犯罪率比无电梯的楼房内的犯罪率约高出一倍。

范宁、卡彭、莫维勒和纽曼所描述的种种不良效果，在什么高度才开始占上风？我们的体会如下：就住宅楼和办公楼两者而言，超过四层高度的，问题就会接踵而至，层出不穷。

在三四层楼上，你仍然能够舒舒服服地走下楼梯，上街去逛逛，你依旧能够凭窗远眺，感到自己置身于街景之中：你能看到街道上的一切细节——熙来攘往的行人，他们清晰的面容，郁郁葱葱的树木和林立的商店。你能从三层楼上大声呼喊，引起下面某人的注意。四层以上，这些联系就中断了。细节模糊不清了。人们谈论下面的景色，仿佛逗乐似的不着边际，他们和地面完全脱节了。四层以上的高楼和大地的联系以及和整个城镇结构的联系变得无足轻重了：高层大楼已成为备有电梯和自助餐厅的独特天地。

high blocks.

More recently,there is the evidence brought forward by Oscar Newman in *Defensible Space*.Newman compared two adjacent housing projects in New York—one high-rise,the other a collection of relatively small three-story walk-up buildings.The two projects have the same overall density,and their inhabitants have roughly the same income.But *Newman found that the crime rate in the high-rise was roughly twice that in the walk-ups*.

At what height do the effects described by Fanning, Cappon,Morville,and Newman begin to take hold?It is our experience that in both housing and office buildings,the problems begin when buildings are more than four stories high.

At three or four stories,one can still walk comfortably down to the street,and from a window you can still feel part of the street scene:you can see details in the street— the people,their faces,foliage,shops.From three stories you can yell out,and catch the attention of someone below.Above four stories these connections break down.The visual detail is lost;people speak of the scene below as if it were a game,from which they are completely detached.The connection to the ground and to the fabric of the town becomes tenuous;the building becomes a world of its own:with its own elevators and cafeterias.

We believe,therefore,that the "fourstory limit" is an appropriate way to express the proper connection between building height and the health of a people.Of course,it is the spirit of the pattern which is most essential.Certainly,a building five stories high,perhaps even six,might work if it were carefully handled.But it is difficult.On the whole,we advocate a

所以，我们认为"不高于四层楼"是一个非常合适的模式，它能恰如其分地表达出建筑的高度和人的身心健康之间的相互联系。诚然，本模式的最精华之处是其精神。无疑，一幢五层高甚至六层高的楼房，如果谨慎从事，处理得当，可具有同等的功能。但这是十分困难的。从整体来看，我们主张在全城镇范围内，除偶然的特殊情况外，普遍采用本模式。

最后，我们替格拉斯哥市的孩子们说几句话。

在格拉斯哥的经济公寓里，向下面街上的孩子，从窗户"扔东西"，如一片面包和果酱，已被公认为司空见惯的了……

果子冻碎块之歌

亚当·麦克诺顿（Adam McNaughton）作

我是摩天大厦的断奶娃，住在二十层楼上，
我真的再也不能走到外面去东玩玩西嚷嚷。
自从我们搬进新居，我一天天消瘦不成样，
我脸色苍白没有血色，我吃得少没好营养。

副歌

噢，你能从二十一层扔出乱七八糟的东西，
我们七百个断奶娃证明干这坏事的就是你。
如果这是黄油、奶酪、果子冻和大圆面包，
它们的碎屑百分之九十九散落到我们这里。
我们给牛津写信，请先生们帮我们说说理，
我们联合告状，把"碎片"排成了一个旅。
我们要向伦敦进军，去要求我们的公民权，
但愿"不再有碎屑乱块从万丈高楼往下飞。"

因此：

在任何一个市区，不管密度如何，都要使大多数建筑保持在四层的高度或低于四层的高度。一些建筑超过这一限制高度是可能的，但它们绝不是供人们居住的住宅楼房。

four-story limit,with only occasional departures,throughout the town.

Finally,we give the children of Glasgow the last word.

To fling a "piece" ,a slice of bread and jam,from a window down to a child in the street below has been a recognised custom in Glasgow's tenement housing...

THE JEELY PIECE SONG

by Adam McNaughton

I'm a skyscraper wean,I live on the nineteenth flair,

On' I'm no'gaun oot tae play ony mair,

For since we moved tae oor new hoose I'm wastin' away,

'Cos I'm gettin' wan less meal ev'ry day.

Refrain

Oh,ye canny fling pieces oot a twenty-storey fiat,

Seven hundred hungry weans will testify tae that,

If it's butter,cheese or jeely, if the breid is plain or pan,

The odds against it reachin'us is ninety-nine tae wan.

We've wrote away tae Oxfam tae try an'get some aid,

We've a' joined thegither an' formed a "piece" brigade,

We're gonny march tae London tae demand oor Civil Rights,

Like "Nae mair hooses ower piece flingin'heights."

Therefore:

In any urban area,no matter how dense,keep the majority of buildings four stories high or less.It is possible that certain buildings should exceed this limit, but they should never be buildings for human habitation.

四层楼房

ᔡᔢᔤ

　　在四层楼极限高度的范围内，个体建筑的精确高度，要根据其所需的楼层的面积、基地的面积和周围建筑的高度，并参照模式**楼层数**（96）才能求算出来。建筑密度的总体变化取决于**密度圈**（29）。从**建筑群体**（95）可以得出，要把大建筑在水平方向上再分割成较小的单元，并把较小的建筑分离开来。**丘状住宅**（39）和**办公室之间的联系**（82）在高度不超过四层楼的限制内，有助于形成多层公寓楼和多层办公楼。最后，不要单从字面上理解本模式。跳出一般原则的偶然例外是颇为重要的——**眺远高地**（62）……

ഇൻ

Within the framework of the four-story limit the exact height of individual buildings,according to the area of floor they need,the area of the site,and the height of surrounding buildings,is given by the pattern NUMBER OF STORIES(96). More global variations of density are given by DENSITY RINGS(29).The horizontal subdivision of large buildings into smaller units,and separate smaller buildings,is given by BUILDING COMPLEX(95).HOUSING HILL(39)and OFFICE CONNECTIONS(82)help to shape multi-storied apartments and offices within the constraints of a four-story limit.And finally,don't take the four-story limit too literally.Occasional exceptions from the general rule are very important—HIGH PLACES(62)...

模式22　停车场不超过用地的9%**

　　……地方交通区的完整性以及地方社区和邻里的宁静，在很大程度上取决于它们所提供的停车场的数量。它们提供的停车场越多，要维护下面的一些模式就越不可能，因为停车场会吸引大量汽车，而汽车必然会侵害地方交通区和邻里——**地方交通区**（11）、**7000人的社区**（12）和**易识别的邻里**（14）。本模式对停车场的分布提出了全面的限制。

❧❧❧

　　当停车场占用地的面积太多时，用地就会遭到毁坏，这是不言而喻的。

22   NINE PER CENT PARKING**

...the integrity of local transport areas and the tranquility of local communities and neighborhoods depend very much on the amount of parking they provide.The more parking they provide,the less possible it will be to maintain these patterns,because the parking spaces will attract cars,which in turn violate the local transport areas andneighborhoods— LOCAL TRANSPORT AREAS(11),COMMUNITY OF 7000 (12),IDENTIFIABLE NEIGHBORHOOD(14).This pattern proposes radical limits on the distribution of parking spaces,to protect communities.

❧☙

Very simply—when the area devoted to parking is too great,it destroys the land.

Very rough empirical observations lead us to believe that it is not possible to make an environment fit for human use when more than 9 per cent of it is given to parking.

Our observations are very tentative.We have yet to perform systematic studies-our observations rely on our own subjective estimates of cases where "there are too many cars" and cases where "the cars are all right." However,we have found in our preliminary observations,that different people agree to a remarkable extent about these estimates.This suggests that we are dealing with a phenomenon which,though obscure,is nonetheless substantial.

An example of an environment which has the threshold

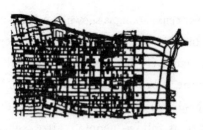

在洛杉矶的商业区，60%以上的用地是供停车使用的
In downtown Los Angels over 60 per cent of the land is given over to the automobile

我们粗略地凭切身的经验观察，认为任何一个环境，停车场占的面积超过9%，从使用角度说，于人就不合适了。

我们的观察是带有试探性的。我们还必须进行系统的研究——我们的观察依赖于我们自己对下列情况的主观估计：什么地方"汽车太拥挤了"，什么地方"汽车不多不少正好"。可是，在我们的初步观察中发现，不同的人们都一致同意要规定一个明白无误的百分比。这表明，我们正在研究的这一现象，虽然还模糊不清，却是极其重要的。

现在我们举出一个环境的例子，见本模式开头的重要照片：照片中所示为俄勒冈大学之一角，其停车场的阈值密度为9%。和我们交谈过的许多人都直观地感受到，现在这里风光秀丽、景色宜人，但是，若有更多的汽车停驻，如画般的校景就会被毁坏掉。

人们不禁要问：这种直觉的可能的功能基础是什么呢？我们推测如下：人们下意识地认识到外界环境是他们社会交际的媒介。正是外界环境，如果具备一切良好的条件，就会成为社会交际的潜在场所，或者成为孤芳自赏之地。

我们猜想，当汽车数量超出一定限度而人们感觉到汽车太多时，真正发生的事情就是他们下意识地感觉到汽车淹没了外界环境，而这环境不再是"他们自己的"了，他

density of 9 per cent parking,is shown in our key photograph:a quadrant of the University of Oregon.Many people we have talked to feel intuitively that this area is beautiful now,but that if more cars were parked there it would be ruined.

What possible functional basis is there for this intuition? We conjecture as follows:people realize, subconsciously, that the physical environment is the medium for their social intercourse.It is the environment which,when it is working properly,creates the potential for all social communion, including even communion with the self.

We suspect that when the density of cars passes a certain limit,and people experience the feeling that there are too many cars,what is really happening is that subconsciously they feel that the cars are overwhelming the environment,that the environment is no longer "theirs," that they have no right to be there,that it is not a place for people,and so on.After all,the effect of the cars reaches far beyond the mere presence of the cars themselves.They create a maze of driveways,garage doors,asphalt and concrete surfaces,and building elements which people cannot use.When the density goes beyond the limit,we suspect that people feel the social potential of the environment has disappeared.Instead of inviting them out,the environment starts giving them the message that the outdoors is not meant for them,that they should stay indoors,that they should stay in their own buildings,that social communion is no longer permitted or encouraged.

We have not yet tested this suspicion.However,if it turns out to be true,it may be that this pattern,which seems to be based on such slender evidence,is in fact one of the most crucial patterns there is,and that it plays a key role in determining

们无权到那里去了，那不是人待的地方了，等等。汽车的影响远远超过其本身单纯存在的影响。汽车创造出了一座由许多纵横交错的车道、各种车库的大门、沥青路面和混凝土路面以及人们无法利用的建筑要素所构成的迷宫。当汽车密度超过限度时，我们猜想人们会感到外界环境作为社交场所的潜在可能已经消失，不复存在了。他们不必离家外出，因为外界的信息告诉他们，外界环境对他们来说已毫无意义，他们应当待在家里，闭门不出。社会交往不再被容许或受到鼓励。

我们尚未验证这种猜测。可是，如果这种猜测被证明是对的，那么，很可能本模式看起来似乎是以微不足道的证据为基础的，其实，这是最关键性的模式之一，而且这一模式在判别外界环境对社会和心理的影响是健康的还是不健康的这方面，起到决定性的作用。

因此，我们推测，从社会和生态方面讲，那些富有人情味的、尚未被停放的汽车所破坏的环境，用于停车场的面积小于9%；绝不允许停车场和汽车库超过用地面积的9%。

重点是要尽量严格地来说明本模式。如果我们容许自己把原来在 A 地的停车场搬迁到相邻的 B 地，以便使 A 地的停车场低于9%，而 B 地的停车场高于9%，那么，本模式就会变得毫无意义了。换句话说，每块土地都必须自己照料好自己；我们绝不允许以牺牲另一块土地为代价来解决这块土地的问题。城镇或社区只有根据这种严格的解释才能实施本模式，具体的实施办法是：在地图的坐标方格上划定独立的停车区——每一区的面积为 1 ～ 10acre——这些独立停车区遍布整个社区，并要坚持在每个停车区内，独立自主、一丝不苟地运用这一原则。

在不同的停车密度区内，9%的原则，对于平衡露天停车和车库停车来说，具有明显的、直接的意义。这一原则

the difference between environments which are socially and psychologically healthy and those which are unhealthy.

We conjecture,then,that environments which are human,and not destroyed socially or ecologically by the presence of parked cars,have less than 9 per cent of the ground area devoted to parking space; and that parking lots and garages must therefore never be allowed to cover more than 9 per cent of the land.

It is essential to interpret this pattern in the strictest possible way.The pattern becomes meaningless if we allow ourselves to place the parking generated by a piece of land A,on another adjacent piece of land B,thus keeping parking on A below 9 per cent,but raising the parking on B to more than 9 per cent.In other words,each piece of land must take care of itself;we must not allow ourselves to solve this problem on one piece of land at the expense of some other piece of land.A town or a community can only implement the pattern according to this strict interpretation by defining a grid of independent "parking zones" -each zone 1 to 10 acres in area- which cover the whole community,and then insisting that the rule be applied,independently,and strictly,inside every parking zone.

The 9 per cent rule has a clear and immediate implication for the balance between surface parking and parking in garages,at different parking densities.This follows from simple arithmetic.Suppose,for example,that an area requires 20 parking spaces per acre.Twenty parking spaces will consume about 7000 square feet,which would be 17 per cent of the land if it were all in surface parking.To keep 20 cars per acre in line with the 9 per cent rule,at least half of them will have to be parked

是通过简单的数学运算而得出的。假设一个区需要每英亩有 20 个停车位。20 个停车位将占地约 7000ft^2，如果它都用作露天停车场，则占用地 17％。为了使每英亩停放 20 辆汽车与 9％的原则相符合，至少其中一半的车辆必须停放在车库内。下面的一份表格提供了不同停车密度的近似数值：

每英亩汽车数/辆	露天停车的百分比（％）	两层停车库内的百分比（％）	三层停车库内的百分比（％）
12	100	—	—
17	50	50	—
23	50	—	50
30	—	—	100

　　关于地下停车场有何看法？我们是否可以认为它是这一原则的一个例外？只要它不违反或限制使用地面用地的原则就行。举例来说，如果停车库设在一片土地的底下，这片土地以前是开阔的空间，上面长着参天大树，郁郁葱葱，那么，现在这地下车库将几乎肯定会改变这片地面空间的性质，因为在它上面再也无法生长绿叶浓荫的大树了。这样的地下停车库是违背土地使用条例的。同样，如果汽车库的结构柱网——60ft 的开间——限制了地面建筑的结构柱网，由此这座建筑无法满足其自身的需要。这样的汽车库也是违背土地使用条例的。地下停车场只有在很少的一些场合，即地面以上土地不受限制的地方才能被容许建造：在主干公路下面或网球场底下。

　　于是，我们就会理解，9％的原则具有巨大的意义。因为地下车库将无法充分满足我们所述的条件，本模式实际上告诉我们，在市区几乎没有一个地方每英亩的停车位会超过 30 个。这将会引起商务中心区的巨大变化。请考虑一下一个典型商业区的一部分。在那里工作的人员中需要两地往返的每英亩就有好几百个；而在今天的条件下，他们

in garages.The table below gives similar figures for different densities:

Cars per acre	Per cent on surface	Per cent in two story garages	Per cent in three story garages
12	100	—	—
17	50	50	—
23	50	—	50
30	—	—	100

What about underground parking? May we consider it as an exception to this rule? Only if it does not violate or restrict the use of the land above.If, for example, a parking garage is under a piece of land which was previously used as open space,with great trees growing on it,then the garage will almost certainly change the nature of the space above,because it will no longer be possible to grow large trees there.Such a parking garage is a violation of the land.Similarly,if the structural grid of the garage—60 foot bays-constrains the structural grid of the building above,so that this building is not free to express its needs,this is a violation too.Under ground parking may be allowed only in those rare cases where it does not constrain the land above at all:under a major road,perhaps,or under a tennis court.

We see then,that the 9 per cent rule has colossal implications. Since underground parking will only rarely satisfy the conditions we have stated,the pattern really says that almost no part of the urban area may have more than 30 parking spaces per acre.This will create large changes in the central business district.Consider a part of a typical downtown area.There may be several hundred commutersper acre working there;and,under

中间许多人的汽车要停放在车库里。如果每英亩的停车位不能超过 30 个是真实的，那么，或者工人们被迫分散工作，或者他们不得不依靠公共交通。总而言之，似乎以环境的社会心理学为基础的这一简单模式使我们得出如同模式**公共交通网**（16）和**分散的工作点**（9）一样意义深远的社会结论。

因此：

在任何一个规定的区内，停车场所占面积不得超过用地的 9％。在一些往往被人忽略的大区内，为了防止出现"成片连串"的停车场，每个城镇和社区都必须把它的土地再划分成"停车区"，每一停车区不大于 10acre，并在每一停车区内应用同一原则。

停车区

每英亩最大停车数为30辆

∞‰

后面两个模式叙述停车场在两种形式中必居其一：或是小型的露天停车场，或有遮挡的停车结构——**有遮挡的停车场**（97）、**小停车场**（103）。如果你同意接受这些模式，那么，9％这一原则将提出一个有效的上限：外界环境的每一处，每英亩 30 个停车位。如今在大街上的停车场，因有许多车道，所以每英亩有 35 个停车位，这是违背上述原则的。而今天那些依赖于汽车的高密度商业发达区同样是违背上述原则的……

today's conditions,many of them park their cars in garages.But if it is true that there cannot be more than 30 parking spaces per acre,then either the work will be forced to decentralize,or the workers will have to rely on public transportation.It seems,in short,that this simple pattern,based on the social psychology of the environment,leads us to the same far reaching social conclusions as the patterns WEB OF PUBLIC TRANS-PORTATION(16)and SCATTERED WORK(9).

Therefore:

Do not allow more than 9 per cent of the land in any given area to be used for parking.In order to prevent the "bunching" of parking in huge neglected areas,it is necessary for a town or a community to subdivide its land into "parking zones" no larger than 10 acres each and to apply the same rule in each zone.

ଽଔଔ

Two later patterns say that parking must take one of two forms:tiny,surface parking lots,or shielded parking structures—SHIELDED PARKING(97),SMALL PARKING LOTS(103). If you accept these patterns the 9 per cent rule will put an effective upper limit of 30 parking spaces per acre,on every part of the environment.Present-day on-street parking,with driveways,which provides spaces for about 35 cars per acre on the ground is ruled out.And those present-day high density business developments which depend on the car are also ruled out...

模式23　平行路

　　……在较早叙述的一些模式中，我们已经建议，城市应当被再分成地方交通区，而这些区的公路允许汽车从环路进出，但坚决禁止汽车穿越其内部——**地方交通区**（11）和**环路**（17）——我们还建议：这些交通区本身应当进一步再被分成社区和邻里，并规定所有主干公路都位于社区和邻里之间的边界内——**亚文化区边界**（13）和**邻里边界**（15）。现在这些公路应如何布局才有助于**地方交通区**（11）所要求的车辆流动并对维护边界有所裨益呢？

23 PARALLEL ROADS

...in earlier patterns,we have proposed that cities should be subdivided into local transport areas,whose roads allow cars to move in and out from the ring roads,but strongly discourage internal movement across the area—LOCAL TRANSPORT AREAS(11),RING ROADS(17) and that these transport areas themselves be further subdivided into communities and neighborhoods,with the provision that all major roads are in the boundaries between communities and neighborhoods SUBCULTURE BOUNDARY(13),NEIGHBORHOOD BOUNDARY(15).Now,what should the arrangement of these roads be like,to help the flow required by LOCAL TRANSPORT AREAS(11),and to maintain the boundaries?

❧❧❧

The net-like pattern of streets is obsolete.Congestion is choking cities.Cars can average 60 miles per hour on freeways,but trips across town have an average speed of only 10 to 15 miles per hour.

Certainly,in many cases,we want to get rid of cars,not help them to go faster.This is fully discussed in LOCAL TRANSPORT AREAS (11).But away from the areas where children play and people walk or use their bikes,there still need to be certain streets which carry cars.The question is:How can these streets be designed to carry the cars faster and without congestion?

It turns out that the loss of speed on present city streets

网状街道的模式现在废弃不用了。交通的堵塞正在使城市窒息。在高速公路上汽车的行驶速度平均每小时为60mi，但在穿越城镇时的车速平均每小时只有 10 ~ 15mi。

在许多情况下，肯定无疑，我们想摆脱汽车，而并不是想方设法使汽车开得更快。这一点在**地方交通区**（11）中已做了充分的论述了。在市区除了有孩子们做游戏的场地和成年人散步或骑自行车的道路外，仍然还需要有汽车通行的一些街道。问题是在于：这些街道需如何设计才能使汽车更快通过而不造成堵塞呢？

现已证明，在今天城市街道上汽车的失速问题主要是由交叉交通所引起的：即左转弯和十字交叉路口。（G.F.Newell，"The Effect of Left Turns on the Capacity of Traffic Intersection"，*Quarterly of Applied Mathematics*，XVII，April 1959，pp.67 ~ 76.）

为了加快交通，必须建造没有十字交叉路口的主干路网并且不采取左转弯穿行。这一点不难做到，如果主干路是交替的单向平行路，相互隔开几百英尺的间距，可以和较小的地方上的路连通，平行路之间的唯一连接是每隔2 ~ 3mi 就有一条横越它们的较大的高速公路。

平行路
Parallel roads

本模式在三篇论文中已被相当详尽地论述过。（"The Pattern of Streets，" C.Alexander，AIP Journal，September 1966；Criticisms by D.Carson and P.Roosen-Runge，and Alexander's reply，in AIP Journal，September 1967.）我们向读者推荐论文原作，它们对平行路的几何细节的来龙

is caused mainly by crossing movements:left-hand turns across traffic and four-way intersections.(G.F.Newell, "The Effect of Left Turns on the Capacity of Traffic Intersection," Quarterly of Applied Mathematics,XVIII,April 1959,pp. 67-76.)

To speed up traffic it is therefore necessary to create a network of major roads in which there are no four-way intersections,and no left-hand turns across traffic.This can easily be done if the major roads are alternating,one-way parallel roads,a few hundred feet apart,with smaller local roads opening off them,and the only connections between the parallel roads given by larger freeways crossing them at two-or three-mile intervals.

This pattern has been discussed at considerable length in three papers("The Pattern of Streets," C.Alexander, AIP Journal,September 1966;Criticisms by D.Carson and P.Roosen-Runge,and Alexander's reply,in AIP Journal, September 1967.)We refer the reader to these original papers for the full derivation of all the geometric details. Our present statement is a radically condensed version.Here we concentrate mainly on one puzzling question—that of detours—because this is for many people the most surprising aspeet of the full analysis.

The pattern of parallel roads—since it contains no major cross streets—creates many detours not present in today's net-like pattern.At first sight it seems likely that these detours will be impossibly large.However,in the papers mentioned above it is shown in detail that they are in fact perfectly reasonable.We summarize the argument below.

It is possible to calculate the probable detour for any

去脉都做了描述。现在我们做一简单扼要的介绍。我们在此主要集中说明一个令人迷惑不解的问题——绕道的问题——因为这是许多人在全面分析中最惊讶不已的方面。

平行路这一模式——因为它不包含主要的交叉街道——会造成今天网状模式中不存在的许多绕道。乍看起来，这些绕道将大得几乎不能成立。不过，在上述论文中已详细说明，实际上绕道是完全合情合理的。我们现将其论据概述如下。

有可能通过这个平行路系统给出一定距离的任何一段旅程，计算出有或然性的绕道，并将它作为交叉路之间距离的一个函数。然后，从对大城市汽车旅行进行的实际研究中可获得任何一次确定的旅程长度的概率。这两类概率最后可能结合起来而得出总平均旅程长度和总平均绕道长度，现列表如下：

旅程长度/mi	1	2	3	4	5	7	10	4.12
旅程长度比例（%） *	28	11	11	9	9	24	8	（总平均旅程长度）
交叉路之间的英里数/mi	平均绕道长度/mi							总平均绕道长度/mi
1	0.12	0.05	0.04	0.03	0.02	0.01	0.01	0.05
2	0.45	0.24	0.15	0.11	0.09	0.07	0.04	0.21
3	0.79	0.58	0.36	0.25	0.20	0.15	0.11	0.41

旅程长度的分布数据引自爱德华·M.霍尔的"圣地亚哥市郊发展的旅行特点"（"Travel Characteristics of Two San Diego Suburban Develop-merits," *Highway Research Board Bulletin* 2039, Washington, D.C., 1958, pp.1-19, Figure 11）。这些数据对整个西方世界的大都市区毫无例外地都具有典型意义。

因此，我们就会明白，即使相隔 2mi 就有交叉公路，但由于没有交叉街道，旅程长度只增加了 5%。同时，旅行的平均速度将从每小时 15mi 增加到每小时 45mi，净增 2 倍。这样节省的大量时间和燃料费用于补偿稍稍延长的

trip of a given length through this proposed parallel road system as a function of the distance between the cross roads. Next,the probability of any given trip length may be obtained from actual studies of metropolitan auto trips.These two types of probabilities can finally be combined to yield an overall mean trip length and overall mean detours as shown below.

Trip Length,miles	1	2	3	4	5	7	10	4.12
Proportion of Trip Lengths %*	28	11	11	9	9	24	8	(Overall Mean Trip Length)
miles between cross roads	Mean Detour,miles							Overall Mean Detour
1	0.12	0.05	0.04	0.03	0.02	0.01	0.01	0.05
2	0.45	0.24	0.15	0.11	0.09	0.07	0.04	0.21
3	0.79	0.58	0.36	0.25	0.20	0.15	0.11	0.41

   Data for distribution of trip lengths was obtained from Edward M.Hall, "Travel Characteristics of Two San Diego Suburban Developments," Highway Research Board Bulletin 2039, Washington,D.C.,1958,pp.1-19,Figure 11.These data are typical for metropolitan areas all over the Western world.

   We see,therefore,that even with cross roads two miles apart,the lack of cross streets only increases trip lengths by 5 per cent.At the same time,the average speed of trips will increase from 15 miles per hour to about 45 miles per hour,a threefold increase.The huge savings in time and fuel costs will more than offset the slight increase in distance.

   Referring back for a moment to the table of detours, it will be noticed that the highest detours occur for the shortest trips.We have argued elsewhere—LOCAL

路程而造成的损失就绰绰有余了。

当我们再回过来参考绕道长度表时，我们就会注意到最高的绕道百分比发生在最短的旅程中。我们在另一模式**地方交通区**（11）中已经论述过，为了使城市环境保持这种性质，必须阻止人们利用汽车做短途旅行，而同时要大力提倡步行、骑自行车、乘公共汽车或者骑马。精确地说，平行路模式具有地方交通区所需要的特征。这就使汽车对较长路程的旅行更为有效，而同时避免汽车做短途的旅行。这样才能给地方交通区提供一种正是它所需要的内部结构，以便维持它的功能。

虽然这一模式乍看起来似乎很奇怪，但事实上它在世界上的许多地方早已存在，并已证明了它存在的价值。举例来说，瑞士的伯尔尼是欧洲没有遭到严重交通堵塞的几个城市之一。当人们观看伯尔尼的地图时，就会看到，伯尔尼的古老中心是由 5 条平行路构成的，而且几乎没有交叉街道。我们认为，在伯尔尼古老中心区一点也没有交通堵塞，正是因为它包含着本模式。今天，在许多大城市内，正在逐步实行具有同样洞察力的明智措施——越来越多地采用单行道：如纽约的交替的单行林荫大街，旧金山商业区的主要的单行道。

伯尔尼的五条主要平行路
Berne's five main parallel streets

因此：

在地方交通区内，根本不要建造交叉的主干路；而要建立一种平行而又交替的单行路系统，以便让车辆通往环路（**17**）。在现存的城镇里，分片分期地创造这种结构，其

TRANSPORT AREAS(11)that to preserve the quality of the city's environment it is necessary to discourage the use of the automobile for very short trips,and to encourage walking,bikes,buses,and horses instead.The pattern of parallel roads has precisely the feature which local transport areas need. It makes longer trips vastly more efficient,while discouraging the very short auto trips,and so provides the local transport area with just the internal structure which it needs to support its function.

Although this pattern seems strange at first sight,it is in fact already happening in many parts of the world and has already proved its worth.For example,Berne,Switzerland,is one of the few cities in Europe that does not suffer from acute traffic congestion.When one looks at a map of Berne,one can see that its old center is formed by five long parallel roads with almost no cross streets.We believe that it has little congestion in the old center precisely because it contains the pattern.In many large cities today,the same insight is being implemented piecemeal—in the form of more and more one-way streets:in New York the alternating one-way Avenues,in downtown San Francisco the one-way major streets.

Therefore:

Within a local transport area build no intersecting major roads at all;instead,build a system of parallel and alternating one-way roads to carry traffic to the RING ROADS(17).In existing towns,create this structure piecemeal,by gradually making major streets one-way and closing cross streets.Keep parallel roads at least 100 yards apart(to make room for neighborhoods between them)and no more than 300 or 400 yards apart.

办法是逐步使主要街道成为单行道并封闭交叉街道。平行路之间的间距至少保持 **100yd**（在平行路之间要为邻里留出空间），至多不超过 **300yd** 或 **400yd**。

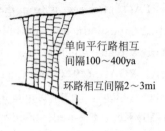

单向平行路相互间隔100～400ya

环路相互间隔2～3mi

෩෨

平行路是**地方交通区**（11）唯一畅通无阻的路。从平行路到公共建筑、住宅团组和私人住宅，使用安全的、狭窄的、并非畅通无阻的慢车道——**区内弯曲的道路**（49）和**绿茵街道**（51）——并使这两种道路与平行路成"丁"字形交叉——**丁字形交点**（50）。使人行道系统与平行路成直角相交，并使人行道高出平行路，而且**小路网络和汽车**（52）和**高出路面的便道**（55）这两个模式必须平行。提供**人行横道**（54），以便让小路穿过平行路。

৪০৫৪

The parallel roads are the only through roads in a LOCAL TRANSPORTAREA(11).For access from the parallel roads to public buildings,house clusters,and individual houses use safe,slow,narrow roads which are not through roads—LOOPED LOCAL ROADS(49),GREEN STREETS(51)—and make their intersections with the parallel roads a "T" —T JUNCTION(50).Keep the pedestrian path system at right angles to the parallel roads,and raised above them where the two must run parallel—NETWORK OF PATHS AND CARS(52),RAISED WALK(55).Provide a ROAD CROSSING(54)where paths cross the parallel roads.

模式24 珍贵的地方*

　　……在每个区域和每个城镇，甚至在每个邻里都有一些特殊的地方，它们象征着这块土地与扎根于此的先人。这些地方或许是天然风景区，或许是过去岁月里留下的历史陈迹。但从某种形式上看，它们却是精华。

ಬಂ

**　　如果人们不能维系居住于其中的物质世界的根，人们便不能维持他们精神上的根及与往昔的联系。**

　　我们在许多社区进行了非正式的实验。令我们感到十分诧异的是，人们不约而同地认为这些地方体现出他们与这块土地和往昔的历史联系。换句话说，仿佛"这些"圣迹就是他们客观上的共同现实似的。

24　SACRED SITES*

...in every region and every town,indeed in every neighborhood,there are special places which have come to symbolize the area,and the people's roots there.These places may be natural beauties or historic landmarks left by ages past. But in some form they are essential.

༃༃ཚ

People cannot maintain their spiritual roots and their connections to the past if the physical world they live in does not also sustain these roots.

Informal experiments in our communities have led us to believe that people agree,to an astonishing extent,about the sites which do embody people's relation to the land and to the past.It seems,in other words,as though "the" sacred sites for an area exist as objective communal realities.

If this is so,it is then of course essential that these specific sites be preserved and made important.Destruction of sites which have become part of the communal consciousness,in an agreed and widespread sense,must inevitably create gaping wounds in the communal body.

Traditional societies have always recognized the importance of these sites.Mountains are marked as places of special pilgrimage;rivers and bridges become holy;a building or a tree,or rock or stone,takes on the power through which people can connect themselves to their own past.

But modern society often ignores the psychological importance

如果确是如此，这些特别珍贵的地方应当保存下来并赋予重要性是不言而喻的。破坏这些地方——一个共同意识的组成部分——从一致同意的广义上讲，势必会在这个共同体内造成创伤。

传统社会始终一贯承认这些地方的重要性。一些山岳被标为朝拜之圣地；一些河流和桥梁也变成圣物。一座建筑、一棵树木、一方岩石或一块石头都具有一种力量，它能使人们同自己过去的历史联系起来。

但是现代社会往往忽视这些地方对人们心理上的重要作用。由于种种政治上和经济上的理由，它们被推土机削平了，被开发利用了，被改变得面目全非了，而未考虑到这些朴实而又重要的感情因素，或者它们早已被置之脑后不屑一顾了。

现在我们提出如下两项措施：

1. 在任何一个大的或小的地理区域内，都要向一大批人征求意见：他们感到什么遗址和什么地方与这一地区有着最密切的联系；什么历史陈迹最能代表过去的重要价值以及体现出他们与这块土地息息相关的命运。随即采取具体办法对这些珍贵的地方加以妥善保护。

2. 一旦这些有意义的地方被选中并且得到保护，就要对它们进行修缮和装饰，以强化它们的公共意义。我们认为最有效的办法就是使人们要步行一段路程才能到达这些地方。这就是"曲折入胜景"的原则，将在模式**圣地**（66）中详细阐述。

只有经过重重的外花园方能进入的花园才具别有洞天的神秘色彩，只有穿过一层层的院落而后方能到达的庙宇才会将其非同一般的重要性留在人们心中，只有翻越那难以攀登的、耸入云霄的峰谷后才能望到，让人倍觉雄伟壮丽之山顶，只有当绰约多姿的少女徐徐揭开面纱才更显得动人。那河畔风光的秀色——湍急的流水、栖息在岸边的

of these sites.They are bulldozed,developed,changed,for political and economic reasons,without regard for these simple but fundamental emotional matters;or they are simply ignored.

We suggest the following two steps.

1.In any geographic area—large or small—ask a large number of people which sites and which places make them feel the most contact with the area;which sites stand most for the important values of the past,and which ones embody their connection to the land.Then insist that these sites be actively preserved.

2.Once the sites are chosen and preserved,embellish them in a way which intensifies their public meaning.We believe that the best way to intensify a site is through a progression of areas which people pass through as they approach the site.This is the principle of "nested precincts," discussed in detail under the pattern HOLY GROUND(66).

A garden which can be reached only by passing through a series of outer gardens keeps its secrecy.A temple which can be reached only by passing through a sequence of approach courts is able to be a special thing in a man's heart.The magnificence of a mountain peak is increased by the difficulty of reaching the upper valleys from which it can be seen;the beauty of a woman is intensified by the slowness of her unveiling;the great beauty of a river bank—its rushes,water rats,small fish,wild flowers— are violated by a too direct approach;even the ecology cannot stand up to the too direct approach—the thing will simply be devoured.

We must therefore build around a sacred site a series of spaces which gradually intensify and converge on the site. The site itself becomes a kind of inner sanctum,at the core.

河鼠、游动着的小鱼、五彩缤纷的野花——如若被人过分直接趋近，就会遭到侵害；以致连这生态环境也难以保持，一切将被吞噬。

因此，我们在这珍贵的地方的四周造出一系列空间，逐步加强，并使之聚合到这一地点来。它本身处于中心位置，就会逐步成为一种内部圣所。如果它很大，比如说一座大山，同样可用上述手法处理，在其周围设一些可以远望它的特殊地点，山还是一座内部的圣所，要到达这些地点，须花费九牛二虎之力，要攀越无数石级。此时望去，大山已不仅是山，也是一座大花园了。比如，游人可从许多不同的角度去欣赏它那非凡的美色。

因此：

珍贵的地方无论是大是小，无论它们位于城镇的中心、邻里或最偏僻的乡村，都要制定法规，对它们加以绝对保护，以免在可见的周围环境内使我们源远流长的根遭到亵渎。

珍贵的地方

保护法案

❀ ❀

给每一个珍贵的地点腾出一片空地或若干空地，让人们在那里休息娱乐，并感到确实存在这样的地点——**僻静区（59）、禅宗观景（134）、树荫空间（171）、园中坐椅（176）**。尤为重要的是要避免开门见山地直接到达这些地点，所以，人们只好徒步走去，并且要经过一系列的门道和门槛，逐步揭示它们的内部奥秘，历尽曲折才入胜景——**圣地（66）**……

And if the site is very large—a mountain—the same approach can be taken with special places from which it can be seen—an inner sanctum,reached past many levels,which is not the mountain,but a garden,say,from which the mountain can be seen in special beauty.

Therefore:

Whether the sacred sites are large or small,whether they are at the center of the towns,in neighborhoods,or in the deepest countryside,establish ordinances which will protect them absolutely-so that our roots in the visible surroundings cannot be violated.

&OCB

Give every sacred site a place,or a sequence of places, where people can relax,enjoy themselves,and feel the presence of the place—QUIET BACKS(59),ZEN VIEW(134),TREE PLACES(171),GARDEN SEAT(176).And above all,shield the approach to the site,so that it can only be approached on foot,and through a series of gateways and thresholds which reveal it gradually-HOLY GROUND(66)...

模式25   通往水域*

……水是无价之宝。在得天独厚的天然景区内有一些**珍贵的地方**（24）。现在我们选出海滩、湖畔和河岸，因为这些地方是不可取代的。为了保护并适当利用这些地方，就需要有一个特殊的模式。

೫೦೮೫

人们具有一种天性，向往着一望无际的万顷碧波，但是人们纷至沓来，会使水质遭到破坏。

要么是公路、高速公路，还有工业，都来破坏滨水区，使它变成一个污秽不堪、险象丛生的无法涉足之地；要么滨水区得到保护，由私人来负责管理。

但是，人们是须臾也离不开水的，水的意义重大，是人类生死攸关的存在。(See,for example,C.G.Jung,Symbols

25   ACCESS TO WATER*

...water is always precious.Among the special natural places covered by SACRED SITES(24),we single out the ocean beaches,lakes,and river banks,because they are irreplaceable. Their maintenance and proper use require a special pattern.

৪০০৪

People have a fundamental yearning for great bodies of water.But the very movement of the people toward the water can also destroy the water.

Either roads,freeways,and industries destroy the water's edge and make it so dirty or so treacherous that it is virtually inaccessible;or when the water's edge is preserved,it falls into private hands.

But the need that people have for water is vital and profound. (See,for example,C.G.Jung,*Symbols of Transformation*,where Jung takes bodies of water which appear in dreams as a consistent representation of the dreamer's unconscious.)

The problem can be solved only if it is understood that people will build places near the water because it is entirely natural;but that the land immediately along the water's edge must be preserved for common use.To this end the roads which can destroy the water's edge must be kept back from it and only allowed near it when they lie at right angles to it.

The width of the belt of land along the water may vary with the type of water,the density of development along it,and the ecological conditions.Along high density development,it may be no more than a simple stone promenade.Along low

of Transformation,where Jungtakes bodies of water which appear indreams as a consistent representation ofthe dreamer's unconscious.)

通往水域之路已被堵死
Access to water is blocked

因为水完全是自然之物，所以，只要人们理解这一点，愿意在靠近水的地方建筑东西，问题就迎刃而解了。但是，沿着滨水区的土地必须保留作为公共用地。为此，凡破坏滨水区的公路只能靠后设置。靠近滨水区的公路只能与之成直角相接。

滨水区的生活方式
Life forms around the water's edge

density development,it may be a common parkland extending hundreds of yards beyond a beach.

Therefore:

When natural bodies of water occur near human settlements,treat them with great respect.Always preserve a belt of common land,immediately beside the water.And allow dense settlements to come right down to the water only at infrequent intervals along the water's edge.

ಜೋಂ

The width of the common land will vary with the type of water and the ecological conditions.In one case,it may be no more than a simple stone promenade along a river bank a few feet wide—PROMENADE(31).In another case,it may be a swath of dunes extending hundreds of yards beyond a beach-THE COUNTRYSIDE(7).In any case,do not build roads along the water within one mile of the water;instead,make all the approach roads at right angles to the edge,and very far apart—PARALLEL ROADS(23).If parking is provided,keep the lots small—SMALL PARKING LOTS(103)...

滨水地带的宽度将随水体的类别、沿水开发区的密度以及生态条件的不同而变化。沿高密度开发区，它可能只有一条简易的石砌的散步道那样宽。沿低密度开发区，它的宽度可能有海滩往上延伸几百码的一片公共用地那么宽。

因此：

当在居民住宅附近有天然水域时，要倍加珍惜它。要始终如一地保护好滨水的公共地带。并只允许稠密的居民住宅偶而有间隔地伸延到滨水区的水边上。

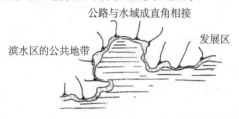

公路与水域成直角相接

滨水区的公共地带

发展区

❧❧

滨水区的公共地带的宽度将随水体类别和生态条件而变化。在一种情况下，它或许只不过是一条沿河岸的简易的石铺的散步道，宽仅几英尺——**散步场所**（31）。在另一种情况下，它或许是跃出海滩向上延伸几百码的一条又长又宽的沙丘——**乡村**（7）。在任何情况下，不要在滨水区离水域 1mi 以内的地方修筑公路；而且要使通往滨水区的所有公路和水体成直角相接，而且各路之间的距离相隔甚远——**平行路**（23）。如果它提供停车场，则停车场一定要小——**小停车场**（103）……

模式26　生命的周期*

　　……一个现实的社区理应使人的感受和生活达到完全的平衡——**7000人的社区**（12）。小而言之，一个良好的邻里也理应如此——**易识别的邻里**（14）。为了使这一愿望成为现实,社区和邻里就必须具备生活所需的丰富多彩的事物,这样才能使每一个人充分感受到其社区生活的广度和深度。

ಶಿ೦ಣ

全世界是一个舞台,
所有的男男女女不过是一些演员:
他们都有下场的时候,也都有上场的时候;
一个人的一生中扮演着好几个角色,
他的表现可以分为七个时期。

26 LIFE CYCLE*

...a real community provides,in full,for the balance of human experience and human life—COMMUNITY OF 7000(12).To a lesser extent,a good neighborhood will do the same—IDENTIFIABLE NEIGHBORHOOD(14).To fulfill this promise,communities and neighborhoods must have the range of things which life can need,so that a person can experience the full breadth and depth of life in his community.

❧

All the world's a stage,
And all the men and women merely players:
They have their exits and their entrances;
And one man in his time plays many parts,
His acts being seven ages.
As,first the infant,
Mewling and puking in the nurse's arms.
And then the whining schoolboy,with his satchel
And shining morning face,creeping like snail
Unwillingly to school.And then the lover,
Sighing like furnace,with a woeful ballad
Made to his mistress'eyebrow.Then the soldier,
Full of strange oaths,and bearded like the pard,
Jealous in honour,sudden and quick in quarrel,
Seeking the bubble reputation
Even in the cannon's mouth.And then the justice,
In fair round belly with good capon lined,

最初是婴孩，在保姆的怀中啼哭呕吐。
然后是背着书包、满脸红光的学童，
像蜗牛一样慢腾腾地拖着脚步，
不情愿地呜咽着上学堂。
然后是情人，像炉灶一样叹着气，
写了一首悲哀的诗歌咏着他恋人的眉毛。
然后是一个军人，满口发着古怪的誓，
胡须长得像豹子一样，
爱惜着名誉，动不动就要打架，
在炮口上寻求着泡沫一样的虚名。
然后是法官，胖胖圆圆的肚子塞满了阉鸡，
凛然的眼光，整洁的胡须，
满嘴都是格言和老生常谈；
他这样扮了他的一个角色。
第六个时期变成了精瘦的趿着拖鞋的龙钟老叟，
鼻子上架着眼镜，腰边悬着钱袋；
他那年轻时候节省下来的长袜子
套在他皱瘪的小腿上显得宽大异常；
他那朗朗的男子的口音，
又变成了孩子似的尖声，像是吹着风笛和哨子。
终结着这段古怪的多事的历史的最后一场，
是孩提时代的再现，全然的遗忘，
没有牙齿，没有眼睛，没有口味，没有一切。

　　莎士比亚（William Shakespeare，1564～1616，英国文艺复兴时期最重要的剧作家和诗人，世界文学史上最伟大的作家之一——译者注）：《大喜欢》，第二幕，第七场。

　　（译文引自朱生豪译、方平校《莎士比亚全集》（三），第139～140页，北京：人民文学出版社，1978年——译者注）

　　为了使人的生活过得美满充实，在其人生的七个时期中，每一时期都要划分得一清二楚，各具特色，绝不雷同。

With eyes severe and beard of formal cut,

Full of wise saws and modern instances;

And so he plays his part.The sixth age shifts

Into the lean and slipper'd pantaloon,

With spectacles on nose and pouch on side;

His youthful hose,well saved,a world too wide

For his shrunk shank;and his big manly voice,

Turning again toward childish treble,pipes

And whistles in his sound.Last scene of all,

That ends this strange eventful history,

Is second childishness and mere oblivion,

Sans teeth,sans eyes,sans taste,sans every thing.

(Shakespeare,As You Like It, Ⅱ . Ⅷ .)

To live life to the fullest,in each of the seven ages,each age must be clearly marked,by the community,as a distinct well-marked time.And the ages will only seem clearly marked if the ceremonies which mark the passage from one age to the next are firmly marked by celebrations and distinctions.

By contrast,in a flat suburban culture the seven ages are not at all clearly marked;they are not celebrated;the passages from one age to the next have almost been forgotten.Under these conditions,people distort themselves.They can neither fulfill themselves in any one age nor pass successfully on to the next. Like the sixty-year-old woman wearing bright red lipstick on her wrinkles,they cling ferociously to what they never fully had.

This proposition hinges on two arguments.

A.The cycle of life is a definite psychological reality. It consists of discrete stages,each one fraught with its own difficulties,each one with its own special advantages.

B.Growth from one stage to another is not inevitable,

对此社区是责无旁贷的。人生从一个时期到另一个时期，如果各有某种庆典活动，则人生的七个时期就被明显地相互区别开来了。

与此相反，在平淡的城郊文化中，七个时期都不祝贺热闹一番；从一个时期过渡到另一个时期，几乎被遗忘得一干二净了。在这种条件下，人们会歪曲自己的形象。他们在任何一个时期，既不能实现自己的愿望，又不能顺利地进入下一个时期。他们恰似年已花甲的老妇，在布满皱纹的脸上搽上了红膏。他们疯狂地抓住过去从未充分享受过的一切。

本模式的主张有两个重要的论据。

A. 生命的周期是一个有明确界限的心理现实。它由各个分立的时期所组成。每一时期都充满着自身的矛盾困难，又各有自己特殊的优势与有利条件。

B. 从一个时期进到另一个时期不是不可避免的。而事实上，除非社区包含一个平衡的生命的周期，否则将不会发生。

A. 生命的周期之现实

每个人都能认识到这一事实：人的一生要经历若干时期，即从幼年到老年。但未必都认识到下面这一点：每个时期都是分立的现实，各有其特定的补偿与艰难；每一时期都有和自己相适应的一些特殊经验。

在这方面最有启发性的一篇文章是埃里克·埃里克森写的《认同感和生命的周期》。(in *Psychological Issues*，Vol.1，No.1，New York : International Universities Press，1959 ; and *Childhoodand Society*，New York : W.W.Norton，1950)

埃里克森描述了一个人如何逐渐发育成人所必须经历的一系列阶段，并指出每一阶段的特征是由一种特定的发育任务——一种成功地解决生活中某一冲突的办法——来表示。他还指出，这一任务必须在此人生理和心理等各方

and,in fact,it will not happen unless the community contains a balanced life cycle.

A.The Reality of the Life Cycle.

Everyone can recognize the fact that a person's life traverses several stages-infancy to old age.What is perhaps not so well understood is the idea that each stage is a discrete reality,with its own special compensations and difficulties;that each stage has certain characteristic experiences that go with it.

The most inspired work along these lines has come from Erik Erikson: "Identity and the Life Cycle," in *Psychological Issues*,Vol.1,No.1,New York:International Universities Press,1959;and *Childhood and Society*,New York:W.W.Norton, 1950.

Erikson describes the sequence of phases a person must pass through as he matures and suggests that each phase is characterized by a specific developmental task—a successful resolution of some life conflict—and that this task must be solved by a person before he can move wholeheartedly forward to the next phase.Here is a summary of the stages in Erikson's scheme,adapted from his charts:

1.*Trust vs.mistrust*:the infant;relationship between the infant and mother;the struggle for confidence that the environment will nourish.

2.*Autonomy vs.shame and doubt*:the very young child; relationship between the child and parents;the struggle to stand on one's own two feet,to find autonomy in the face of experiences of shame and doubt as to one's capacity for self-control.

3.*Initiative vs.guilt*:the child;relationship to the family,the ring of friends;the search for action,and the integrity of one's

面都已具备条件进入下一阶段之前解决。下面是埃里克森方案中的内容提要，是由他的图表改编而成的：

1. 信任对不信任：婴儿；婴儿与母亲的关系；为一种信心而斗争：周围环境将养育他。

2. 自主性对羞怯和怀疑：幼儿；幼儿与双亲的关系；为自己的双脚能站立起来而斗争；在面临感到羞怯和怀疑自我克制能力之时寻求自主性。

3. 主动精神对内疚：儿童；与家庭和一小圈子朋友的关系；探索行动和行为的完满性；动手干活和渴望学习，但受到恐惧心理和自己越轨行为时内疚的抑制。

4. 勤奋对自卑感：少年；与邻里和学校的关系；适应社会上使用的各种工具；单独一人或和其他人一起都能把事办好的意识，相对于失败和不能胜任工作的经验。

5. 认同感对认同感的扩散：青年；与同辈的关系以及和"别的集团"的关系；探索成年人的生活方式；面对混乱和疑虑探索自己个性的连续性；某种活动的暂停；腾出时间去寻找世界的信条和纲领，并与之发生联系。

6. 亲密感对孤独感：初出茅庐的成年人；亲密伙伴、性欲、工作；在与他人相处中完全把自己托付给别人，在别人身上失去或发现自己，反对孤独和回避他人。

7. 开创力对迟钝性：成年人；个人和劳动分工之间的关系；建立一个美满和谐的家庭；努力去建业、管理和创造，反对无所作为和停滞不前的悲观情绪。

8. 完整性对绝望：老年；人和他周围世界的关系，以及和他的同类——人类的关系；智慧的成果；对自己和人类的热爱；以自己一生凝聚起来的力量坦然地面对死亡；相对于活着无用的绝望。

B. 但是，经过生命的周期的发展并非注定不变。

这种发展取决于一个平衡的社区是否存在，即取决于

acts;to make and eagerly learn,checked by the fear and guilt of one's own aggressions.

4.*Industry vs.inferiority*:the youngster;relationship to the neighborhood,the school;adaptation to the society's tools;the sense that one can make things well,alone,and with others,against the experience of failure,inadequacy.

5.*Identity vs.identity diffusion*:youth,adolescence; relationship to peers and "outgroups" and the search for models of adult life;the search for continuity in one's own character against confusion and doubt;a moratorium;a time to find and ally oneself with creeds and programs of the world.

6.*Intimacy vs.isolation*:young adults;partners in friendship, sex,work;the struggle to commit oneself concretely in relations with others;to lose and find oneself in another,against isolation and the avoidance of others.

7.*Generativity vs.stagnation*:adults;the relationship between a person and the division of labor,and the creation of a shared household;the struggle to establish and guide,to create,against the failure to do so,and the feelings of stagnation.

8.*Integrity vs.despair*:old age;the relationship between a person and his world,his kind,mankind;the achievement of wisdom;love for oneself and one's kind;to face death openly,with the forces of one's life integrated;vs.the despair that life has been useless.

B.But growth through the life cycle is not inevitable.

It depends on the presence of a balanced community,a community that can sustain the give and take of growth.Persons at each stage of life have something irreplaceable to give and to take from the community,and it is just these transactions which help a person to solve the problems that beset each stage.Consider the case of a young couple and their new child.

社区是否能维持这种在生长过程中的给予与索取。处于生命的每一时期的人都有某种不可取代的东西给予社区或从社区索取，恰恰是这样一些交往的活动才帮助他解决了在每一时期内存在的各种问题。请考虑一对年轻夫妇和他们的新生儿的情况吧。他们和新生儿之间的联系完全是相互的。当然，新生儿要"依靠"父母对他的精心照料和无微不至的爱，这种关怀和爱是解决与婴儿期相适应的信任的冲突所必需的。但同时，新生儿给予双亲以养儿育女的经验，这种经验有助于他们应付成年时期所特有的开创力的冲突。

如果我们抽象地谈论：我们认为父母的一方"具有"这样那样的个性，在新生儿呱呱坠地之后，他们的个性仍然保持一成不变，并一直影响这个可怜的小东西，那么我们就会曲解这种情况。为了这个弱小的和不断变化的小生命，全家人忙得不可开交。婴儿控制并培育他们的家庭，正如他们被家庭控制一样；事实上，我们可以说，家庭培育婴儿，而婴儿也培育着家庭。从生物学角度看，不管接受什么样的反应，从发展观点看，不管预先确定什么样的时间表，都应当被看作改变相互调整方式的一系列的潜在可能性。(埃里克森，《认同感和生命的周期》，第69页)

相似的相互调整的模式发生在老年人和青年人之间；青年和未成熟的成年人之间；在儿童和婴儿之间；在十三四岁的少年和十八九岁的青年之间；在年轻男子和老态龙钟的妇女之间；在妙龄少女和眼花耳聋的老叟之间；等等。而这些模式之所以成为现实可行是由于现行的社会风俗习惯以及外界环境中的这样一些地方——学校、托儿所、家庭、咖啡馆、寝室、运动场、车间、工作室、花园、墓地……

可是，我们认为，通过生命的周期而正常发展的环境的平衡条件已不复存在。每个人在任何时候都越来越不可能同整个生命的周期保持接触。取代具有一个平衡的生命

The connection between them is entirely mutual.Of course,the child "depends" on the parents to give the care and love that is required to resolve the conflict of trust that goes with infancy. But simultaneously,the child gives the parents the experience of raising and bearing,which helps them to meet their conflict of generativity,unique to adulthood.

We distort the situation if we abstract it in such a way that we consider the parent as "having" such and such a personality when the child is born and then,remaining static,impinging upon a poor little thing.For this weak and changing little being moves the whole family along.Babies control and bring up their families as much as they are controlled by them;in fact,we may say that the family brings up a baby by being brought up by him.Whatever reaction patterns are given biologically and whatever schedule is predetermined developmentally must be considered to be a series of potentialities for changing patterns of mutual regulation.[Erikson,ibid.p.69.]

Similar patterns of mutual regulation occur between the very old and the very young;between adolescents and young adults,children and infants,teenagers and younger teenagers,young men and old women,young women and old men,and so on.And these patterns must be made viable by prevailing social institutions and those parts of the environment which help to maintain them—the schools,nurseries,homes,cafes,bedrooms,sports fields, workshops,studios,gardens,graveyards...

We believe,however,that the balance of settings which allow normal growth through the life cycle has been breaking down.Contact with the entire cycle of life is less and less available to each person,at each moment in time.In place of natural communities with a balanced life cycle we have retirement villages,bedrooms suburbs,teenage culture,ghettos

周期的天然社区有：退休村、郊区"睡城"、青少年文化协会、失业者的贫民窟、学院城镇、公墓、工业园区。在这样一些条件下，人们要解决在生命的周期中每一阶段随之而来的冲突的可能性已微乎其微。

为了再创造一个具有平衡的生命周期的社区，首先要明确发展社区的主导思想：每一建筑项目，无论是增添住宅，还是修筑新的公路，盖医院，都可以被看作或有助于或有碍于地方社区的适当平衡。我们猜想，在《俄勒冈实验》（本丛书的第三卷）的第五章中所论及的社区修整图，在协助鼓励社区发展平衡的生命周期方面，能够起到特别有益的作用。

但是，本模式只是所需要做的工作的一种说明而已。在这一方面每个社区都应当审时度势，寻找出自己的相对"平衡"的途径，并规定自己朝正确的方向发展。这是一个令人颇感兴趣而又十分重要的问题。这需要做大量的开拓、实验工作和理论探讨。如果埃里克森是正确的，如果这方面的工作跟不上，要发展信任感、自主性、主动精神、勤奋精神、认同感、亲密感、开创力、完整性，就会全然化成泡影。

时期	重要的环境	经过的世俗礼仪
1.婴儿 信任感	家庭，有栏杆的儿童小床，托儿所，花园	诞生地，安顿在家里……从有栏杆的儿童小床出来安顿出一块地方
2.幼儿 自主性	自己的空间，夫妻的空间，儿童的空间，公共用地，相互沟通的游戏场所	散步，安顿出一块地方，特殊的生日
3.儿童 主动 精神	游戏的场所，自己的空间，公共用地，邻里，动物	在城镇的最初几次的冒险活动……参加

of unemployed,college towns, mass cemeteries,industrial parks. Under such conditions,one's chances for solving the conflict that comes with each stage in the life cycle are slim indeed.

To recreate a community of balanced life cycles requires, first of all,that the idea take its place as a principal guide in the development of communities.*Each building project,whether the addition to a house,a new road,a clinic,can be viewed as either helping or hindering the right balance for local communities.* We suspect that the community repair maps,discussed in *The Oregon Experiment*,Chapter V(Volume 3 in this series),can play an especially useful role in helping to encourage the growth of a balanced life cycle.

But this pattern can be no more than an indication of work that needs to be done.Each community must find ways of taking stock of its own relative "balance" in this respect,and then define a growth process which will move it in the right direction.This is a tremendously interesting and vital problem;it needs a great deal of development,experiment,and theory.If Erikson is right,and if this kind of work does not come,it seems possible that the development of trust,autonomy,initiative,industry,identity,intimacy,generativity, integrity may disappear entirely.

STAGE	IMPORTANT SETTINGS	RITES OF PASSAGE
1.INFANT Trust	Home,crib,nursery, garden	Birth place,setting up the home...out of the crib,making a place
2.YOUNG CHILD Autonomy	Own place,couple's realm, children's realm, commons, connected play	Walking,making a place,special birthday

时期	重要的环境	经过的世俗礼仪
4.少年勤奋	儿童之家，学校，自己的空间，冒险性的游戏场地，俱乐部，社区	青春期的礼仪，私人的入口，付分内应付之款
5.青年认同感	住所，青少年协会，寄宿舍，学徒，城镇，区域	毕业典礼，结婚，工作，建造住宅
6.未成熟的成年亲切感	家务，夫妻的空间，工作小组，家庭，学习网	生儿育女，创造社会财富……建造住宅
7.成年开创力	工作社区，家庭的市政大厅，自己的房间	特殊的生日，集会，工作变动
8.老人完整性	分散的工作点，别墅，家庭，独立区域	死亡，葬礼，墓地

因此：

确信整个生命的周期是在每个社区表现出来并趋于平衡的。把这种平衡的生命周期的理想确定为社区发展的重要指南。这意味着：

1. 每个社区都包含着从婴儿到老人的生命周期的每一时期的平衡；包含着对生命的所有这些时期所需的环境的历史记录。

2. 社区包含对环境的历史记录，它们最好地标志着人生从一个时期到另一个时期所经历的世俗礼仪。

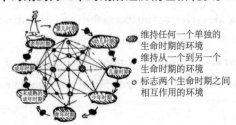

维持任何一个单独的生命时期的环境

维持从一个到另一个生命时期的环境

标志两个生命时期之间相互作用的环境

STAGE	IMPORTANT SETTINGS	RITES OF PASSAGE
3.CHILD Initiative	Play space,own place, common land, neighborhood, animals	First ventures in town...joining
4.YOUNGSTER Industry	Children's home,school, own place,adventure play, club,community	Puberty rites, private entrance paying your way
5.YOUTH Identity	Cottage,teenage society, hostels,apprentice,town and region	Commencement, marriage,work, building
6.YOUNG ADULT Intimacy	Household,couple's realm, small work group,the family, network of learning	Birth of a child, creating social wealth... building
7.ADULT Generativity	Work community,the family town hall, a room of one's own	Special birthday, gathering, change in work
8.OLD PERSON Integrity	Settled work, cottage,the family, independent regions	Death,funeral, grave sites

Therefore:

Make certain that the full cycle of life is represented and balanced in each community.Set the ideal of a balanced life cycle as a principal guide for the evolution of communities.This means:

1.That each community include a balance of people at every stage of the life cycle,from infants to the very old;and include the full slate of settings needed for all these stages of life;

2.That the community contain the full slate of settings which best mark the ritual crossing of life from one stage to the next.

为世人提供举行仪式的非常具体的场所是**圣地**（66）。其他一些特意为维持生命的七个时期和举行过渡仪式而设计的专门模式是**户型混合**（35）、**老人天地**（40）、**工作社区**（41）、**地方市政厅**（44）、**市区内的儿童**（57）、**分娩场所**（65）、**墓地**（70）、**家庭**（75）、**自己的家**（79）、**师徒情谊**（83）、**青少年协会**（84）、**店面小学**（85）、**儿童之家**（86）、**出租房间**（153）、**青少年住所**（154）、**老人住所**（155）、**固定工作点**（156）、**夫妻用床**（187）。

⳾⳾⳾

The rites of passage are provided for, most concretely, by HOLY GROUND(66). Other specific patterns which especially support the seven ages of man and the ceremonies of transition are HOUSEHOLD MIX(35), OLD PEOPLE EVERYWHERE(40), WORK COMMUNITY(41), LOCAL TOWN HALL(44), CHILDREN IN THE CITY (57), BIRTH PLACES(65), GRAVE SITES(70), THE FAMILY(75), YOUR OWN HOME(79), MASTER AND APPRENTICES(83), TEENAGE SOCIETY(84), SHOPFRONT SCHOOLS (85), CHILDREN'S HOME(86), ROOMS TO RENT(153), TEENAGER'S COTTAGE(154), OLD AGE COTTAGE(155), SETTLED WORK(156), MARRIAGE BED(187).

模式27　男人和女人

　　……正如社区或邻里对于不同年龄的人的各种活动必须有一个恰如其分的平衡一样——**7000人的社区**（12）、**易识别的邻里**（14）、**生命的周期**（26）——它们各自也必须调整自己的活动，并使之适应于男人和女人的平衡，向他们提供同等数量的、能反映生活中男性方面的和女性方面的工作。

❧❦

　　20世纪70年代的城镇按男女性别的不同发生了分化。妇女在郊区工作，男子在工作场所工作；幼儿园的活妇女干，职业学校的事男子做；超级市场由妇女经营管理，五金商店由男人营业。

　　因为在生活中没有任何一个领域纯粹是男人干的事或纯粹是女人干的事，所以，在男女分离极端严重的社会中，

27   MEN AND WOMEN

...and just as a community or neighborhood must have a proper balance of activities for people of all the different ages-COMMUNITY oF 7000(12),IDENTIFIABLE NEIGHBORHOOD(14),LIFE CYCLE(26)-so it must also adjust itself and its activities to the balance of the sexes,and provide,in equal part,the things which reflect the masculine and feminine sides of life.

⋙⋘

The world of a town in the 1970'S is split along sexual lines.Suburbs are for women,workplaces for men; kindergartens are for women,professional schools for men; supermarkets are for women,hardware stores for men.

Since no aspect of life is purely masculine or purely feminine,a world in which the separation of the sexes is extreme,distorts reality,and perpetuates and solidifies the distortions.Science is dominated by a masculine,and often mechanical mentality;foreign diplomacy is governed by war,again the product of the masculine ego.Schools for young children are swayed by the world of women,as are homes.The house has become the domain of woman to such a ridiculous extreme that home builders and developers portray an image of homes which are delicate and perfectly "nice," like powder rooms.The idea that such a home could be a place where things are made or vegetables grown,with sawdust around the front door,is almost inconceivable.

人们往往会曲解现实，并使这种畸形的社会现象永久化和凝固化。男性统治科学，常见的是机械思维；外交是由战争主导的，同样是男性利己主义的产物。妇女支配儿童学校，同时支配家庭。住宅成为妇女活动的领域，已经达到荒谬可笑的程度——开发商和建筑商把家的形象描绘得十分可爱：精巧别致，完美"无缺"，像化妆室一般。家可能是制造东西、栽培蔬菜的地方，门前屋后散落着锯屑——这样一种家的概念几乎是不可思议的了。

　　本模式和其他若干模式能否解决上述问题，暂时还不清楚。我们只能稍稍提示建筑的式样、如何使用土地和那些能使问题妥善解决的机构。只有某些社会事实得到确认并能对环境施加全面影响之时，人们才能理解本文插图的几何图形。总而言之，只有男性和女性两者都能够共同影响城镇生活的每一部分时，我们才会茅塞顿开：什么样的空间模式将最好地与这种社会秩序共处。

　　因此：

　　要确信，环境的每一处——每一幢住宅、开阔的空间、邻里和工作社区——都是由男性和女性两者的本能混合而形成的。对于从厨房到钢厂的任何规模的任何工程项目而言，均要切记男性和女性的平衡。

女性的精神

男性的精神

ଚ୨୦୫

　　没有一个大的住宅区没有男人工作的车间；没有一个工作社区不向妇女提供兼职工作和儿童看护服务——**分散的工作点**（9）。在男性和女性已趋于平衡的每一地区内，要确信男女双方都要有空间以便使自己获得充分的发展，他们都有这种权利，将他们双方区别并分离开来——**个人居室**（141）……

The pattern or patterns which could resolve these problems are,for the moment,unknown.We can hint at the kinds of buildings and land use and institutions which would bring the problem into balance.But the geometry cannot be understood until certain social facts are realized,and given their full power to influence the environment.*In short,until both men and women are able to mutually influence each part of a town's life,we shall not know what kinds of physical patterns will best coexist with this social order.*

Therefore:

Make certain that each piece of the environment—each building,open space,neighborhood,and work community-is made with a blend of both men's and women's instincts. Keep this balance of masculine and feminine in mind for every project at every scale,from the kitchen to the steel mill.

⁖⁗

No large housing areas without workshops for men;no work communities which do not provide for women with part-time jobs and child care-SCATTERED WORK(9).Within each place which has a balance of the masculine and feminine,make sure that individual men and women also have room to flourish,in their own right,distinct and separate from their opposites—A ROOM OF ONE'S OWN (141)...

both in the neighborhoods and the communities,and in between them,in the boundaries,encourage the formation of local centers:

28.ECCENTRIC NUCLEUS
29.DENSITY RINGS

在邻里和社区的内部以及在它们两者之间，在边界内，鼓励形成以下的地方中心：

28. 偏心式核心区

29. 密度圈

30. 活动中心

31. 散步场所

32. 商业街

33. 夜生活

34. 换乘站

30.ACTIVITY NODES

31.PROMENADE

32.SHOPPING STREET

33.NIGHT LIFE

34.INTERCHANGE

模式28　偏心式核心区*

……迄今为止，我们对城市总体的建筑高度已作出了限制，随后又对城市的平均密度也作出了限制——**不高于四层楼**（21）。如果我们也假设，每个 30 万人口的城市都包含若干中心，各中心的位置按**城市的魅力**（10）中的原则分布，则城市的建筑密度将随着离开这些中心而逐渐下降：最高的密度靠近中心，最低的密度远离中心。这意味着任何一个单独的 **7000 人的社区**（12）将具有一个总密度，这一总密度由社区离开最近的商业区的距离的远近而定。于是问题就接踵而至了：在社区内，在地方上密度应当如何变化？密度应当具有什么样的几何模式？根据**亚文化区边界**（13）的原理来看，上述问题是极端错综复杂的。因为亚文化区边界要求在社区四周，而不是在社区的几何中心设置服务机构。本模式和下一个模式将说明和本文内容一致的地方密度分布。

❦

地方密度的随机性质搞乱了我们社区的个性，并在土地利用的模式中制造了混乱。

让我们先从考虑城镇居住密度的典型轮廓开始。就居住密度而言，存在着一种总的偏离倾向：向着中心区密度高，向着郊区密度低。但在这总的偏离倾向中没有可识别的结构，即没有清晰可见的在城市中能反复见到的重复性模式。现在试把这一点同一条山脉的轮廓进行比较。在一条山脉中，有许多易识别的结构；我们看见它那连绵起伏的山岭和山谷、山麓、山凹，以及在那地质演变过程中天然突起的峰峦；而这种结构，在整条山脉内从一处到另一

28   ECCENTRIC NUCLEUS*

...so far,we have established an overall height restriction on the city,with its attendant limitation on average density-FOUR-STORY LIMIT(21).If we assume,also,that the city contains major centers for every 300,000 people,spaced according to the rules in MAGIC OF THE CITY(10),it will then follow that the overall density of the city slopes off from these centers:the highest density near to them,the lowest far away.This means that any individual COMMUNITY OF 7000(12)will have an overall density,given by its distance from the nearest downtown.The question then arises:How should density vary locally,within this community;what geometric pattern should the density have?The question is complicated greatly by the principle of SUBCULTURE BOUNDARY(13),which requires that communities are surrounded by their services,instead of having their services at their geometric centers.This pattern,and the next,defines a local distribution of density which is compatible with this context.

ജ്ഞ

The random character of local densities confuses the identity of our communities,and also creates a chaos in the pattern of land use.

Let us begin by considering the typical configuration of the residential densities in a town.There is an overall slope to the densities:they are high toward the center and lower toward the outskirts.But there is no recognizable structure within this

处，一再重复出现。

当然，这仅仅是一种类比。但是，这种类比会使人联想出这样一个问题：如果一个城镇的密度结构具有这样的随机性质，难道是合乎自然的和合理的？假如在密度模式中有某种更能看得见的、前后连贯的结构，有某种自成系统的变化，难道城镇的境况就不会变得更好？

当城镇的地方密度在现存的杂乱无章的、互不连贯的形式中变化时，什么情况将会发生呢？在有潜在可能支持密集活动的高密度区，实际上做不到这一点，因为分布得太广了。而在有潜在可能保持寂静和安宁的低密度区，当它们集中时，有可能做到这点，但也因为过于分散而收效甚微。结论是：城镇既没有十分密集的活动区，也没有异常的安静区。由于我们已有许多论据来表明，一个城镇向它的居民提供两类地区是何等的重要：密集的活动地区和幽静而令人满意的安静地区——**珍贵的地方**（24）、**活动中心**（30）、**散步场所**（31）、**僻静区**（59）、**池塘**（71）——所以，这种密度的随机性，的确伤害着城市生活。

我们确实认为，如果一个城镇包含一个相互连贯的密度模式，那么，它就会好得多。现在我们就将可能自然地影响密度模式的各种因素作一系统的说明，希望表明什么样的前后一贯的模式是切合实际的、有用的。这一论证分下列五个步骤来进行。

1. 我们有理由假定，在每一个 7000 人的社区中，至少有一个由地方服务行业形成的中心。这个中心将是我们称之为**商业街**（32）的典型中心。我们在**商业网**（19）中已经指明，每 10000 人的地区就有一条商业街。

2. 我们通过在**亚文化区边界**（13）中所提供的论据知道，这个活动中心因为是一个服务中心，所以应当出现在亚文化区之间的边界内，应当有助于亚文化区之间的边界的形成，应当坐落在边界区之内——不是位于社区的内部，

overall slope:no clearly visible repeating pattern we can see again and again within the city.Compare this with the contours of a mountain range.In a mountain range,there is a great deal of recognizable structure;we see systematic ridges and valleys,foothills,bowls,and peaks which have arisen naturally from geological processes;and all this structure is repeated again and again,from place to place,within the whole.

Of course,this is only an analogy.But it does raise the question:Is it natural,and all right,if density configurations in a town are so random;or would a town be better off if there was some more visible coherent structure,some kind of systematic variation in the pattern of the densities?

What happens when the local densities in a town vary in their present rambling,incoherent fashion? The high density areas,potentially capable of supporting intense activity cannot actually do so because they are too widely spread.And the low density areas,potentially capable of supporting silence and tranquility when they are concentrated,are also too diffusely scattered.The result:the town has neither very intense activity,nor very intense quiet.Since we have many arguments which show how vital it is for a town to give people both intense activity,and also deep and satisfying quiet—SACRED SITES(24),ACTIVITY NODES(30),PROMENADE(31),QUIET BACKS (59),STILL WATER(71)—it seems quite likely,then,that this randomness of density does harm to urban life.

We believe,indeed,that a town would be far better off if it did contain a coherent pattern of densities.We present a systematic account of the factors which might naturally influence the pattern of density—in the hope of showing what kind of coherent pattern might be sensible and useful.The argument has five steps.

而是位于社区之间。

3. 我们还知道,这个中心必须恰好位于边界内最接近较大的城镇或城市中心的那个地方。这一结论是从一系列惹人注目但尚未被充分认识的研究成果中得出来的。这些研究成果表明,购物中心的顾客会集区,不是像人们天真地想象的那样呈圆圈形,而是呈半圆形,而这个半圆形弧度是在离开城市中心的一侧,因为居民总是去市中心的购物中心采购东西而绝不会去市郊的购物中心采购东西。

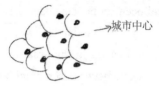

布伦南的顾客会集区
Brenna's catch basins

这一现象由布伦南在战后对英格兰的乌未罕普顿进行了详细的研究后而最早发现的 (T.Brennan, *Midland City*, London : Dobson 1948)。从那时以来,已有几位著作家证实了这一现象,并对它作了进一步的研究,其中蜚声世界的是特伦斯·李的一篇文章《被感知的距离作为市内方向的函数》(*Environment and Behavior*, June 1970, 40~51)。李业已指明,引起这一现象的原因不单单是人们对通往市中心的公路和小路更为熟悉,更加频繁地使用它们,而且也是因为他们对距离的感知随方向而起变化:人们在心理上觉得通往市中心去的道路的距离比离开市中心的道路的距离要短得多。

既然我们肯定要使社区与它的顾客会集"中心"相符合,那么重要的是,这个中心的位置应当是偏心的——事实上,这个中心就位于社区朝向较大城市中心的那个点上。当然,这与我们已经探讨过的概念是一致的,即中心应当位于社区的边界内。

1.We may assume,reasonably,that some kind of center,formed by local services,will occur at least once in every community of 7000.This center will typically be the kind we have called a SHOPPING STREET(32).In WEB OF SHOPPING(19) we have shown that shopping streets occur about once for every 10,000 persons.

2.From the arguments presented in SUBCULTURE BOUNDARY (13),we know that this center of activity,since it is a service,should occur in the boundary between subcultures, should help to form the boundary between subcultures,and should therefore be located in the area of the boundarynot inside the community,but between communities.

3.We know,also,that this center must be in just that part of the boundary which is closest to the center of the larger town or city.This follows from a dramatic and little known series of results which show that catch basins of shopping centers are not circles,as one might naively suppose,but half-circles,with the half-circle on that side of the center away from the central city,because people always go to that shopping center which lies toward the center of their city,never to the one which lies toward the city's periphery.

This phenomenon was originally discovered by Brennan in his post-war studies of Wolverhampton(T.Brennan,*Midland City*,London:Dobson,1948).It has,since then,been confirmed and studied by several writers,most notably Terence Lee, "Perceived Distance as a Function of Direction in the City," *Environment and Behavior*,June 1970,40-51.Lee has shown that the phenomenon is not only caused by the fact that people are simply more familiar with the roads and paths that lie toward the center,and use them more often,but that

4. 即使中心位于社区的一侧,从而形成社区的边界,我们也可以认为,中心需要稍稍凸入社区。这一看法是根据下列事实得出的:即服务行业必须位于社区的边界内,而非位于社区的中央,人们仍有某种需要去形成一个他们自己的社区的心理中心。这一心理中心起码要位于朝向社区的引力几何中心的某一处。如果我们使边界向着社区的几何中心凸入,那么,这条轴线将自然而然地形成一个中心——而且,进而言之,根据上面所给出的数据,它的顾客会集区将几乎完全符合社区的要求。

向内的凸入区
The inward bulge

5. 最后,虽然我们知道,这个中心多半必须位于边界区内,但是我们还不确定这个中心需要多大的规模。城市边缘地区的总密度是低的,所以中心将是小型的。城市中心的总密度是较高的,所以中心也将较大,因为较大的人口密度能够维持更多的服务行业。在这两种情况下,中心将位于边界之内。如果中心太大,以致无法容纳于一个点上,它将势必沿着边界延伸,但仍未超越边界,结果,在边界内,根据它在较大城市中的位置,就形成一个月牙形区,一个局部的或长或短的马蹄形区。

their very perception of distance varies with direction,and that distances along lines toward the center are seen as much shorter than distances along lines away from the center.

Since we certainly want the community to correspond with the catch basin of its "center" it is essential,then,that the center be placed off-center—in fact,at that point in the community which lies toward the center of the larger city.This is,of course,compatible with the notion discussed already,that the center should lie in the boundary of the community.

4.Even though the center lies on one side of the community,forming a boundary of the community,we may also assume that the center does need to bulge into the community just a little.This follows from the fact that,even though services do need to be in the boundary of the community,not in its middle,still,people do have some need for the psychological center of their community to be at least somewhere toward the geometric center of gravity.If we make the boundary bulge toward the geometric center,then this axis will naturally form a center—and,further,its catch basin,according to the data given above,will correspond almost perfectly with the community.

5.Finally,although we know that the center needs to be mainly in the boundary,we do not know exactly just how large it needs to be.At the edge of the city,where the overall density is low—the center will be small.At the center of the city,where the overall density is higher,it will be larger,because the greater density of population supports more services.In both cases,it will be in the boundary.If it is too large to be contained at one point,it will naturally extend itself along the boundary,but still within the boundary,thus forming a lune,a partial horseshoe,long or short,according to

局部的马蹄形区
A partial horseshoe

这些规则是相当简单的。如果我们遵循它们，就将发现一条优美的鳞状梯度变化的曲线，它颇似鱼的鳞片。如果城市逐渐地获得这种高度连贯的结构，则我们可以确信，稠密区和非稠密区就会泾渭分明，活动区和安静区也就一清二楚了。每一个区都具有鲜明特色，互不混同，而且人人都可以去。

因此：

根据下列规则，鼓励人口密度的增大和集聚，以便形成一个清晰的峰谷轮廓：

1. 把城镇视为许多个 **7000** 人的社区的集合体。这些社区，按照各自人口的总密度，将位于直径为 **1/4mi** 的范围内或位于直径为 **2mi** 的范围内。

2. 在每一社区的边界内，标出最接近于最近的主要城市中心的那一个地点。这一地点将成为密度峰，并将成为"偏心"核的核心。

3. 容许高密度区从边界向社区的引力中心凸入，从而扩大朝着中心的偏心核。

4. 继续发展这种高密度，以便在马蹄形的边界附近形成一个脊——马蹄形区的长度取决于总平均密度值——也就是在城市的那个地方，即向着区域中心呈马蹄形凸入的那个地方，形成一个脊，马蹄形区就会根据其在区域中的位置形成一条梯度曲线。靠近主要商业区的马蹄形区几乎是完整的；离商业区较远的马蹄形区只不过是半完整的；离中心最远的马蹄形区则缩小为一个点。

its position in the greater city.

These rules are rather simple.If we follow them,we shall find a beautiful gradient of overlapping imbricated horseshoes,not unlike the scales of a fish.If the city gradually gets this highly coherent structure,then we can be sure that the articulation of dense areas,and areas of little density,will be so clear that both activity and quiet can exist,each intense,unmixed,and each available to everyone.

Therefore:

Encourage growth and the accumulation of density to form a clear configuration of peaks and valleys according to the following rules:

1.Consider the town as a collection of communities of 7000.These communities will be between 1/4 mile across and 2 miles across,according to their overall density.

2.Mark that point in the boundary of each conununity which is closest to the nearest major urban center.This point will be the peak of the density,and the core of the "eccentric" nucleus.

3.Allow the high density to bulge in from the boundary,toward the center of gravity of the community,thus enlarging the eccentric nucleus toward the center.

4.Continue this high density to form a ridge around the boundary in horseshoe fashion-with the length of the horseshoe dependent on the overall mean gross density,at that part of the city,and the bulge of the horseshoe toward the center of the region,so that the horseshoes form a gradient,according to their position in the region.Those close to a major downtown are almost complete;those further away are only half complete;and those furthest from centers are shrunken to a point.

低密度区
高密度区
偏心核

商业区

&cb&

　　假如存在这样一个总的结构，并依照下一个模式——**密度圈**（29）中所提出的计算值，我们就能求算出距高密度脊的不同距离内的平均密度值；使主要商业街和散步场所朝向马蹄形的稠密区——**活动中心**（30）、**散步场所**（31）、**商业街**（32）；使安静区朝向马蹄形的开阔区——**珍贵的地方**（24）、**僻静区**（59）、**池塘**（71）……

Given this overall configuration,now calculate the average densities at different distances from this ridge of high density,according to the computations given in the next pattern—DENSITY RINGS (29);keep major shopping streets and promenades toward the dense part of the horseshoe—ACTIVITY NODES(30),PROMENADE(31),SHOPPING STREET(32);and keep quiet areas toward the open part of the horseshoe-SACRED SITES(24),QUIET BACKS(59),STILL WATER (71)...

模式29 密度圈*

……在**偏心式核心区**（28）中，我们已经提出密度"峰谷"轮廓的一般形式，这种轮廓是同**亚文化区的镶嵌**（8）和**亚文化区边界**（13）相关的。现在，我们假定**7000人的社区**（12）中的商业活动中心的位置，根据**偏心式核心区**（28）的规定及区域内的总密度来选定。那么，我们就会面临一个问题：在密度峰周围的不同距离内，为住宅团组和工作社区确定地方密度。本模式为设计地方密度的梯度曲线提出了一条规则。具体地说，密度的梯度曲线可以十分清楚地、形象化地用图表示出来：从主要的活动中心起，在不同的距离内画出圆圈若干个，继而对每一圆圈确定不同的密度值，在有先后顺序的圆圈中，各种不同的密度值就会形成一条密度的梯度曲线。这种梯度曲线因各社区的不同情况而异：它既取决于社区在某一区域内的位置，又取决于社区人民的文化背景。

❧❧❧

为了寻找刺激和方便，人们都想住在商店和服务机构附近。可是，他们为了寻找宁静和绿化区，就要离开这些地方。这两种愿望的准确平衡因人而异。但是，总括起来说，正是这两种愿望的平衡决定邻里内的住宅密度的梯度曲线。

为了使住宅密度的梯度曲线更为精确，让我们立即同意利用三同心半圆法来分析在主要活动中心周围等径向宽度内的密度值。〔我们把这些圈画成半圆，而不是完整的圈，因为实际经验早已表明，一个规定的地方中心的顾客汇集区是半圆，位于离开城市的一侧——请参阅在**偏心式核心区**（28）中的论述，以及布伦南和李所提供的参考文献。

29   DENSITY RINGS*

...in ECCENTRIC NUCLEUS(28)we have given a general form for the configuration of density "peaks"and"valleys,"with respect to the MOSAIC OF SUBCULTURES(8)and SUBCULTURE BOUNDARIES (13).Suppose now that the center of commercial activity in a COMMUNITY OF 7000(12) is placed according to the prescriptions of ECCENTRIC NUCLEUS(28),and according to the overall density within the region.We then face the problem of establishing local densities,for house clusters and work communities,at different distances around this peak.This pattern gives a rule for working out the gradient of these local densities.Most concretely,this gradient of density can be specified,by drawing rings at different distances from the main center of activity and then assigning different densities to each ring,so that the densities in the succeeding rings create the gradient of density.The gradient will vary from community to community-both according to a community's position in the region,and according to the cultural background of the people.

⊗⊃⊂⊗

People want to be close to shops and services,for excitement and convenience.And they want to be away from services,for quiet and green.The exact balance of these two desires varies from person to person,but in the aggregate it is the balance of these two desires which determines the gradient of housing densities in a neighborhood.

可是，即使你不接受这种发现，并希望去假设这些圈是完整的，下面的分析在实质上仍然是不变的。〕现在我们把密度的梯度曲线规定为三个密度值为一组，每一密度的梯度曲线，表示这一组中三个半圆各自的密度值。

→城市中心

等宽圈
Rings of egual thickness

密度的梯度曲线
A density gradient

设想某个实际存在的邻里的三个半圆的密度值分别为 D_1、D_2 和 D_3。现在假设，一个新来的人搬进了这个邻里。正如我们所说，在这条规定的密度的梯度曲线内，他将会在这样一个半圆内选择住地：这块住地恰好处于他的两种愿望的平衡之中：一方面，他喜欢绿化区和安静区，另一方面，他也喜欢商店和公共服务机构。这意味着，每个人实际上都面临着一种选择：在三个可供选择的密度和距离的组中任选一个：

圈1：密度值 D_1，至商店的距离约为 R_1。
圈2：密度值 D_2，至商店的距离约为 R_2。
圈3：密度值 D_3，至商店的距离约为 R_3。

现在，当然，每个人都将作出不同的选择——各人根据自己对于密度和距离两者的平衡所持的偏爱态度而定。仅仅是为了论证的缘故，让我们设想一下，在该邻里中的全体居民都被要求作出上述选择（暂且不提哪些住宅是可以使用的）。有人将会选择圈1，有人将会选择圈2，有人将会选择圈3。假定 N_1 选择圈1，N_2 选择圈2，N_3 选择圈3。因为这三个圈都有专门的、明确规定的地区，所以，已经选择了这三个地区的人数可以转换成假想的密度值。换句话说，如果我们（在想象中）根据他们的选择把他们分

In order to be precise about the gradient of housing densities,let us agree at once,to analyze the densities by means of three concentric semi-circular rings,of equal radial thickness,around the main center of activity.

[We make them semi-circles,rather than full circles,since it has been shown,empirically,that the catch basin of a given local center is a half-circle,on the side away from the city—see discussion in ECCENTRIC NUCLEUS(28)and the references to Brennan and Lee given in that pattern.However,even if you do not accept this finding,and wish to assume that the circles are full circles,the following analysis remains essentially unchanged.]We now define a density gradient,as a set of three densities,one for each of the three rings.

Imagine that the three rings of some actual neighborhood have densities D_1,D_2,D_3.And assume,now,that a new person moves into this neighborhood.As we have said,within the given density gradient,he will choose to live in that ring,where his liking for green and quiet just balances his liking for access to shops and public services.This means that each person is essentially faced with a choice among three alternative density-distance combinations:

Ring 1.The density D_1,.with a distance of about R_1 to shops.

Ring 2.The density D_2,with a distance of about R_2 to shops.

Ring 3.The density D_3,with a distance of about R_3 to shops.

Now,of course,each person will make a different choice-according to his own personal preference for the balance of density and distance.Let us imagine,just for the sake of argument,that all the people in the neighborhood are asked to make this choice(forgetting,for a moment,which houses are available).Some will choose ring 1,some ring 2,and some ring 3.Suppose that N_1 choose ring 1,N_2 choose ring 2,and N_3 choose ring 3.Since the three rings have specific,well-defined areas,the numbers of people who have chosen the three

配到三个圈内，结果，我们能得出将出现在三个圈内的假想密度值。

现在我们突如其来地面临着两种引人注目的可能性：

Ⅰ.这些新的密度值不同于实际的密度值。

Ⅱ.这些新的密度值与实际的密度值一样。

情况Ⅰ发生的可能性较大，但这是不稳定的，因为人们的选择将趋向于改变这些密度值。情况Ⅱ发生的可能性较小，但这是稳定的，因为这意味着进行自由选择的人们将会一起再创造完全相同的密度模式，在这种密度模式之内，他们已经作出这些选择。这种区别是根本性的区别。

如果我们假设，一个规定的邻里具有一个规定的总面积，它必须容纳一定数量的人口（这一定数量的人口由该区域的一个点的平均密度值给出），那么，从这种意义上讲，只有一种稳定的密度结构。现在我们来描述一种计算过程，通过它就可以获得这种稳定的密度结构。

在解释这一计算过程之前，我们必须解释清楚这种稳定的密度结构是何等的重要和必不可少。

在今天的世界上，密度的梯度曲线通常是不稳定的，就我们所指的意义而言，大多数人被迫居住在安静和活动失去平衡的环境中，这是违背他们的愿望和要求的，因为在不同的距离内可利用的住宅和公寓的总数是不恰当的。结果发生的情况是：付得起钱的富翁们就能找到合乎他们所要求的动静平衡的住宅和公寓，不那么富裕的人和贫穷的人只好快快离去，另觅栖身之所。所有这一切都因中产阶级的"地租"经济学而变得合法了。"地租"经济学是这样一种观念：即对离活动中心不同距离之内的土地要求不同的地价，因为多多少少总有人想要住在这些地方。但实际上，级差地租这一事实是一种经济机制，它萌发于一个不稳定的密度轮廓之中，去抵消这一轮廓的不稳定性。

areas,can be turned into hypothetical densities.In other words,if we(in imagination)distribute the people among the three rings according to their choices,we can work out the hypo—thetical densities which would occur in the three rings as a result.

Now we are suddenly faced with two fascinating possibilities:

I.These new densities are different from the actual densities.

II.These new densities are the same as the actual densities.

Case Ⅰ is much more likely to occur.But this is unstable— since people's choices will tend to change the densities.Case II,which is less likely to occur,is stable—since it means that people,choosing freely,will together recreate the very same pattern of density with-in which they have made these choices. This distinction is fundamental.

If we assume that a given neighborhood,with a given total area,must accommodate a certain number of people(given by the average density of people at that point in the region),then there is just one configuration of densities which is stable in this sense.We now describe a computational procedure which can be used to obtain this stable density configuration.

Before we explain the computational procedure,we must explain how very fundamental and important this kind of stable density configuration is.

In today's world,where density gradients are usually not stable,in our sense,most people are forced to live under conditions where the balance of quiet and activity does not correspond to their wishes or their needs,because the total number of available houses and apartments at different distances is inappropriate.What happens,then,is that the rich,who can afford to pay for what they want,are able to find houses and apartments with the balance that they want;the not so rich and poor are forced to take the leavings.All this is made legitimate by the middleclass economics of "ground

我们想要指出，在具有稳定密度轮廓（稳定是指我们所说的那种稳定）的邻里中，在离活动中心不同的距离内，土地的价格就没有必要有所不同，因为在每一密度圈内可利用的住宅总数将完全符合希望住在这些距离内的人数。因为在每一密度圈中都要提供数量相等的住宅，地租或地价，在每一圈内就可能是相同的了，而且，每一个人，无论贫富，可能会有他所需要的动静平衡。

我们现在来看一个规定的邻里的稳定密度的计算问题。这种稳定性取决于十分微妙的人的心理力量。迄今为止，我们知道这些力量不能用数学方程式以心理上精确的方法表示，因此，至少在目前，不可能为这种稳定的密度提出一个数学模型。而眼下我们宁可利用人人都能够作出的他们所需要的动静平衡的选择这一事实，并将人们的选择编排成简单的游戏，作为计算源。一言以蔽之，我们已经设计出了一种游戏，它能使人们在几分钟内求出这种稳定的密度结构。这种游戏在实质上模拟真实系统的行为。我们认为，这种游戏比任何一种数学计算更为可靠。

密度梯度曲线的智力游戏

1. 先画一幅三个同心的半圆的地图。如果你接受**偏心式核心区（28）**的论据，就画成半圆——不然，你就画个完整的圆。把半圆展开，使之适合于最高密度的马蹄形区，并把半圆的中心标为马蹄形区的中心。

2. 如果半圆的总半径为 R，则三个半圆的平均半径由下式给出，分别为

$$R_1 = R/6$$
$$R_2 = 3R/6$$
$$R_3 = 5R/6$$

3. 制作一游戏盘，上面示出三个同心圈。每个同心圈均有用街区标出的半径，所以人们能容易理解，1000ft=3个街区。

rent" -the idea that land at different distances from centers of activity,commands different prices,because more or less people want to be at those distances.But actually the fact of differential ground rent is an economic mechanism which springs up,within an unstable density configuration,to compensate for its instability.

We want to point out that in a neighborhood with a stable density configuration(stable in our sense of the word),the land would not need to cost different prices at different distances,because the total available number of houses in each ring would exactly correspond to the number of people who wanted to live at those distances.With demand equal to supply in every ring,the ground rents,or the price of land,could be the same in every ring,and everyone,rich and poor,could be certain of having the balance they require.

We now come to the problem of computing the stable densities for a given neighborhood.The stability depends on very subtle psychological forces;so far as we know these forces cannot be represented in any psychologically accurate way by mathematical equations,and it is therefore,at least for the moment,impossible to give a mathematical model for the stable density.Instead,we have chosen to use the fact that each person can make choices about his required balance of activity and quiet,and to use people's choices,within a simple game,as the source of the computation.In short,we have constructed a game,which allows one to obtain the stable density configuration within a few minutes.This game essentially simulates the behavior of the real system,and is,we believe,far more reliable than any mathematical computation.

DENSITY GRADIENTS GAME

1.First draw a map of the three concentric half rings.Make it a half-circle—if you accept the arguments of ECCENTRIC NUCLEUS(28)—otherwise a full circle Smooth this half—

4. 规定这个邻里的总人口。它和定居于该区的总平均密度值是相同的。它将不得不和该区域的总的密度模式大致符合。现在让我们假定该社区的总人口为 N 个家庭。

5. 找出 10 个与该社区的居民大致相似的人——在文化习惯背景等方面相似的人。如果可能，他们应当是这个实际社区居民中的 10 个人。

6. 给做游戏者观看各区的一组照片，这些照片表明不同的人口密度（以大概 1acre 的家庭数目为单位）典型的最佳实例，并把它们在游戏过程中陈列出来，以便做游戏的人在作出选择时能够利用它们。

7. 给每个做游戏的人一个薄平的圆片，他可以把它放到盘上三个圈的任何一个圈内。

8. 现在，为了开始做游戏，先要决定在三个圈的每一个圈内总人口的百分比是多少。你从什么样的百分比开始选择，这是无关紧要的——因为随着游戏的继续，这些百分比很快会得到调整——但是，为了简便起见，请在每一圈内选择一个 10% 的倍数，即在圈 1 为 10%，在圈 2 为 30%，在圈 3 为 60%。

9. 现在请把这些百分比转换成纯粹 1acre 的实际的家庭密度值。因为在整个游戏过程中，你将不得不换算许多次，所以设计出一张能把百分比直接转换成密度值的表格是可取的。你可以通过对 N 和 R 插入数值的方法编制出这种表格，并为你的社区把这些 N 值和 R 值选入下列公式。这些公式是以面积和人口的简单的数学运算为依据的。R 用数百码来表示——粗略地用街区示出。密度值用大概 1acre 的家庭数目来表示。用 1 和 10 之间的一个数乘以每个圈的密度值，并按该圈的百分比求算。结果，如果圈 3 为 30%，则密度值 3 倍于公式中所列的该项数值，即 $24N/5\pi R^2$。

circle to fit the horseshoe of the highest density-mark its center as the center of that horseshoe.

2.If the overall radius of the half-circle is R,then the mean radii of the three rings are R_1,R_2,R_3 given by:

$$R_1=R/6$$
$$R_2=3R/6$$
$$R_3=5R/6$$

3.Make up a board for the game,which has the three concentric circles shown on it,with the radii marked in blocks,so people can understand them easily,i.e.,1000 feet=3 blocks.

4.Decide on the total population of this neighborhood. This is the same as settling on an overall average net density for the area.It will have to be roughly compatible with the overall pattern of density in the region.Let us say that the total population of the community is N families.

5.Find ten people who are roughly similar to the people in the community -vis-à-vis cultural habits,background,and so on.If possible,they should be ten of the people in the actual community itself.

6.Show the players a set of photographs of areas that show typical best examples of different population densities(in families per gross acre),and leave these photographs on display throughout the game so that people can use them when they make their choices.

7.Give each player a disk,which he can place on the board in one of the three rings.

8.Now,to start the game,decide what percentage of the total population is to be in each of the three rings.It doesn't matter what percentages you choose to start with—they will soon tight themselves as the game gets under way-but,for the sake of simplicity,choose multiples of 10 per cent for each ring,i.e.,10 per cent in ring 1,30 per cent in ring 2,60 per cent in ring 3.

9.Now translate these percentages into actual densities of families per net acre.Since you will have to do this many times

<u>10%</u>

圈 1	$8N/\pi R^2$
圈 2	$8N/3\pi R^2$
圈 3	$8N/5\pi R^2$

10. 一旦你已经发现合适的密度值，就将它们从公式中分别抄到三张纸条上，并把三张纸条分别放到游戏盘上它们合适的圈内。

11. 这些纸条说明该社区的暂时的密度结构。每个圈离中心都有某种代表性的距离。而且每一圈都有一种密度值。现在先请 10 人仔细查看代表这些密度值的照片，然后再次确定三个圈中的哪一个会给他们提供"安静区和绿化区"与"通往商店"这两者之间最佳的平衡。最后再请他们把薄平的圆片——放到他们所选中的圈内。

12. 当所有 10 个薄平的圆片都放在游戏盘上后，这就规定了一种新的人口分布。或许，这种分布与你所开始做游戏时的不同。现在，把你原先规定的和现在 10 个薄平圆片规定的这两个百分比折半，再以四舍五入法把百分比舍入最接近的 10%，就可粗略地编出一组新百分比了。

旧百分比	10 人的薄平圆片		新百分比
10%	3=30%	→	20%
30%	4=40%	→	30%
60%	3=30%	→	50%

正如你所见，新百分比不是另外两个百分比的真正的折半——但是你能获得尽量接近你想获得的数值，而且还都是 10% 的倍数。

13. 现在请回到步骤 9，并一再反复通过 9、10、11、12，直到那一回合 10 个薄平圆片所规定的百分比与你原先规定的百分比相同为止。如果你把这些最终得到的稳定的百分比转换成密度值，你就会发现这个社区的稳定的密度结构。请停止吧,大家为这个游戏的整个回合而举杯祝贺吧。

during the course of the game,it is advisable to construct a table which translates percentages directly into densities.You can make up such a table by inserting the values for N and R which you have chosen for your community into the formulae below.The formulae are based on the simple arithmetic of area,and population.R is expressed in hundreds of yardsroughly in blocks.The densities are expressed in families per gross acre.Multiply each ring density by a number between 1 and 10,according to the per cent in that ring. Thus,if there are 30 per cent in ring 3,the density there is 3 times the entry in the formulae,or $24N/5\pi R^2$.

10%

Ring 1	$8N/\pi R^2$
Ring 2	$8N/3\pi R^2$
Ring 3	$8N/5\pi R^2$

10.Once you have found the proper densities,from the formulae,write them on three slips of paper,and place these slips into their appropriate rings,on the game board.

11.The slips define a tentative density configuration for the community.Each ring has a certain typical distance from the center.And each ring has a density.Ask people to look carefully at the pictures which represent these densities,and then to decide which of the three tings gives them the best balance of quiet and green,as against access to shops.Ask each person to place his disk in the ring he chooses.

12.When all ten disks are on the board,this defines a new distribution of population.Probably,it is different from the one you started with.Now make up a new set of percentages,half-way between the one you originally defined,and the one which people's disks define,and,again,round off the percentages to the nearest 10 per cent. Here is an example of the way you can get new percentages.

在我们的许多实验中，我们已经发现，这种游戏确确实实可以十分迅速地得到密度的稳定状态。10个人在几分钟之内，就能够规定一个稳定的密度分布。我们在下列的表格中提供了一组游戏的结果。

不同规模的社区稳定的密度分布
（这些数据是为半圆的社区提供的）

半径 以街区为 单位	人口 以家庭为 单位	密度值，以大概1acre的 家庭数为单位		
		圈1	圈2	圈3
2	150	15	9	5
3	150	7	5	2
3	300	21	7	5
4	300	7	3	2
4	600	29	7	4
6	600	15	4	2
6	1200	36	9	3
9	1200	18	5	1

重点是要承认，就现状来看，本表格中所列出的密度值并没有被明智地利用。这些数据将随着该邻里的精确的几何形状以及在不同的亚文化区中的不同的文化观点而变化。为此，我们认为重要的是，一个规定的社区的居民想要应用本模式，可以玩玩这种游戏，以便为他们自己的境况找出一条稳定的密度梯度曲线。为了便于说明，我们在上面列举的数值多于任何一个模式。

因此：

一旦社区核心十分清楚地选定了位置，就要环绕核心规定逐步递减地方住宅密度圈。如果你不能避免这一点，则可从前面所述的表格中选择各密度值。但是，如果你能尽力而为，那么玩一玩密度圈游戏就是更好的选择，你可以从将要住进这个社区的那些人的直觉中获得这些密度值。

Old percentages	People's disks		New percentages
10%	3=30%	→	20%
30%	4=40%	→	30%
60%	3=30%	→	50%

As you see,the new ones are not perfectly half-way between the other two-but as near as you can get,and still have multiples of ten.

13.Now go back to step 9,and go through 9,10,11,12 again and again,until the percentages defined by people's disks are the same as the ones you defined for that round.If you turn these last stable percentages into densities,you have found the stable density configuration for this community.Stop,and have a drink all round.

In our experiments,we have found that this game reaches a stable state very quickly indeed.Ten people,in a few minutes,can define a stable density distribution.We have presented the results of one set of games in the table which follows below.

STABLE DENSITY DISTRIBUTIONS FOR DIFFERENT SIZED COMMUNITIES

These figures are for semi-circular communities.

Radius in blocks	Population in families	Density in families per gross acre		
		Ring 1	Ring 2	Ring 3
2	150	15	9	5
3	150	7	5	2
3	300	21	7	5
4	300	7	3	2
4	600	29	7	4
6	600	15	4	2
6	1200	36	9	3
9	1200	18	5	1

⊱⊰

　　在密度圈内，鼓励以住宅团组的形式建造住宅——
8～15户的自我管理的合作社，其总体规模随密度的大
小而变化——**住宅团组**（37）。根据不同圈内的密度值，
把这些住宅建成独立的住宅——**住宅团组**（37）、**联排式
住宅**（38），或密度更高的住宅团组——**丘状住宅**（39）。
将公共场所——**散步场所**（31）、**小广场**（61）——建
设在那些高密度区内，并使它们生气勃勃——**行人密度**
（123）……

It is essential to recognize that the densities given in this table cannot wisely be used just as they stand.The figures will vary with the exact geometry of the neighborhood and with different cultural attitudes in different subcultures.For this reason,we consider it essential that the people of a given community,who want to apply this pattern,play the game themselves,in order to find a stable gradient of densities for their own situation.The numbers we have given above are more for the sake of illustration than anything else.

Therefore:

Once the nucleus of a community is clearly placed─define rings of decreasing local housing density around this nucleus.If you cannot avoid it,choose the densities from the foregoing table. But,much better,if you can possibly manage it,play the density rings game,to obtain these densities,from the intuitions of the very people who are going to live in the community.

৪৩৫৫

Within the rings of density,encourage housing to take the form of housing clusters—self-governing cooperatives of 8 to 15 households,their physical size varying according to the density HOUSE CLUSTER(37).According to the densities in the different rings,build these houses as free-standing houses—HOUSE CLUSTER (37),ROW HOUSES(38),or higher density clusters of housing—HOUSING HILL(39). Keep public spaces—PROMENADE(31),SMALL PUBLIC SQUARES(61)—to those areas which have a high enough density around them to keep them alive—PEDESTRIAN DENSITY(123)...

模式30　活动中心**

　　……本模式形成一些重要的生活中心，它们有助于产生**易识别的邻里**（14）、**散步场所**（31）、**小路网络和汽车**（52）和**步行街**（100）。为了理解本模式的作用，设想一个社区及其边界是在**7000人的社区**（12）、**亚文化区边界**（13）、**易识别的邻里**（14）、**邻里边界**（15）、**偏心式核心区**（28）和**密度圈**（29）的影响之下不断发展的。随着上述模式的发展，一些"星状中心"开始形成，一些最重要的小路都在那里交会。这些星状中心是社区潜在的重要之点。它们的发展及其周围小路的发展都需要加以引导，以便形成真正的社区交叉路。

※⁂

　　独立地分散在全市的社区设施无益于城市生活。

30   ACTIVITY NODES**

...this pattern forms those essential nodes of life which help to generate IDENTIFIABLE NEIGHBORH OOD(14),PROMENADE(31),NETWORK OF PATHS AND CARS(52),and PEDESTRIAN STREET (100). To understand its action,imagine that a community and its boundary are growing under the influence of COMMUNITY OF 7000(12),SUBCULTURE BOUNDARY(13),IDENTIFIABLE NEIGHBORHOOD (14),NEIGHBORHOOD BOUNDARY (15), ECCENTRIC NUCLEUS(28),and DENSITY RINGS(29).As they grow,certain "stars" begin to form, where the most important paths meet.These stars are potentially the vital spots of a community.The growth of these stars and of the paths which form them need to be guided to form genuine community crossroads.

⊰⊱

Community facilities scattered individually through the city do nothing for the life of the city.

One of the greatest problems in existing communities is the fact that the available public life in them is spread so thin that it has no impact on the community.It is not in any real sense available to the members of the community. Studies of pedestrian behavior make it clear that people seek out concentrations of other people,whenever they are available(for instance,Jan Gehl, "Mennesker til Fods

现有的社区中存在着的最大问题之一就是可以进行的公共活动太分散，结果对社区毫无影响。就任何现实的意义而言，对此社区的成员是无裨益的。许多学者通过对行人行为的研究才弄清楚：人们无论何时都想到可以找到其他人们集中去的地点（for instance，Jan Gehl，"Mennesker til Fods（Pedestrians）"，*Arkitekten*，No.20，1968）。

为了在社区内创建居民集中常去的地点，各种设施必须密集地围在很小的广场四周，这些小广场能起到中心的作用：有组织地使社区内所有的步行者都通过这些中心。这种中心要求具有以下四个特点：

第一，每个中心必须使其周围的社区内的行人主路接近自己。行人主路都应当会合于广场，行人小道会合于行人主路，从而形成基本的星状模式。做到这一点要比人们所想象的困难得多。当我们试图在一城镇内建立这种模式时，遇到了重重困难。我们下面的例子可资佐证。我们现在展示下列的一张平面图——这是我们为秘鲁的住宅建筑而设计的一个方案——在这一平面图内行人主路全都会集到为数不多的几个广场。

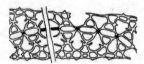

会集于活动中心的公共小路
Public paths converge on centers of action

这不是一幅很理想的平面图——它太刻板，且拘泥于形式。完全有可能以较松散的方式进行设计。在任何情况下，正确处理小路、社区设施和广场三者之间的关系是极为重要的，也是很难的。这一点作为城市的主要特征，从一开始就必须认真加以对待。

第二，为使人们的活动集中，重点是广场要相当小，小到出乎人们的意料之外。一个面积约为45ft×60ft的广场，

(Pedestrians)," *Arkitekten*, No.20,1968).

To create these concentrations of people in a community, facilities must be grouped densely round very small public squares which can function as nodes—with all pedestrian movement in the community organized to pass through these nodes.Such nodes require four properties.

First,each node must draw together the main paths in the surrounding community.The major pedestrian paths should converge on the square,with minor paths funneling into the major ones,to create the basic starshape of the pattern.This is much harder to do than one might imagine. To give an example of the difficulty which arises when we try to build this relationship into a town,we show the following plan—a scheme of ours for housing in Peru—in which the paths are all convergent on a very small number of squares.

This is not a very good plan—it is too stiff and formal. But it is possible to achieve the same relationship in a far more relaxed manner.In any case the relationship between paths,community facilities,and squares is vital and hard to achieve.It must be taken seriously,from the very outset,as a major feature of the city.

Second,to keep the activity concentrated,it is essential to make the squares rather small,smaller than one might imagine. A square of about 45×60 feet can keep the normal pace of public life well concentrated.This figure is discussed in detail under SMALL PUBLIC SQUARE(61).

Third,the facilities grouped around any one node must be chosen for their symbiotic relationships.It is not enough merely to group communal functions in socalled

它能使正常的公共生活井然有序地集中起来。这一数据在**小广场**（61）中还要加以详细论述。

第三，在任何一个中心的周围，各种设施必须顾及它们的共生关系。仅仅把所谓的社区中心的共同功能集中起来是远远不够的。举例来说，教堂、电影院、幼儿园、警察分局都是社区的设施，但它们不会相互支持。不同的人们，在不同的时间，怀着不同的心情，走进不同的机构。没有一个点能把他们集中在一起。为了创造活动的密集性，设在任何一个中心周围的机构都必须以合作的方式协同发挥作用，在一天的同一时间里必须吸引同样的人们。例如，集中举办晚间娱乐活动，凡想前来欢度夜晚的人都可以参加其中的任何一项娱乐活动，这样，活动的总集中量就会增加——参阅**夜生活**（33）。当幼儿园、小公园和花园都集中起来，拖儿带女的年轻夫妇就可以利用其中的任何一个，这些地方的整体吸引力就增加了。

第四，这些活动中心应该相当均匀地分布在整个社区，使每一幢住宅或每一个工作场所与一个中心的距离都不超过几百码。以此方式，在一个小范围内就可形成一个"繁华区和安静区"的鲜明对照——最终，大片没有生气的"死区"就可能避免。

不同规模的中心
Nodes of different size

community centers.For ex-ample,church,cinema,kindergar ten,and police station are all community facilities,but they do not support one another mutually.Different people go to them,at different times,with different things in mind. There is no point in grouping them together.To create intensity of action,the facilities which are placed together round any one node must function in a cooperative manner,and must attract the same kinds of people,at the same times of day.For example,when evening entertainments are grouped together,the people who are having a night out can use any one of them,and the total concentration of action increases—see NIGHT LIFE(33). When kindergartens and small parks and gardens are grouped together,young families with children may use either,so their total attraction is increased.

Fourth,these activity nodes should be distributed rather evenly across the community,so that no house or workplace is more than a few hundred yards from one.In this way a contrast of "busy and quiet" can be achieved at a small scale—and large dead areas can be avoided.

Therefore:

Create nodes of activity throughout the community,spread about 300 yards apart.First identify those existing spots in the community where action seems to concentrate itself.Then modify the layout of the paths in the community to bring as many of them through these spots as possible.This makes each spot function as a "node" in the path network.Then,at the center of each node,make a small public square,and surround it with a combination of community facilities and shops which are mutually supportive.

因此：

在整个社区创立许多活动中心，相互散开的距离约为300yd。首先，要鉴别社区内的那些现存的点，这些点的活动是否真能起到集中的作用。其次，修改社区内的小路的布局设计，以便使尽可能多的小路通过这些点。这就使在小路网络中的每一个点起到一个"中心"的作用。最后，在每个中心的中央形成一个小广场，并在它的四周设置一连串的社区设施和商店，它们相互支持，互为补充。

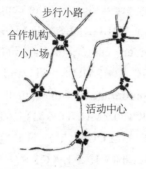

步行小路

合作机构

小广场

活动中心

⊰⊱

将那些最稠密的中心同一条较宽的、更为重要的供人们散步的道路连接起来——**散步场所**（31）；为晚间活动设立特殊的中心——**夜生活**（33）；无论何时修建新的小路时，都要确有把握地让它们通过这些中心，以便它们进一步强化生活——**小路和标志物**（120）；并且改变小路的宽度：接近中心的路面是宽的，而离开中心的路面较窄——**公共性的程度**（36）。在每一中心的中央，建造一个小广场——**小广场**（61）；在每一广场的四周，有许多彼此自我强化的、间杂适中的机构——**工作社区**（41）、**像市场一样开放的大学**（43）、**地方市政厅**（44）、**保健中心**（47）、**分娩场所**（65）、**青少年协会**（84）、**店面小学**（85）、**个体商店**（87）、**临街咖啡座**（88）、**啤酒馆**（90）、**饮食商亭**（93）……

Connect those centers which are most dense, with a wider,more important path for strolling—PROMENADE (31);make special centers for night activities—NIGHT LIFE(33);whenever new paths are built,make certain that they pass through the centers,so that they intensify the life still further—PATHS AND GOALS(120);and differentiate the paths so they are wide near the centers and smaller away from them—DEGREES OF PUBLICNESS(36). At the heart of every center,build a small public square-SMALL PUBLIC SQUARES (61),and surround each square with an appropriate mix of mutually self-reinforcing facilities—WORK COMMUNITY(41),UNIVERSITY AS A MARKETPLACE(43),LOCAL TOWN HALL(44),HEALTH CENTER(47),BIRTH PLACES(65),TEENAGE SOCIETY(84),SHOPFRONT SCHOOL(85),INDIVIDUALLY OWNED SHOPS(87),STREET CAFE(88),BEER HALL(90),FOOD STANDS(93)...

模式31　散步场所**

……现在假设，有一个市区被划分成亚文化区和社区，并且各自都有边界。在**亚文化区的镶嵌**（8）中，每个亚文化区和每个**7000人的社区**（12）都有一个散步场所作为自己的要素。每个散步场所有助于在它自己的范围内形成**活动中心**（30），而川流不息的行人是活动中心赖以生存的必不可少的条件。

&&&

每个亚文化区都需要有一个居民进行公共活动的中心，即一个你可以去那里观看他人并被他人所观看的地方。

31   PROMENADE**

...assume now that there is an urban area,subdivided into subcultures and communities each with its boundaries.Each subculture in the MOSAIC OF SUBCULTURES(8),and each COMMUNITY OF 7000 (12)has a promenade as its backbone. And each promenade helps to form ACTIVITY NODES(30) along its length,by generating the flow of people which the activity nodes need in order to survive.

❧❧

Each subculture needs a center for its public life:a place where you can go to see people,and to be seen.

The promenade, "paseo" , "passegiata" ,evening stroll,is common in the small towns of Italy,Spain,Mexico,Greece,Yugoslavia,Sicily,and South America.People go there to walk up and down,to meet their friends,to stare at strangers,and to let strangers stare at them.

Throughout history there have been places in the city where people who shared a set of values could go to get in touch with each other.These places have always been like street theaters:they invite people to watch others,to stroll and browse,and to loiter:

In Mexico,in any small town plaza every Thursday and Sunday night with the band playing and the weather mild,the boys walk this way,the girls walk that,around and around,and the mothers and fathers sit on iron-scrolled benches and watch.(Ray Bradbury, "The girls walk this way;the boys walk that way..." *West,Los* Angeles Times Sunday Magazine,April 5,1970.)

散步场所、"公共大道"、"散步的地方"、黄昏时散心闲逛处，所有这些名称，在意大利、西班牙、墨西哥、希腊、南斯拉夫、西西里和南美的许多小城镇中，都是一样通用的。人们到散步场所漫步，会见亲朋好友，凝视那些素不相识的陌生人，而那些陌生人也目不转睛地凝视着他们。

从历史上看，在城市里都有一些供具有共同价值观念的人们彼此接触会面的地方。这些地方往往有点像街头演出场所：人们会被纷纷吸引过去看别人的热闹，或是悠闲地散步，或是漫然地看店内和摊上的货物，怡然自得：

在墨西哥的任何一个小城镇的集市广场上，每逢星期四和星期日晚，都有乐队演出，高奏美妙动听的乐曲。时值风和日丽，男女老少成群结队而来。小伙子们往这边来闲逛溜达，姑娘们往那边去信步漫游，来往穿梭，络绎不绝。父母们端坐在饰有涡卷形花纹的铁凳子上，观看这洋溢着欢歌笑语的快乐场面。(Ray Bradbury, "The girls walk this way ; the boys walk that way···" West,Los Angeles Times Sunday Magazine,April 5, 1970.)

在所有这些场合，散步场所之美简单地表现为：具有共同生活方式的人们聚集在一起，相互交往，从而使他们的社区更加巩固。

散步场所在实际上难道仅仅是拉丁人的风俗习惯吗？我们的实验表明，情况并非如此。事实是那种有散步情景的散步场所在一座城市内并非普遍的，尤其在一些城市延伸扩大地区更不普遍。卢斯·拉西昂纳罗在加州大学伯克利分校建筑系时进行了实验，结果表明，无论何处都确实存在着公众交往的可能的地点，只要它离居民很近，人们一定会去寻找它的。拉西昂纳罗访问了住在旧金山几个地区的 37 人，他们分别住在离一个散步场所不同的距离内，结果发现，20min 内就能到达的居民都去散步，超过 20min 才能到达的居民就不怎么去了。

In all these places the beauty of the promenade is simply this:people with a shared way of life gather together to rub shoulders and confirm their community.

Is the promenade in fact a purely Latin institution? Our experiments suggest that it is not.The fact is that the kinds of promenades where this strolling happens are not common in a city,and they are especially uncommon in a sprawling urban region.But experiments by Luis Racionero at the Department of Architecture at the University of California,Berkeley,have shown that wherever the possibility of this public contact does exist,people will seek it,as long as it is close enough.Racionero interviewed 37 people in several parts of San Francisco,living various distances from a promenade,and found that people who lived within 20 minutes used it,while people who lived more than 20 minutes away did not.

	Use the promenade	Do not use the promenade
People who live less than 20 minutes away	13	1
People who live more than 20 minutes away	5	18

It seems that people,of all cultures,may have a general need for the kind of human mixing which the promenade makes possible;but that if it is too far,the effort to get there simply outweighs the importance of the need.In short,to make sure that all the people in a city can satisfy this need,there must be promenades at frequent intervals.

Exactly how frequent should they be? Racionero establishes 20 minutes as the upper limit,but his survey

	利用散步场所	未利用散步场所
20min内能到达的居民	13	1
超过20min才能到达的居民	5	18

看来，具有各种不同的文化背景的人似乎都有一种基本的需要，即人的交往。散步场所使这一需要成为可能。但是，如果散步场所离得太远，所要花的力气已经大大超过人们这种需要本身的重要性了。简而言之，在市内，每隔一定的距离就有一个散步场所，这样，你就会确信，全体居民的这种需要能够得到满足。

这些散步场所相互间隔的距离究竟应该多大呢？拉西昂纳罗提出了一个上限值——20min能到达。但是，他的调查报告并未研究散步场所的使用频率。我们知道，散步场所离得越近，人们就越频繁地使用它。我们推测，在10min或更少的时间内就能到达散步场所，人们将会经常使用它——甚至每周一两次。

散步场所的散步者会集区和该散步场所实际铺设地面的总面积之间的关系是极其关键的。我们将在**行人密度**（123）中指明，在有铺装地面的地方，若每 $150 \sim 300\text{ft}^2$ 内不到 1 人，看起来将显得死气沉沉，毫无吸引力。因此，极为重要的是要确有把握做到下面这一点：那些有可能外出到散步场所闲逛的人的数量，通常要达到足以保持在散步场所长度内所必需的行人密度。为了核查这种关系，我们计算如下：

一次 10min 的步行所经过的路程总计达 1500ft 左右（每分钟 150ft），这一路程大概就是散步场所本身的长度。这就意味着散步场所散步人的会集区具有略似下图的形状：

does not investigate frequency of use.We know that the closer the promenade is,the more often people will use it.We guess that if the promenade is within 10 minutes or less,people will use it often—perhaps even once or twice a week.

The relation between the catch basin of the promenade,and the actual physical paved area of the promenade itself,is extremely critical.We show in PEDESTRIAN DENSITY(123), that places with less than one person for every 150 to 300 square feet of paved surface,will seem dead and uninviting.It is therefore essential to be certain that the number of people who might,typically,be out strolling on the promenade,is large enough to maintain this pedestrian density along its length.To check this relation,we calculate as follows:

A 10-minute walk amounts to roughly 1500 feet(150 feet per minute),which is probably also about the right length for the promenade itself.This means that the catch basin for a promenade has a shape roughly like this:

This area contains 320 acres.If we assume an average density of 50 people per gross acre,then there are 16,000 people in the area.If one-fifth of this population uses the promenade once a week,for an hour between 6 and 10 p.m.,then at any given moment between those hours,there are some 100 people on the promenade.If it is 1500 feet long,at 300 square feet per person,it can therefore be 20 feet wide,at the most,and would be better if it were closer to 10 feet wide.It is feasible,but only just.

We see then,that a promenade 1500 feet long,with the catch basin we have defined and the population density stated,should be able to maintain a lively density of

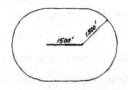

一个散步场所和散步人的会集区
A promenade and its catch basin

这一散步人会集区的面积有 320acre。我们现在假设 1acre 的平均密度约为 50 人，那么，在这一面积内就有 16000 人。如果其中 1/5 的人每周利用它一次，每次一小时（下午 6 时至晚间 10 时），那么，在该时段内的任何一个时刻在散步场所约有 100 人。如果散步场所长 1500ft，每人占 300ft^2，则它的宽度至多可达 20ft，若其接近 10ft 则更好。这是切实可行且恰到好处的。

于是我们看到，一个 1500ft 长的散步场所具有我们已经说明的集中区和人口密度，如果它的宽度不超过 20ft，应当能维持一个充满生气的活动密度。我们要强调指出，除非行人密度足够大，否则散步场所将不会发挥效能，我们还要强调，必须经常进行这种计算，以便核查它的可行性。

前面列举的数据是说明性的。这些数据为散步场所及其散步者会集区的人口确立了一个粗略的数量级。但是，我们还看见过一些成功的例子，如供 2000 居民用的散步场所（秘鲁的一个渔村）和供 200 万人用的散步场所（巴塞罗那的散步广场）。两者虽然在性质上迥然不同，但各有好处。这个仅供 2000 人会集区用的小散步场所有效地发挥着作用，因为当地的文化习俗对散步的要求非常强烈，更加频繁地使用它的居民的百分比较高，而散步的行人密度则比我们想象的要小——它的景色是如此的绚丽多彩，人们赞赏它，即使那里并不那么人头济济。而那巨大的散步广场却发挥着一个城市的作用。人们愿意长途驱车前往，他

activity,provided that it is not more than about 20 feet wide. *We want to emphasize that a promenade will not work unless the pedestrian density is high enough,and that a calculation of this kind must always be made to check its feasibility.*

The preceding figures are meant to be illustrative.They establish a rough order of magnitude for promenades and their catch basin populations.But we have also seen successful promenades for populations of 2000(a fishing village in Peru);and we have seen apromenade for 2,000,000(Las Ramblas in Barcelona).They both work,although they are very different in character.The small one with its catch basin of 2000 works,because the cultural habit of the paseo is so strong there,a higher percentage of the people use it more often,and the density of people on the promenade is less than we would imagine—it is so beautiful that people enjoy it even if it is not so crowded.The large one works as a citywide event.People are willing to drive a long distance to it-they may not come as often,but when they do,it is worth the ride-it is exciting-packed -teeming with people.

We imagine the pattern of promenades in a city to be just as varied-a continuum ranging from small local promenades serving 2000 people to large intense ones serving the entire city—each different in character and density of action.

Finally,what are the characteristics of a successful promenade?Since people come to see people and to be seen,a promenade must have a high density of pedestrians using it.It must therefore be associated with places that in themselves attract people,for example,clusters of eating places and small shops.

们不会常去，但一旦去了，就觉得是值得的。这是多么激动人心的场面：人山人海，热闹非凡。

我们把市内散步场所这一模式恰恰设想成变化的——它是一个连续的统一体，从小的地方散步场所（供 2000 人使用）直到巨大的活动频繁的散步广场（供整个城市的居民使用）都包括在它的范围之内——各种散步场所在性质上不同，活动密度也不同。

最后，一个令人称心如意的散步场所要有哪些特征呢？因为人们去散步场所是为了观看他人和被他人观看的。所以，它一定要有高密度的步行人来往其间。由此可见，它必须和那些引人入胜的地方联结起来，比如，与星罗棋布的小吃店和小商店融为一体。

巴黎的一个散步场所
A promenade in Paris

简而言之，即使去散步场所的真正理由是观看他人和被他人观看，人们也会发现，如果他们心中有一处"目的地"，散起步来，就会感到更加轻松愉快。这样的目的地可以是实实在在的，如可口可乐商店或咖啡座，也可以是部分想象的——"让我们在街区溜达溜达。"但是，散步场所

Further,even though the real reasons for coming might have to do with seeing people and being seen,people find it easier to take a walk if they have a "destination." This destination may be real,like a coke shop or cafe,or it may be partly imaginary, "let's walk round the block." But the promenade must provide people with a strong goal.

It is also important that people do not have to walk too far between the most important points along the promenade. Informal observation suggests that any point which is more than 150 feet from activity becomes unsavory and unused. In short,good promenades are part of a path through the most active parts of the community;they are suitable as destinations for an evening walk;the walk is not too long,and nowhere on it desolate:no point of the stroll is more than 150 feet from a hub of activity.

A variety of facilities will function as destinations along the promenade:ice cream parlors,coke shops,churches,public gardens,movie houses,bars,volleyball courts.Their potential will depend on the extent to which it is possible to make provisions for people to stay:widening of pedestrian paths,planting of trees,walls to lean against,stairs and benches and niches for sitting,opening of street fronts to provide sidewalk cafes,or displays of activities or goods where people might like to linger.

Therefore:

Encourage the gradual formation of a promenade at the heart of every community,linking the main activity nodes,and placed centrally,so that each point in the community is within 10 minutes' walk of it.Put main points of attraction at the two ends,to keep a constant movement up and down.

必须向行人提供一个有强烈吸引力的目的地。

　　沿散步场所散步，离最热闹的点不要太远了，这一点同样是重要的。非正式的研究表明，任何一处地点离开人群活动 150ft 以上，就会令人厌烦、变成人们不怎么去的地方。总之，良好的散步场所是通往社区最活跃地区的小路的一部分，是傍晚散步合适的目的地。散步不要走得太远。在散步场所没有一处是荒凉寂寞的：因为没有一个散步点离开活动中心超过 150ft。

　　在散步场所四周兴建的各种设施和商店将作为散步的目的地而发挥效用：如冷饮店、可口可乐商店、教堂、公园、电影院、酒吧间和排球场等。它们潜力的发挥将取决于在多大程度上为可能吸引住散步的人而预作的种种安排：如扩展人行道的路面；植树栽花；增设可让人靠着的墙面、台阶、凳子、坐凳；临街开敞的店面，经营人行道旁的露天咖啡座，或展示各种活动或商品。这些正是散步者愿意徘徊的地方。

　　因此：

　　鼓励在每个社区的中央逐步形成一个散步场所，并和主要的活动中心连接起来。散步场所在社区中心选址，以便社区各点的居民能在 10min 以内步行至此。为了保持恒定的人流，把引人入胜的主要的点设置在散步场所的两端。

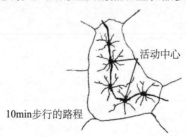

活动中心

10min步行的路程

No matter how large the promenade is,there must be enough people coming to it to make it dense with action,and this can be precisely calculated by the formula of PEDESTRIAN DENSITY(123).The promenade is mainly marked by concentrations of activity along its length—ACTIVITY NODES(30);naturally,some of these will be open at night—NIGHT LIFE(33);and somewhere on the promenade there will be a concentration of shops—SHOPPING STREET(32). It might also be appropriate to include CARNIVAL(58)and DANCING IN THE STREET(63)in very large promenades. The detailed physical character of the promenade is given by PEDESTRIAN STREET(100)and PATH SHAPE(121)...

　　不管散步场所大小如何，都必须有足够数量的人来此地，使这里的活动密集频繁。这一点可以通过**行人密度**（123）的公式精确地计算出来。散步场所的主要标志是：沿着它的长度是活动的集中区——**活动中心**（30）；很自然，其中一些中心是在晚间开放的——**夜生活**（33）；在散步场所的某一处将是商店集中区——**商业街**（32）。在巨大的散步广场，包括**狂欢节**（58）和**街头舞会**（63）也许是合适的。关于散步场所的外部空间性质将在**步行街**（100）和**小路的形状**（121）中进行详细的描述……

模式32　商业街*

……本模式有助于完善**城市的魅力**（10）和**散步场所**（31）。而且，每当商业街建成之日，就是**商业网**（19）行将形成之时。

❦❧

购物中心的形成和发展取决于交通的畅通程度：它们必须位于主要的交通干线附近。可是顾客并未从交通中受惠：因为他们需要安静、舒适和方便，他们想要从周围地区的步行道通往购物中心。

这种简单明了的冲突几乎从未被有效地解决。我们有商业大道，这里有各种商店林立在主要交通干线两侧。这对驾驶汽车来购买东西的顾客是方便的，但对步行来购买商品的顾客是不方便的。商业大道没有步行区的那些特征。

32   SHOPPING STREET*

...this pattern helps to complete the MAGIC OF THE CITY(10)and PROMENADE(31).And,each time a shopping street gets built,it will also help to generate the WEB OF SHOPPING(19).

ଊୠ

Shopping centers depend on access:they need locations near major traffic arteries.However,the shoppers themselves don't benefit from traffic:they need quiet,comfort,and convenience,and access from the pedestrian paths in the surrounding area.

This simple and obvious conflict has almost never been effectively resolved.On the one hand,we have shopping strips. Here the shops are arranged along the major traffc arteries.This is convenient for cars,but it is not convenient for pedestrians. A strip does not have the characteristics which pedestrian areas need.

On the other hand,we have those "pre-automobile" shopping streets in the center of old towns.Here the pedestrians'needs are taken into account,at least partially.But,as the town spreads out and the streets become congested,they are inconvenient to reach;and again the cars dominate the narrow streets.

The modern solution is the shopping center.They are

商业大道：供驱车前来的顾客选购物品
Shopping strip—for cars

另外，我们在旧城镇的中心有那样一些"在出现汽车之前就存在的"商业街。这些商业街对步行前来选购物品的顾客的需要加以考虑，至少部分地加以考虑。但是，随着城镇的扩展和街道变得日益拥挤，要前往这些地方颇为不便。结果，汽车又重新统治着这些狭窄的街道。

旧式的商业街：对驱车购物者和步行购物者都是不方便的
Old shopping street—inconvenient for cars and people

在现代，解决的办法是建立购物中心。它们通常位于主要交通干线两侧或靠近主要交通干线的地方，所以它们对驱车购物者来说是方便的；它们也常常有步行购物者的商店区——所以，至少在理论上，它们对步行购物者来说也是舒适和方便的。但是，它们通常孤零零地位于大停车场的中央，和周围地区的步行道路结构断绝了联系。总而言之，你是无法走到这样的商店区的。

usually located along,or near to,major traffic arteries,so they are convenient for cars;and they often have pedestrian precincts in them—so that,in theory at least,they are comfortable and convenient for pedestrians.But they are usually isolated,in the middle of a vast parking lot,and thereby disconnected from the pedestrian fabric of the surrounding areas.In short,you cannot walk to them.

To be convenient for traffic,and convenient for people walking,and connected to the fabric of the surrounding town,the shops must be arranged along a street,itself pedestrian,but opening off a major traffic artery,perhaps two,with parking behind,or underneath,to keep the cars from isolating the shops from surrounding areas.

We observed this pattern growing spontaneously in certain neighborhoods of Lima,Peru:a wide road is set down for automobile traffic,and the shops begin to form themselves,in pedestrian streets that are perpendicular offshoots off this road.

This pattern is also the form of the famous Stroget in Copenhagen.The Stroget is the central shopping spine for the city;it is extremely long—almost a mile—and is entirely pedestrian,only cut periodically by roads which run at right angles to it.

Therefore:

Encourage local shopping centers to grow in the form of short pedestrian streets,at right angles to major roads and opening off these roads-with parking behind the shops, so that the cars can pull directly off the road,and yet not harm the shopping street.

新的购物中心：只供驱车购物者选购物品
New shopping center—only for cars

为了交通方便，为了行人的方便，也为了与周围的城镇内的住宅保持联系，各种商店必须沿街开设，而这条街本身就是步行街，但是要靠近一条主要的交通干线或两条主要的交通干线。在商店区的后面或下面设有停车场，以便汽车不致把周围地区与商店隔离开来。

我们在秘鲁首都利马的一些邻里中观察到了这种自发形成的模式：一条宽阔的公路是为汽车交通而修筑的，商店在一些步行街两侧，形成自己的格局。这些步行街都是与这条公路相垂直的支线。

在秘鲁利马自发形成的商业街
Shopping streets growing spontaneously in Lima,Peru

本模式也是哥本哈根著名的斯特洛格特步行街的形式。斯特洛格特是该市中央的购物中心。这是一条极长的（几乎长达 1mi）、地地道道的步行街，每隔一定的距离，就有公路与它成直角相交。

Treat the physical character of the street like any other PEDESTRIAN STREET(100)on the NETWORK OF PATHS AND CARS(52),at right angles to major PARALLEL ROADS(23);have as many shops as small as possible— INDIVIDUALLY OWNED SHOPS(87);where the shopping street crosses the road,make the crossing wide,giving priority to the pedestrians—ROAD CROSSING(54);parking can easily be provided by a single row of parking spaces in an alley lying behind the shops—all along the backs of the shops,off the alley,with the parking spaces walled,and perhaps even given canvas roofs,so that they don't destroy the area—SHIELDED PARKING(97),CANVAS ROOFS (244).Make sure that every shopping street includes a MARKET OF MANY SHOPS(46),and some HOUSING IN BETWEEN(48)...

因此：

鼓励地方的购物中心以短步行街的形式逐步发展起来，它们要和主干路成直角相交，而不在这些公路上——停车场设在商店区的后面，这样，可以直接把汽车从公路开到这些停车场，而又对商业街没有丝毫妨碍。

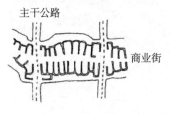

根据**小路网络和汽车**（52），像处理任何其他的**步行街**（100）一样处理商业街外部空间的特点，并使商业街与主干的**平行路**（23）成直角相交。在商业街上开设尽可能多而小的店面——**个体商店**（87）；在商业街与公路成十字相交的地方，要使交叉道路宽阔，并向行人提供优先权——**人行横道**（54）；为了便于停放汽车，在商店区后面的小胡同内设一排停车场——它们全都设在商店僻静区内离小胡同不远的地方，把停车场用围墙围起来，甚至提供帆布顶篷，使停车场不破坏这一地区——**有遮挡的停车场**（97）、**帆布顶篷**（244）。一定要使每条商业街都包括**综合商场**（46）和某种**住宅与其他建筑间杂**（48）……

模式33 夜生活*

  ……每个社区都有某种公开的夜生活——**城市的魅力**
（10）、**7000 人的社区**（12）。如果社区内有一个散步场
所，夜生活就沿着散步场所进行，至少有一部分在那里进
行——**散步场所**（31）。本模式描述晚间集中活动的情景。

⋙⋘

  城市里的大部分活动要到夜里才告结束。那些仍然开
门营业的场所如果不集中在一起，对于城市的夜生活就不
会有所裨益。
  本模式是从下列 7 个观点得出的：

33   NIGHT LIFE*

...every community has some kind of public night life—MAGIC OF THE CITY(10),COMMUNITY OF7000(12).If there is a promenade in the community,the night life is probably along the promenade,at least in part-PROMENADE(31).This pattern describes the details of the concentration of night time activities.

❧❀☙

Most of the city's activities close down at night;those which stay open won't do much for the night life of the city unless they are together.

This pattern is drawn from the following seven points:

1.People enjoy going out at night;a night on the town is something special.

2.If evening activities such as movies,cafes,ice cream parlors,gas stations,and bars are scattered throughout the community,each one by itself cannot generate enough attraction.

3.Many people do not go out at night because they feel they have no place to go.They do not feel like going out to a specific establishment,*but they do feel like going out.*An evening center,particularly when it is full of light,functions as a focus for such people.

4.Fear of the dark,especially in those places far away from

1. 人们乐意在夜晚出门。城镇的夜晚别有一番情趣。

2. 如果晚间的活动点，如电影院、咖啡座、冷饮店、加油站和酒吧，分散在整个社区，每个点本身都不能产生足够的吸引力。

夜间，一个单独的酒吧显得冷冷清清

One bar by itself is a lonely place at night

3. 许多人在晚上闭门不出，原因是他们感到无处可去。他们不是想要到某一特定的设施去，而只是想要出门。一个晚间活动中心，尤其是当它灯火通明时，就会成为吸引这些人的集中点了。

4. 人们的共同心理是害怕黑暗，尤其在远离自己家后院的那些地方，更是令人毛骨悚然，这一点是不难理解的。在我们的进化过程中，夜晚一直是安静的并受到保护的良宵，而不是任意游荡的时辰。

一组夜间活动场所构成了街道生活

A cluster of night spots creates life in the street

one's own back yard,is a common experience,and quite simple to understand.Throughout our evolution night has been a time to stay quiet and protected,not a time to move about freely.

5.Nowadays this instinct is anchored in the fact that at night street crimes are most prevalent in places where there are too few pedestrians to provide natural surveillance,but enough pedestrians to make it worth a thief's while,in other words,dark,isolated night spots invite crime.A paper by Shlomo Angel, "The Ecology of Night Life" (Center for Environmental Structure,Berkeley, 1968),shows the highest number of street crimes occurring in those areas where night spots are scattered.Areas of very low or very high night pedestrian density are subject to much less crime.

6.It is difficult to estimate the exact number of night spots that need to be grouped to create a sense of night life. From observation,we guess that it takes about six, minimum.

7.On the other hand,massive evening centers,combining evening services which a person could not possibly use on the same night,are alienating.For example,in New York the Lincoln Center for the Performing Arts makes a big splash at night,but it makes no sense.No one is going to the ballet and the theater and aconcert during one night on the town.And centralizing these places robs the city as a whole of several centers of night life.

All these arguments together suggest small,scattered centers of mutually enlivening night spots,the services grouped to form

5. 现在，我们凭直觉就能知道这样的事实：夜间，街头犯罪最猖獗的场所就是行人稀少、难以提供自然监视的地方，但这些地方确有足够的行人可供窃贼作案。换言之，灯光暗淡的、孤零零的夜间活动场所是滋生事端之地。斯洛莫·安吉尔在他的《夜生活的生态学》(*Center for Enviromnental Structure*, Bekeley, 1968) 一文中指出，在夜间活动场所分散的地方，街头犯罪率较高。夜里，行人密度很高和很低的地方，犯罪率就较低。

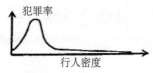

孤零零的夜间活动场所是肇事犯罪之地
Isolated night spots invite crime

6. 很难估计为创造一种名副其实的夜生活而需要集中的夜间活动场所的确切数字。根据观察，我们揣测，最小的数字是 6 个左右。

7. 另外，把各种晚间服务机构（一个人在同一个夜里是光顾不过来的）联合成大规模的夜市活动中心，是会使人在感情上疏远的不当之举。例如，纽约的林肯表演艺术中心，入夜后灯火辉煌，光彩夺目，但是它不合情理。每当夜幕降临城镇之时，没有人能在一个晚上去那里又看芭蕾舞，又去影剧院，又去听音乐会。把这些地方都集中于一处，实际上把原来有可能形成城市中几个夜生活中心的可能给剥夺了。

所有这些论据统统表明，数个小而分散的夜生活中心相映成趣，围在广场四周的各种服务机构会使广场呈现出一派喜气洋洋的气氛，那里灯光明亮，可以徘徊徜徉，人们都会兴致勃勃地去消磨时光，一连几小时都会感到有兴味。下面是一些实例——几组相互维持夜生活活动的商店

cheery squares,with lights and places to loiter,where people can spend several hours in an interesting way.Here are some examples of small groups of mutually sustaining night activities.

A movie theater,a restaurant and a bar,and a bookstore open till midnight;a smoke shop.

A laundromat,liquor store and cafe;and a meeting hall and beer hall.

Lodge hall,bowling alley,bar,playhouse.

A terminal,a diner,hotels,nightclubs,casinos.

Therefore:

Knit together shops,amusements,and services which are open at night,along with hotels,bars,and all-night diners to form centers of night life:well-lit,safe,and lively places that increase the intensity of pedestrian activity at night by drawing all the people who are out at night to the same few spots in the town.Encourage these evening centers to distribute themselves evenly across the town.

❧❧

Treat the physical layout of the night life area exactly like any other ACTIVITY NODE(30),except that all of its establishments are open at night.The evening establishments might include LOCAL TOWN HALL (44),CARNIVAL(58),DANCING IN THE STREET(63), STREET CAFE(88),BEER HALL(90), TRAVELER'S INN(91)...

和机构。它们分别为：

第一组：一个电影院、一个餐厅、一个酒吧间和一个书店，营业直至午夜；还有一个烟馆。

第二组：一间自动洗衣店、一个酒店和一个咖啡座；还有一个会议厅和一个啤酒馆。

第三组：一个旅店大厅、一个保龄球场、一个酒吧和一个小剧场。

第四组：一个公共汽车的终点站、一个餐车式饭店、几家旅馆、几个夜总会和几个娱乐场。

因此：

要把晚间营业的商店、娱乐场所和服务机构同旅馆、酒吧以及通宵营业的餐车式饭店联结在一起，以便形成夜生活的活动中心，那里灯火通明，安全舒适，生动活泼，兴高采烈。这样就可把夜里外出的游人都吸引到该城镇内寥寥可数的几个夜间活动场所去，从而增加夜间行人活动的热烈气氛。鼓励这些夜生活活动中心均匀地分布在整个城镇。

连成一片的夜间服务设施

⁂

处理夜生活区的总体布局，如同任何一个**活动中心**（30）一样，所不同的是它的全部商店和机构都要在夜间营业。其中可以包括**地方市政厅**（44）、**狂欢节**（58）、**街头舞会**（63）、**临街咖啡座**（88）、**啤酒馆**（90）和**旅游客栈**（91）······

模式34　换乘站

　　……本模式阐明形成**公共交通网**（16）的一些要点。在每一交通区的中心保证能有换乘站从而有助于完善**地方交通区**（11）。在换乘站，人们能够由自行车或地方小公共汽车换乘长途运输线上的车辆。这些长途运输线把各个运输区相互连接起来。

≬≮

　　换乘站在公共交通中起核心作用。除非换乘站运转正常，否则公共交通系统将无法维持。

34   INTERCHANGE

...this pattern defines the points which generate the WEB OF PUBLIC TRANSPORTATION(16).It also helps to complete LOCAL TRANSPORT AREAS(11)by guaranteeing the possibility of interchanges at the center of each transport area,where people can change from their bikes,or local mini-buses,to the long distance transit lines that connect different transport areas to one another.

⋈⋈

Interchanges play a central role in public transportation. Unless the interchanges are working properly,the public transportation system will not be able to sustain itself.

Everyone needs public transportation sometimes.But it is the steady users who keep it going.If the steady users do not keep it going,then there is no system for the occasional user.To maintain a steady flow of users,interchanges must be extremely convenient and easy to use:1.Workplaces and the housing for people who especially need public transportation must be distributed rather evenly around interchanges.2.The interchanges must connect up with the surrounding flow of pedestrian street life.3.It must be easy to change from one mode of travel to another.

In more detail:

1.Workers are the bread and butter of the transportation system.If the system is to be healthy,all the workplaces in town must be within walking distance of the interchanges.

每个人都不时地需要公共交通。而正是这些稳定的乘客才使公共交通得以维持下去。如果他们不来乘车，偶尔来的乘客是不可能享有这种公共交通系统的。为了保持稳定的客流量，换乘站必须使乘客感到十分方便，容易利用：1. 那些特别需要公共交通的人的工作地点和住宅必须相当匀称地分布在换乘站的四周。2. 换乘站必须和周围的步行街的人流沟通。3. 从一种旅行方式改变到另一种旅行方式必须很方便。

下面是更为详细的说明：

1. 工人是交通系统赖以生存的衣食父母。如果这一系统是健全的，城镇内的所有的工作点都必须位于离换乘站步行可达的距离之内。除此之外，换乘站四周的工作点多多少少应当是均匀分布的——**分散的工作点**（9）。当它们都集中在一两个换乘站周围时，高峰时刻的人流涌入列车，挤得水泄不通，就整体而言，该系统就会降低效率。

再者，在换乘站周围的地区内，应划出一些地方来供完全依赖公共交通的人——尤其是老人——建造住宅之用。年迈的老人依赖公共交通系统，他们构成该系统常客中的一大部分。为了满足他们的需要，在换乘站周围的地区内应划定地段，以便兴建适合于他们居住的住宅——**老人天地**（40）。

2. 必须让从家里或工作地点徒步前来乘车的旅客感到换乘站既方便又安全。如果换乘站肮脏不堪，工作人员玩忽职守，呈现出衰颓败落的景象，人们将弃之不用。这意味着换乘站必须和地方行人的生活保持连续的节奏。停车场必须设置在换乘站的同一侧，这样，乘客到换乘站去就不必跨越它了。在换乘站必须开设数量上足够多的商店和商亭，以便保持一个稳定的客流量进出或经过换乘站。

Furthermore,the distribution of workplaces around interchanges should be more or less even—see SCATTERED WORK(9). When they are concentrated around one or two,the rush hour flow crowds the trains,and creates inefficiencies in the system as a whole.

Furthermore,some of the area around interchanges should be given over to houses for those people who rely entirely on public transportation—especially old people.Old people depend on public transportation;they make up a large proportion of the system's regular users.To meet their needs,the area around interchanges must be zoned so that the kind of housing that suits them will develop there—OLD PEOPLE EVERYWHERE(40).

2.The interchange must be convenient for people walking from their homes and jobs,and it must be safe.People will not use an interchange if it is dingy,derelict,and deserted. This means that the interchange must be continuous with local pedestrian life.Parking lots must be kept to one side,so that people do not have to walk across them to get to the station.And there must be enough shops and kiosks in the interchange,to keep a steady flow of people moving in and out of it and through it.

3.If the system is going to be successful,there must be no more than a few minutes'walk—600 feet at the most—between points of transfer.And the distance should decrease as the trips become more local:from bus to bus,100 feet maximum;from rapid transit to bus,200 feet maximum;from train to rapid transit,300 feet maximum.In rainy climates the connecting paths should be almost entirely covered—ARCADES(119). What's more,the most important transfer connections should

3. 如果该系统将来获得圆满成功，那旅客必定在几分钟内就能步行到换乘站：换车的两点之间的距离至多为 600ft。随着旅游更具有地方性质，换车的距离也应相应缩短：从公共汽车到公共汽车的最大间距为 100ft；从快速运输线到公共汽车的最大间距为 200ft；从列车到快速运输线的最大间距为 300ft。在多雨的气候环境下，和换乘站连接的各条小路十之八九都应有防雨设备——**拱廊**（119）。而且，最重要的换乘站枢纽不应包含交叉街道：如有必要，让公路在地面下通过或架设桥梁，以便换车顺利进行。

至于换乘站的组织细则，参阅《快速运输站的 390 项要求》[see "390 Requirements for Rapid Transit Stations," Center for Environmental Strueture, 1964, Partly published in "Relational Complexes in Architecture"（Christopher Alexander, Van Maren King, Sara Ishikawa, Michael Baker, *Architectural Record*, September 1966, PP.185 ～ 190）]。

因此：

在交通网的每一个换乘站都应遵守如下原则：

1. 特别需要公共交通的人的工作地点和住宅应位于换乘站的周围。

2. 使换乘站的内部和外部的行人网络成为一个连续的整体，并通过如下措施来保持这种连续性：在换乘站内建造许多小商店和商亭；停车场设在换乘站的同一侧。

3. 无论在什么地方，只要有可能，要使不同交通方式间的换车距离缩小到 **300ft**，而绝对最大值则为 **600ft**。

not involve crossing streets:if necessary sink the roads or build bridges to make the transfer smooth.

For details on the organization of interchanges,see "390 Requirements for Rapid Transit Stations," Center for Environmental Structure,1964, partly published in "Relational Complexes in Architecture" (Christopher Alexander,Van Maren King,Sara Ishikawa,Michael Baker,*Architectural Record*,September 1966,pp.185-190).

Therefore:

At every interchange in the web of transportation follow these principles:

1.Surround the interchange with workplaces and housing types which specially need public transportation.

2.Keep the interior of the interchange continuous with the exterior pedestrian network,and maintain this continuity by building in small shops and kiosks and by keeping parking to one side.

3.Keep the transfer distance between different modes of transport down to 300 feet wherever possible,with an absolute maximum of 600 feet.

80G3

Recognize that the creation of workplaces around every interchange contributes to the development of SCATTERED WORK(9).Place HOUSING HILLS(39),OLD PEOPLE EVERYWHERE(40),and WORK COMMUNITIES(41) round the interchange;treat the outside of the interchange as an ACTIVITY NODE(30)to assure its continuity with the pedestrian network;treat the transfers as ARCADES(119)where

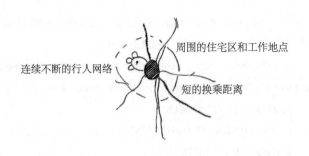

连续不断的行人网络

周围的住宅区和工作地点

短的换乘距离

❧❦

　　承认在每一换乘站周围建立工作点有助于发展**分散的工作点**（9）。将**丘状住宅**（39）、**老人天地**（40）和**工作社区**（41）都布置在换乘站的周围；把换乘站以外的地区作为**活动中心**（30）来处理，以便确保换乘站和行人网络的连续性；把换车的过道作为**拱廊**（119）来处理，上面设有必要的防雨顶盖；为每一换乘站提供一个**公共汽车站**（92），以便利用**小公共汽车**（20）网络……

　　在这些中心周围，根据人群面对面相处的原则，以住宅团组的形式发展住宅建筑：

　　35. 户型混合

　　36. 公共性的程度

　　37. 住宅团组

　　38. 联排式住宅

　　39. 丘状住宅

　　40. 老人天地

necessary to keep them under cover;give every interchange a
BUS STOP(92)on the MINI-BUS(20)network...

around these centers,provide for the growth of housing in
the form of clusters,based on face to face human groups:

35.HOUSEHOLD MIX

36.DEGREES OF PUBLICNESS

37.HOUSE CLUSTER

38.ROW HOUSES

39.HOUSING HILL

40.OLD PEOPLE EVERYWHERE

模式35 户型混合*

……在一个地区内, 在形成或破坏**易识别的邻里** (14)、**住宅团组** (37)、**工作社区** (41) 或通常最重要的**生命的周期** (26) 的性质方面所起的作用, 几乎没有什么别的东西可与户型混合相比拟。问题在于: 一个平衡良好的邻里应当包含何种混合呢?

৪০৫৪

在生命的周期中, 没有一个时期是自我满足的。

在生命的周期中, 人们需要从和他们自己处于不同时期的人那里获得支持和认可, 同时也需要从和他们自己处于同一时期的人那里获得支持。

但是, 人们某些会产生分离的需要远远超过混合的需要。现在的住宅模式倾向于不同户型的相互隔离。现在有大片的两居室住宅, 不同的工作室加单居室的公寓, 以及

35   HOUSEHOLD MIX*

...the mix of households in an area does almost more than anything else to generate,or destroy,the character of an IDENTIFIABLE NEIGHBORHOOD(14),of a HOUSE CLUSTER(37),of a WORK COMMUNITY(41),or,most generally of all,of a LIFE CYCLE(26).The question is,what kind of mix should a wellbalanced neighborhood contain?

❧⳯

No one stage in the life cycle is self-sufficient.

People need support and confirmation from people who have reached a different stage in the life cycle,at the same time that they also need support from people who are at the same stage as they are themselves.

However,the needs which generate separation tend to overwhelm the need for mixture.Present housing patterns tend to keep different types of households segregated from each other.There are vast areas of two-bedroom houses,other areas of studio and one-bedroom apartments,other areas of three-and four-bedroom houses.This means that we have corresponding areas of single people,couples,and small families with children,segregated by type.

The effects of household segregation are profound.In the pattern LIFE CYCLE(26),we have suggested that normal growth through the stages of life requires contact,at each stage,with people and institutions from all the other ages of man.Such contact is completely foiled if the housing mix in

不同的三居室住宅和四居室住宅。这意味着，我们有相应的单人、夫妻和拖儿带女的不同类型、相互隔离的小家庭。

户型隔离的影响深远。在模式**生命的周期**（26）中，我们业已建议，通过生命不同时期的正常成长，在每一时期，都需要和所有的年龄不同的人和各种机构接触。如果在某一邻里中，户型混合被曲解为仅仅涉及人生的一两个时期，那么这种接触将完全落空。另外，当生命的周期的平衡与某一邻里中可用的各种住宅类型有着良好的关系时，接触的可能性就变得具体了。每个人都能发现，他在和他的街坊四邻面对面的相处之中，至少可以接触到处于生命各个时期的人。青年小伙子看见一对对情投意合的伴侣，老人关注着活蹦乱跳的娃娃，独居者能得到大家庭的关怀，少年们视年富力强的中年人为榜样，等等。这一切都是媒介，人们通过这种媒介才会感觉到他们自己的生命之路。

年龄相仿和生活方式相似的人愿意彼此挨近居住的这种需要一定会抵消掉户型混合的需要。如果我们对这两种需要相提并论，那么户型混合恰如其分的平衡到底是怎样的呢？

这种恰如其分的户型混合的平衡可以直接从区域的统计数字引出来。第一，就区域整体而言，要确定每一种户型的百分比；第二，利用这同一的百分比去指导在该邻里内逐步发展户型混合。例如，如果大城市区的住户中40％是家庭，25％是夫妻户，20％是单人户，10％是集体户，那么我们就可以期待在每一邻里内的住宅具有大抵相同的平衡。

最后，让我们试问一下：户型混合应当运用于多大的一个群体？我们可以试图在每一所住宅内（显然是荒谬可笑的），或在每一个由12幢住宅构成的住宅团组内，或在每一个邻里中，或仅仅在每一个城镇中（这最后一种情况

one's neighborhood is skewed toward one or two stages only. On the other hand,when the balance of life cycles is well related to the kinds of housing that are available in a neighborhood,the possibilities for contact become concrete.Each person can find in the face-to-face life of his neighborhood at least passing contact with people from every·stage·of life.Teenagers see young couples,old people watch the very young,people living alone draw sustenance from large families,youngsters look to the middle-aged for models,and so on:it is all a medium through which people feel their way through life.

This need for a mix of housing must be offset against the need to be near people similar in age and way of life to oneself. Taking these two needs together,what is the right balance for the housing mix?

The right balance can be derived straightforwardly from the statistics of the region.First,determine the percentage of each household type for the region as a whole;second,use the same percentages to guide the gradual growth of the housing mix within the neighborhood.For example,if 40 percent of a metropolitan region's households are families,25 percent are couples,20 percent are individuals,and 10 percent group households,then we would expect the houses in each neighborhood to have roughly the same balance.

Let us ask,finally,how large a group should the mix be applied to?We might try to create a mix in every house (obviously absurd),or in every cluster of a dozen houses,or in every neighborhood,or merely in every town(this last has almost no significant effect).We believe that the mix will only work if it exists in a human group small enough to have some internal political and human intercourse—this could be a

几乎毫无意义）形成一种混合。我们认为，如果这种混合存在于一个小得足以具有某种内部的、政治和人情交往的群体中时——这可能是由 12 个家庭组成的一个住宅团组或一个 500 人的邻里——仅仅在此时，这种混合才会发挥积极的作用。

因此：

鼓励在每一邻里内、每一住宅团组内发展户型混合，以便单人住宅、夫妻住宅、拖儿带女的家庭和集体住户相邻共处。

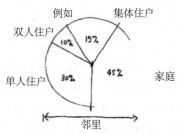

꧁꧂

尤其要深信不疑的是，在每一邻里中要为老人预作安排——**老人天地**（40），即使有这种户型混合，年幼的儿童仍将有足够的游戏伙伴——**相互沟通的游戏场所**（68）；建立不同的户型混合细则，根据合适的、更加详细的模式，以便加强这种混合——**家庭**（75）、**小家庭住宅**（76）、**夫妻住宅**（77）和**单人住宅**（78）……

cluster of a dozen families,or a neighborhood of 500 people.

Therefore:

Encourage growth toward a mix of household types in every neighborhood,and every cluster,so that one-person households,couples,families with children,and group households are side by side.

৪০৪৪

Make especially sure there are provisions for old people in every neighborhood—OLD PEOPLE EVERYWHERE(40),and that even with this mix,young children will have enough playmates—CONNECTED PLAY(68);and build the details of the different kinds of households,according to the appropriate more detailed patterns to reinforce the mix—THE FAMILY(75),HOUSE FOR A SMALL FAMILY(76),HOUSE FOR A COUPLE(77),HOUSE FOR ONE PERSON(78)...

模式36　公共性的程度**

……在邻里内——**易识别的邻里**（14），很自然，一些地方活动相当集中——**活动中心**（30），另一些地方活动比较缓慢，而还有一些地方，活动则居于两者之间——**密度圈**（29）。根据这一梯度，把住宅团组和通往住宅团组的小路区别开来是十分重要的。

৪০৫৪

人是各不相同的，其最基本的区别之一是他们以不同的方式在邻里内为自己的住宅选址。

一些人想住在活动频繁的地方。另一些人想住在清静的地方。这一点符合人类基本的个性类型，它们可以被称为"外向型"和"内向型"，或可称为"喜爱社区型"和

36   DEGREES OF PUBLICNESS**

...within the neighborhoods—IDENTIFIABLE
NEIGHBORHOOD(14)—there are naturally some areas where
life is rather concentrated ACTIVITY NODES(30),others where
it is slower,and others in between—DENSITY RINGS(29).
It is essential to differentiate groups of houses and the paths
which lead to them according to this gradient.

❧❦☙

**People are different,and the way they want to place
their houses in a neighborhood is one of the most basic
kinds of difference.**

Some people want to live where the action is.Others
want more isolation.This corresponds to a basic human
personality dimension,which could be called the "extrovert-
introvert" dimension,or the "community loving-privacy
loving" dimension.Those who want the action like being
near services,near shops,they like a lively atmosphere
outside their houses,and they are happy to have strangers
going past their houses all the time.Those who want more
isolation like being away from services and shops,enjoy a
very small scale in the areas outside their houses,and don't
want strangers going past their houses.(See for example,Nancy
Marshall, "Orientations Toward Privacy:Environmental and
Personality Components," James Madison College,Michigan
State University,East Lansing,Michigan.)

The variation of different people along the extrovert-

"喜爱私密型"。凡是喜欢热热闹闹活动的人都愿意住在服务机构和商店的附近，他们喜欢自家的住宅外充满着生动活泼的气氛，他们乐意看到从早到晚有许多陌生人川流不息地经过自己的家门。凡喜欢清静的人都愿意远离服务机构和商店，他们只对自己宅外的一小块地方感到兴趣，不愿意看到许多陌生人路过自己的家门。(See for example, Nancy Marshall, "Orientations Toward Privacy: Environmental and Personality Components," James Madison College, Michigan State University, East Lansmg, Michigan.)

弗兰克·亨德里克斯和马尔科姆·麦克内厄对于"外向型"和"内向型"的人做了极为生动的描述 (See "Concepts of Environmental Quality Standards Based on Life Styles," report to the American Public Health Association, February 12, 1969, PP.11-15)。作者鉴别了几种不同的人，并根据他们在外向活动和内向活动方面所花的时间的相对量分别作了说明。弗朗西斯·洛伊特尔对此问题作了进一步的说明 (See "Environment Attitudes and Social Life ill Santa Clara County," Santa Clara County Plannillg Department, San Jose, California, 1967)。他逐一调查了 3300 户，询问他们希望住在离不同的社区服务机构多远的地方。调查的结果表明：其中 20% 的住户希望住在离商业中心区不到 3 个街区的地方；60% 的住户希望住在离商业中心区 4～6 个街区的地方：20% 的住户希望住在离商业中心区 6 个街区以外的地方（在圣克拉拉县平均的街区为 150yd）。这种正确无误的距离只适用于圣克拉拉县。但是，全部的结果以绝对的优势支持了我们的论点，即人们是以上述方式在起着变化，同时表明，就居住的位置和特点而言，人们的需要是十分不同的。

为了确保不同的人能寻找到会使他们各自的特殊需要

introvert dimension is very well described by Frank Hendricks and Malcolm MacNair in "Concepts of Environmental Quality Standards Based on Life Styles," report to the American Public Health Association,February 12,1969,pp.11-15.The authors identify several kinds of persons and characterize each by the relative amount of time spent in extroverted activities and in introverted activities.Francis Loetterle has shed further light on the problem in "Environment Attitudes and Social Life in Santa Clara County," Santa Clara County Planning Department,San Jose,California,1967.He asked 3300 households how far they wanted to be from various community services.The results were:20 percent of the households interviewed wanted to be located less than three blocks from commercial centers;60 percent wanted to be located between four and six blocks away;20 percent wanted to be located more than six blocks away(mean block size in Santa Clara County is 150 yards).The exact distances apply only to Santa Clara. But the overall result overwhelmingly supports our contention that people vary in this way and shows that they have quite different needs as far as the location and character of houses is concerned.

To make sure that the different kinds of people can find houses which satisfy their own particular desires,we suggest that each cluster of houses,and each neighborhood should have three kinds of houses,in about equal numbers:those which are nearest to the action,those which are half-way between,and those which are almost completely isolated.And,to support this pattern we need,also,three distinct kinds of paths:

1.Paths along services,wide and open for activities and crowds,paths that connect activities and encourage busy

得到满足的住宅，我们建议，每个住宅团组和每个邻里都应当有三类住宅，数量大致相等：一类是最靠近活动中心的那些住宅，另一类几乎完全是孤零零的那些住宅，还有一类是介乎于上述两者之间的那些住宅。而且，为了支持这一模式，我们还需要三种不同的道路。

1. 沿途有服务机构的道路既要宽广，又要便于行人的往来和活动连成一片，并鼓励频繁的穿行交通。

2. 远离服务机构的道路既狭窄又弯曲，阻拦穿行交通，其中有许多道路成直角相交，并有尽端。

3. 中间型道路将最远而又安静的道路与最中心的繁华道路连接起来。

本模式在设计住宅团组和邻里方面都是同等重要的。过去当我们协助一批人设计他们的住宅团组时，首先要求每一个人根据"外向型"和"内向型"来考虑，他偏爱在什么地方选址。于是出现了三批人：4个"外向型"者希望尽可能住在行人的和社区的活动中心，4个"内向型"者希望尽可能住在远离活动中心的僻静的地方，其余4个人希望住在介乎两者之间的地方。他们利用本模式设计出来的基地平面图列举如下，图中一并标出了这三批人所选中的地址。

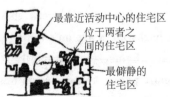

在一个住宅团组内：私密性强的、公共性强的和居于两者之间的住宅区

In one house cluster:private homes,public homes,and in-between

因此：

在三种住宅之间要有鲜明的区别：一些在僻静的后池

through traffic.

2.Paths remote from services,narrow and twisting,to discourage through traffic,with many at right angles and dead ends.

3.Intermediate types of paths linking the most remote and quiet paths to the most central and busy ones.

This pattern is as important in the design of a cluster of a few houses as it is in the design of a neighborhood. When we were helping a group of people to design their own cluster of houses,we first asked each person to consider his preference for location on the basis of extrovert-introvert. Three groups emerged:four "extroverts" who wished to be as near the pedestrian and community action as possible, four "introverts" who desired as much remoteness and privacy as possible,and the remaining four who wanted a bit of both. The site plan they made,using this pattern,is shown below,with the positions which the three kinds of people chose.

Therefore:

Make a clear distinction between three kinds of homes—those on quiet backwaters,those on busy streets, and those that are more or less in between.Make sure that those on quiet backwaters are on twisting paths,and that these houses are themselves physically secluded;make sure that the more public houses are on busy streets with many people passing by all day long and that the houses themselves are relatively exposed to the passers-by.The in between houses may then be located on the paths half-way between the other two.Give every neighborhood about equal numbers of these three kinds of homes.

塘处；另一些在繁华的街道边；还有一些或多或少介于两者之间。要保证那些位于僻静处的住宅周围有弯弯曲曲的小路，而且其外部空间也是隐蔽的；要保证那些更靠近活动中心的处于繁华街道区内，从早到晚都有许多行人从旁通过，而且对于过往行人是比较暴露的；处于中间状态的住宅可以位于上述两者中间的小路两侧。在每一邻里内建造大致等量的三种类型的住宅。

最靠近活动中心的住宅区

位于两者之间的住宅区

最僻静的住宅区

❧❧❧❧

利用本模式有助于区别邻里中的住宅和住宅团组的住宅。在邻里内部，把较高密度的住宅团组沿比较繁华的街道两侧布置——**丘状住宅**（39）、**联排式住宅**（38），而较低密度的住宅团组布置在僻静的地方——**住宅团组**（37）、**联排式住宅**（38）。实际上，繁华街道本身应当或是**步行街**（100），或是沿主干公路的高出**路面的便道**（55）；偏僻的地方应当或是**绿茵街道**（51），或是具有不同的**小路的形状**（121）的狭窄道路。凡在充满生机的街道上，都要保证足够高的住宅密度，以便形成这种蓬勃的生机——**行人密度**（123）……

Use this pattern to help differentiate the houses both in neighborhoods and in house clusters.Within a neighborhood, place higher density clusters along the busier streets—HOUSING HILL(39),ROW HOUSES(38),and lower density clusters along the backwaters—HOUSE CLUSTER(37),ROW HOUSES(38).The actual busy streets themselves should either be PEDESTRIAN STREETS(100)or RAISED WALKS(55)on major roads;the backwaters GREEN STREETS(51),or narrow paths with a distinct PATH SHAPE(121).Where lively streets are wanted,make sure the density of housing is high enough to generate the liveliness-PEDESTRIAN DENSITY(123)...

模式37　住宅团组**

　　……在邻里内部的基本的组织单元——**易识别的邻里**（14）——是一个由 12 幢住宅构成的团组。通过不断改变住宅团组的密度和布局, 本模式也有助于形成**密度圈**（29）、**户型混合**（35）和**公共性的程度**（36）。

⊰⊱

　　居民在各自的住宅内不会感到舒适，除非住宅能形成一个住宅团组，并有一块各户共有的公共用地在其间。

37 HOUSE CLUSTER**

...the fundamental unit of organization within the neighborhood—IDENTIFIABLE NEIGHBORHOOD(14)—is the cluster of a dozen houses.By varying the density and composition of different clusters,this pattern may also help to generate DENSITY RINGS(29),HOUSEHOLD MIX(35),and DEGREES OF PUBLICNESS(36).

∞⊗

People will not feel comfortable in their houses unless a group of houses forms a cluster,with the public land between them jointly owned by all the householders.

When houses are arranged on streets,and the streets owned by the town,there is no way in which the land immediately outside the houses can reflect the needs of families and individuals living in those houses.The land will only gradually get shaped to meet their needs if they have direct control over the land and its repair.

This pattern is based on the idea that the duster of land and homes immediately around one's own home is of special importance.It is the source for gradual differentiation of neighborhood land use,and it is the natural focus of neighborly interaction.

Herbert Gans,in The Levittowners(New York: Pantheon, 1967),has collected some powerful evidence for this tendency. Gans surveyed visiting habits on a typical block tract development.Of the 149 people he surveyed,all of them

当住宅排列在街道的两侧，而街道是属于城镇时，紧挨这些住宅外侧的一片土地就无法反映出居住在这些住宅内的家庭和个人的需要。如果家庭和个人能直接管理这片土地，并对它进行修整，这片土地的面貌才会逐渐成形，从而满足他们的需要。

本模式所依据的思想如下：紧邻住宅周围的土地和住宅团组具有特殊的重要性。它是对邻里内的土地利用逐步划分的原始资料，而且它也是邻居间相互交往的天然集中点。

赫伯特·甘斯在《莱维特镇的居民们》（New York：Pantheon，1967）一书中已针对上述趋势收集了一些强有力的证据。甘斯就一个典型的街区发展调查了居民串门访问的习惯。他共调查了149人，人人都以某种方式对邻居进行定期的访问。这一有趣的发现就是这一访问模式的形态学。

试考虑下面的一张示意图——这样的示意图对区内几乎每一幢住宅而言都能画出来。图中每一侧都有一幢住宅，在街道对面有一幢或两幢住宅，有一幢住宅在正后面，隔着一道花园的篱笆。

整个邻里93%的住户进行互访的范围局限于这一空间团组。

在一个典型的街区，每一住宅都位于它自己的住宅团组的中心
On a typical block each home is at the center of its own cluster

当被问及"你最常去访问的是谁"时，91%的人异口同声地说，他们访问最多的是街对面的那一家和紧邻的隔壁人家。

这一发现的好处就在于它表明了住宅团组空间布局具

were engaged in some pattern of regular visiting with their neighbors.The interesting finding is the morphology of this visiting pattern.

Consider the following diagram—one like it can be made for almost every house in a tract.There is a house on either side,one or two across the street,and one directly behind,across a garden fence.

Ninety-three percent of all the neighborhood visiting engaged in by the subjects is confined to this spatial cluster.

And when asked "Whom do you visit most?" 91 percent said the people they visit most are immediately across the street or next door.

The beauty of this finding is its indication of the strength of the spatial cluster to draw people together into neighborly contact.*The most obvious and tribal-like cluster—the homes on either side and across the street—forms roughly a circle,and it is there that most contact occurs.*And if we add to this shape the home immedi-ately behind,although it is separated by private gardens and afence,we can account for nearly all the visiting that goes on in the Levittown neighborhood.

We conclude that people continue to act according to the laws of a spatial cluster,even when the block layout and the neighborhood plan do their best to destroy this unit and make it anonymous.

Gans'data underscore our intuitions:people want to be part of a neighborly spatial cluster;contact between people sharing such a cluster is a vital function.And this need stands,even when people are able to drive and see friends all over the city.

What about the size of the cluster? What is the appropriate size?In Gans'investigations each home stands at the center of

有吸引人们共同进行邻里交往的力量。最明显的、部落式的住宅团组——在每一侧或街道对面都有一些住宅——大致形成一个圆圈，而且，正是在这个圈内不断发生着大部分的接触。如果我们把这一图形加到后面那家住宅上，虽然它被私人花园或篱笆分隔开，我们就能估计出在莱维特镇邻里中几乎全部邻里间互访的原因。

由此我们得出结论，人们继续根据空间团组规律来活动，即使当街区布置和邻里规划都力图破坏这一单元并使它失去个性特征时也还是如此。

甘斯的数据注重强调我们的直觉：居民都希望成为邻里空间团组的一部分。分享这样一个住宅团组的居民们相互进行接触正是一项极重要的功能。而这一点需要坚持，即使当他们能够在全市驱车去探亲访友时也不例外。

你对住宅团组的规模有何看法呢？它的适当规模究竟该是多大呢？据甘斯的调查报告称，每一幢住宅都位于由5幢或6幢其他住宅所构成的住宅团组的中心。但是，这当然不是对住宅团组的一种天然限制，因为莱维特镇街区是非常局限的。就我们的经验而论，当住宅的选址与住宅团组模式协调一致时，天然的限制完全是由该团组的非正式性和一致性之间的平衡所引起的。

如果每一住宅团组由8～12幢住宅构成，由此看来，它的效果最佳。每个家庭派出一名代表，人数刚好够围坐在一张普通会议桌旁，彼此可以面对面地直接交谈，从而就有可能作出关于如何利用他们所共有的土地的明智决定。由8幢或10幢住宅组成的住宅团组的居民可以在厨房的餐桌旁见面聊天，可以在街道上或花园里交换新闻，一般来说，不必费心就能和整个团组保持接触。当一个住宅团组超过10幢或12幢住宅时，这种平衡就显得勉强了。因此，我们规定了一个能自然而然形成团组的户数的上限值为12。当然，团组的平均规模可以小些，户数可以为6或8；就

a cluster of five or six other homes.But this is certainly not a natural limit for a housing cluster since the Levittown block layouts are so confining.In our experience,when the siting of the homes is attuned to the cluster pattern,the natural limit arises entirely from the balance between the informality and coherence of the group.

The clusters seem to work best if they have between 8 and 12 houses each.With one representative from each family,this is the number of people that can sit round a common meeting table,can talk to each other directly,face to face,and can therefore make wise decisions about the land they hold in common.With 8 or 10 households,people can meet over a kitchen table,exchange news on the street and in the gardens,and generally,without much special attention,keep in touch with the whole of the group.When there are more than 10 or 12 homes forming a cluster,this balance is strained. We therefore set an upper limit of around 12 on the number of households that can be naturally drawn into a cluster.Of course,the average size for clusters might be less,perhaps around 6 or 8;and clusters of 3,4,or 5 homes can work perfectly well.

Now,assuming that a group of neighbors,or a neighborhood association,or a planner,wants to give some expression to this pattern,what are the critical issues?

First,the geometry.In a new neighborhood,with houses built on the ground,we imagine quite dramatic clusters,with the houses built around or to the side of common land;and with a core to the cluster that gradually tapers off at the edges.

In existing neighborhoods of free-standing houses,the pattern must be brought into play gradually by relaxing zoning

连户数为 3、4 或 5 的住宅团组也有很好的效果。

现在假定一群邻居或一个邻里协会或一位规划人员想要对本模式作若干说明，关键性的问题究竟是哪一些呢？

首先是几何形状。在一个新的邻里中，在地面上建起了住宅，我们想象那一定是一些富有戏剧性的、建造在公共用地四周或一侧的住宅团组。它有一个核心，在边缘处逐渐减少。

一个由12幢住宅组成的住宅团组
A cluster of 12 houses

在现存的独立式住宅的邻里中，本模式必须通过放宽土地区划条令和允许居民把现行的地图坐标方格之外的住宅团组逐步连接起来而渐渐发挥作用——参阅**公共用地**（67）和**家庭**（75）。利用**联排式住宅**（38）和**丘状住宅**（39）来贯彻本模式甚至也是可能的。在这种情况下，联排式住宅的轮廓和公寓楼房的翼楼共同形成住宅团组。

在一切情况下，住宅团组共用的公共土地是住宅团组的一个重要的组成部分。公共用地起到一个焦点的作用，它从空间上把住宅团组连成一体。这片公共用地可小至一条道路，大至一方绿地。

另外，务必注意不要使住宅团组过于密集或自给自足，否则，它们会排斥较大的社区，或者看起来太闭关自守和与世隔绝了。住宅团组必须有开敞的尽端，并与其他团组重叠。

ordinances,and allowing people to gradually knit together clusters out of the existing grid—see COMMON LAND(67) and THE FAMILY(75).It is even possible to implement the pattern with ROW HOUSES(38)and HOUSING HILLS(39). In this case the configuration of the rows,and the wings of the apartment building,form the cluster.

In all cases common land which is shared by the cluster is an essential ingredient.It acts as a focus and physically knits the group together.This common land can be as small as a path or as large as a green.

On the other hand,care must be taken not to make the clusters too tight or self-contained,so that they exclude the larger community or seem too constricting and claustrophobic. There needs to be some open endedness and overlapping among clusters.

Along with the shape of the cluster,the way in which it is owned is critical.*If the pattern of ownership is not in accord with the physical properties of the cluster,the pattern will not take hold.*Very simply,the cluster must be owned and maintained by its constituent households.The households must be able to organize themselves as a corporation,capable of owning all the common land they share.There are many examples of tiny,user-owned housing corporations such as this. We know several places in our region where such experiments are under way,and places where they have been established for many years.And wehave heard,from visitors to the Center,of similar developments in various parts of the world.

We advocate a system of ownership where the deed to one home carries with it part ownership in the cluster to which the home belongs;and ideally,this in turn carries with it part

土耳其农村的重叠式住宅团组
Overlapping clusters in a Turkish village

随着住宅团组空间形状的确定，关键的问题就是以何种方式占有住宅团组。如果占有方式和住宅团组的外界环境特性不相符合，则本模式将无法成立。事情非常简单，住宅团组必须由构成它的住户来占有和维持。这些住户一定能把他们自己组织起来，建立一个公司，这样，他们才能拥有他们共用的全部公共土地。像这种小型的、用户占有的住宅公司的例子多得不胜枚举。就我们所知，在我们的区域内的一些地方正在进行着这种实验，另一些地方住宅公司已建立了许多年。我们还从参观访问我们环境结构中心的来宾那里获悉，在世界的许多地方都在经历着相似的发展过程。

我们提倡一种所有制，即住宅团组内的每一户都有部分所有权，并发给证书。再将这种部分所有制推广到由若干住宅团组构成的邻里中去，那是最理想的了。这样一来，每个所有者就自动地成为公共用地的不同级别的股东。而且，从住宅团组的住户开始的每一个级别都是一个政治单位，有权控制它自身的发展并负责修缮事宜。

在这样一种所有制下，无论在低密度的还是高密度的邻里内，住宅建筑就能逐渐找到自己的道路即一种具有永久性的模式：住宅团组。住宅团组将会提高邻里生活的质量。而现在我们从日益衰落的邻里中所感受到的只是暗淡。

人类未公开的秘密就是希望他的伙伴确认他的生命和存在的价值，同时他也希望能同样确认他们的……不仅在家庭、晚会和小酒店里，而且在亲切的邂逅相遇中，也许

ownership in the neighborhood made up of several clusters.In this way,every owner is automatically a shareholder in several levels of public land.And each level,beginning with the homes in their clusters,is a political unit with the power to control the processes of its own growth and repair.

Under such a system,the housing,whether in low or high density neighborhoods,can gradually find its way toward an abiding expression of the cluster.And the clusters themselves will come to support a quality of neighborhood life that,from our broken down neighborhoods now,we can only dimly perceive.

The unavowed secret of man is that he wants to be confirmed in his being and his existence by his fellow men and that he wishes them to make it possible for him to confirm them,and...not merely in the family,in the party assembly or in the public house,but also in the course of neighborly encounters,perhaps when he or the other steps out of the door of his house or to the window of his house and the greeting with which they greet each other will be accompanied by a glance of well-wishing,a glance in which curiosity,mistrust,and routine will have been overcome by a mutual sympathy:the one gives the other to understand that he affirms his presence. This is the indispensable minimum of humanity.(Martin Buber,Gleanings,New York:Simon and Schuster,1969, p.94.)

Therefore:

Arrange houses to form very rough,but identifiable clusters of 8 to 12 households around some common land and paths.Arrange the clusters so that anyone can walk through them,without feeling like a trespasser.

在他或另一人跨出他家的大门或走近他家的窗户时，他们会彼此寒暄，互致问候，并互相投以善意的祝愿的目光。这目光会驱散好奇心、不信任感和例行公事的习气而代之以相互的同情气氛：这就是人相互确认对方存在的价值。这是不可缺少的最低限度的人性。(Martin Buber, Gleanings，New York：Simon and Schuster，1969，p.94)

因此：

在某一公共用地的四周或道路两侧安排建造住宅，以便形成非常粗略但易识别的由 8 ~ 12 幢住宅构成的住宅团组。住宅团组的布局以每个人都能步行通过它而无侵入私人土地之感为准绳。

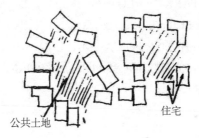

公共土地　　　　　　　　　　住宅

※○※

把本模式应用于每英亩至多约 15 幢住宅的低密度区；在较高的密度区，要以**联排式住宅**（38）或**丘状住宅**（39）所提供的附加结构来修正住宅团组。在住宅之间始终要有公共的土地——**公共用地**（67）和一个共同使用的普通车间——**家庭工间**（157）。合理布置道路——**内部交通领域**（98）——并且，在设计道路时要考虑到繁忙热闹的和清静的两类道路，甚至在住宅团组之内——**公共性的程度**（36）；使汽车停在**小停车场**（103），并使住宅团组内的住宅分别适合于未来住户的需要——**家庭**（75）、**小家庭住宅**（76）、**夫妻住宅**（77）、**单人住宅**（78）、**自己的家**（79）……

৪৩৫৪

Use this pattern as it is for low densities,up to about
15 houses per acre;at higher densities,modify the cluster
with the additional structure given by ROW HOUSES(38)or
HOUSING HILL(39).Always provide common land between
the houses—COMMON LAND (67)and a shared common
workshop-HOME WORKSHOP(157).Arrange paths clearly—
CIRCULATION REALMS(98)-and lay these paths out in such
a way that they create busier paths and backwaters,even within
the cluster—DEGREES OF PUBLICNESS (36);keep parking
in SMALL PARKING LOTS(103),and make the houses in
the cluster suit the households which will live there—THE
FAMILY(75),HOUSE FOR A SMALL FAMILY(76),HOUSE
FOR A COUPLE (77),HOUSE FOR ONE PERSON(78),YOUR
OWN HOME(79)...

模式38　联排式住宅*

　　……在社区的某些地方，**住宅团组**（37）内分离而又孤立的住宅和花园将不会产生积极的效果，因为它们尚未密集到足以在**密度圈**（29）和**公共性的程度**（36）中形成密度较大的地方。为了促成这些较大的模式，建造联排式住宅势在必行。

38   ROW HOUSES*

...in certain parts of a community,the detached homes and gardens of a HOUSE CLUSTER(37)will not work,because they are not dense enough to generate the denser parts of DENSITY RINGS(29)and DEGREES OF PUBLICNESS(36).To help create these larger patterns,it is necessary to build row houses instead.

ഓᏣᏇ

At densities of 15 to 30 houses per acre,row houses are essential.But typical row houses are dark inside,and stamped from an identical mould.

Above 15 houses per acre,it is almost impossible to make houses freestanding without destroying the open space around them;the open space which is left gets reduced to nothing more than shallow rings around the houses.And apartments do not solve the problem of higher densities;they keep people off the ground and they have no private gardens.

Row houses solve these problems.But row houses,in their conventional form,have problems of their own.Conventional row houses all conform,approximately,to the following diagram.The houses have a short frontage and a long depth,and share the party wall along their long side.

Because of the long party walls,many of the rooms are poorly lit.The houses lack privacy—there is nowhere in the houses or their yards that is very far from a party wall.The

每英亩的住宅密度为 15 ~ 30 幢时，联排式住宅就举足轻重了。但是，典型的联排式住宅内部是昏暗的，而且像是从同一个模子里造出来的。

每英亩的住宅密度超过 15 幢，要使住宅成为独立式的，而同时又不毁坏周围的开阔空间，这几乎是不可能的；留下来的外部空间缩小成绕宅的小环。公寓解决不了较高的密度问题。公寓没有私人花园，并使居民脱离地面。

联排式住宅却能解决上述问题。但联排式住宅在通常的形式中有着自身的矛盾。通常的联排式住宅大都符合下列的示意图。这些住宅正面狭窄而进深很大，并在进深方向共用隔墙。

因为共用隔墙长，许多房间光线不充足。这种住宅缺乏私密性——在住宅或庭院内没有一处是远离共用隔墙的。小庭院造得更加糟糕。它们位于住宅的短的一端，所以只有一小部分的室内空间可与花园邻接。住宅内几乎没有任何余地可做富有个性的变动，导致联排式住宅的平台往往是乏味的。

典型的联排式住宅模式
Typical row house pattern

上述联排式住宅的四个问题可以通过沿小路两侧建造像村舍那样的狭长形住宅而获得解决。在这种情况下，就有足够的余地从一幢住宅到另一幢住宅做精巧的变动——每一幢的平面可能是十分不同的；安排平面时让阳光照进宅内是容易办到的。

这种住宅有 30% 的固定周边，70% 的自由变动的周

small yards are made even worse by the fact that they are at the short ends of the house,so that only a small part of the indoor space can be adjacent to the garden.And there is almost no scope for individual variation in the houses,with the result that terraces of row houses are often rather sterile.

These four problems of row houses can easily be solved by making the houses long and thin,along the paths,like cottages. In this case,there is plenty of room for subtle variations from house to house-each plan can be quite different;and it is easy to arrange the plan to let the light in.

This kind of house has 30 percent of its perimeter fixed and 70 percent free for individual variations.A house in a conventional terrace of row houses has 70 percent of its perimeter fixed and only 30 percent open to individual variations.So the house can take on a variety of shapes,with a guarantee of a reasonable amount of privacy for its garden and for most of the house,an increase in the amount of light into the house,and an increase in the amount of indoor space that can be next to outdoor areas.

These advantages of the long thin row house are so obvious,it is natural to wonder why they aren't used more often.The reason is,of course,that roads do not permit it.So long as houses front directly onto roads,it is imperative that they have the shortest frontage possible,so as to save the cost of roads and services—the cost of roads is a large part of any housing budget.But in the pattern we propose,we have been able to avoid this difficulty altogether,by making the houses front only onto paths—which don't cost much—

边。而通常有平台的联排式住宅内，70%的周边是固定的，30%的周边可自由变动。所以这种住宅可取不同的形状，只要确保它的花园和宅内大部分空间具有适度的私密性，增加宅内采光量，并扩大紧邻室外的室内空间。

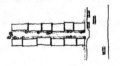

沿小路两侧的狭长形联排式住宅
Houses long and thin along the path

卷曲形联排式住宅
及其变化
Crinkling and variation

狭长形联排式住宅的优点十分明显。人们想要了解这种住宅未被广泛采用的原因是理所当然的。诚然，理由是道路不许可。只要住宅直接面向道路，就强制性地要求住宅的正面尽量狭窄，以便节省修建道路和服务管线的费用——筑路费总是占任何住宅建筑预算中的一大部分。但在我们建议的本模式中，我们已有可能完全避免上述困难，办法是：住宅只面向小路——小路的建筑造价不高——然后再把小路与公路连接起来，相互成直角相交，如在**小路网络和汽车**（52）中所规定的那样。

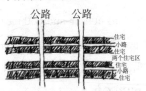

离开联排式住宅的公路
Roads away from houses

最后，略微谈一下密度问题。正如我们从下列的草图中所见到的，在一块 30×20 的土地上，利用约 $1300ft^2$ 的总面积（包括小路、住宅和花园）建造一幢 $1200ft^2$ 的二层住宅是可能的，甚至利用绝对最小值 $1000ft^2$ 的面积也有可能建成这样一幢住宅。

and it is then these paths which connect to the roads,at right angles,in the way prescribed by NETWORK OF PATHS AND CARS (52).

Finally,a word on density.As we see from the sketch below,it is possible to build a two-story house of 1200 square feet on an area 30×20,using a total area(path,house,garden)of about 1300 square feet,and it is even possible to manage with an absolute minimum of 1000 square feet.

It is therefore possible to build row houses at a density of 30 per net acre.Without parking,or with less parking,this figure could conceivably be even higher.

Therefore:

For row houses,place houses along pedestrian paths that run at right angles to local roads and parking lots,and give each house a long frontage and a shallow depth.

❧❧❧

Make the individual houses and cottages as long and thin along the paths as possible—LONG THIN HOUSE(109);vary the houses according to the different household types—THE FAMILY(75),HOUSE FOR A SMALL FAMILY(76),HOUSE FOR A COUPLE(77),HOUSE FOR ONE PERSON(78);build roads across the paths,at right angles to them—PARALLEL ROADS(23),NETWORK OF PATHS AND CARS(52),with small parking lots off the roads—SMALL PARKING LOTS(103).In other respects build row houses in clusters—HOUSE CLUSTER(37),BUILDING COMPLEX(95)...

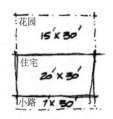

每幢住宅占地1300ft²

1300 square feet of land per house

由此可见，以每一标准英亩 30 幢的密度建设联排式住宅是可能的。如果没有停车场或停车场较小，这个数字还可考虑得高一些。

因此：

联排式住宅要坐落在人行道两侧，而这些人行道与地方公路和停车场连接，成直角相交；要使每一住宅的正面宽而进深浅。

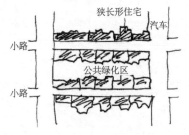

❧❧❧

尽可能将私人住宅和别墅建在人行道两侧，呈狭长形结构——**狭长形住宅**（109）；根据不同的住户类型，建造不同的住宅——**家庭**（75）、**小家庭住宅**（76）、**夫妻住宅**（77）、**单人住宅**（78）；建造与小路成直角相交的公路——**平行路**（23）、**小路网络和汽车**（52），还有偏离公路的小停车场——**小停车场**（103）。在其他方面，如在住宅团组内建造联排式住宅——**住宅团组**（37）、**建筑群体**（95）……

模式39　丘状住宅

……在社区**密度圈**（29）的内圈所要求的较高密度区，以及任何住宅密度超过每英亩30幢或住宅高达四层楼——**不高于四层楼**（21）的地方，住宅团组就变得像一座座小丘似的。

❧◆❧

每一城镇都有一些令人称心如意的中心区，那里每英亩至少有30~50户人家将长期居住下去。但是，达到这种密度的公寓住宅几乎都是非个性化的。

39 HOUSING HILL

...at the still higher densities required in the inner ring of the community's DENSITY RINGS(29),and wherever densities rise above 30 houses per acre or are four stories high—FOUR-STORY LIMIT(21),the house clusters become like hills.

৪০৫৪

Every town has places in it which are so central and desirable that at least 30-50 households per acre will be living there.But the apartment houses which reach this density are almost all impersonal.

In the pattern YOUR OWN HOME(79),we discuss the fact that every family needs its own home with land to build on,land where they can grow things,and a house which is unique and clearly marked as theirs.A typical apartment house,with flat walls and identical windows,cannot provide these qualities.

The form of the HOUSING HILL comes essentially from three requirements.First,people need to maintain contact with the ground and with their neighbors,far more contact than high-rise living permits.Second,people want an outdoor garden or yard.This is among the most common reasons for their rejecting apartment living.And third,people crave for variation and uniqueness in their homes,and this desire is almost always constrained by high-rise construction,with its regular facades and identical units.

我们将在模式**自己的家**（79）中论述以下事实：每个家庭都要有自己的家和建房的土地。在土地上能种植花草树木；而住宅要与众不同，表明自己的特色。一幢典型的公寓住宅，只有平直的墙壁和一模一样的门窗，不具备上述的品格。

模式**丘状住宅**（39）主要取决于三种需要。第一，人们需要同地面和邻居保持接触，需有比身居高楼所允许的多得多的这类接触。第二，人们希望有一个户外花园或庭院。这一点是反对住公寓住宅者老生常谈的理由了。第三，人们渴望自己的住宅形式有变化，新颖别致。可是，这些愿望总是受到高层结构的制约，千篇一律的立面和毫无二致的单元。

1. 同地面和邻居的联系。D.M. 范宁提出了有关这方面的最强有力的证据（"Families in Flats"，British Medical Journal，November 1967，PP.382 ～ 386）。范宁指出精神错乱的发生和身居高楼之间有着直接的相互关系。这些发现已在**不高于四层楼**（21）中作了详细说明。看来，身居高楼会使人有离群索居、囿于公寓而一筹莫展之感。电梯、过道和长长的楼梯把家庭生活和随便的街道生活隔断了。两者完全脱节了。居民决定外出参加某种公共活动，都会感到拘谨和难于应付，除非有某项特殊任务急需外出处理，否则宁可独自一人待在家里。

范宁在研究中还发现，住在高层公寓内的家庭相互间缺乏交流的程度达到令人惊讶的地步。妇女和儿童格外感到孤独和与世隔绝。妇女们除上街采购东西外，不会感到有什么必要带着轻松的心情离开公寓到外面去。她们和孩子们恰似身陷囹圄，同地面和邻居断绝了联系。

看来，仿佛地面、住宅间的公共用地，是使居民有可能彼此接触交往的媒介物。居住地面上，绕宅的庭院和邻居的庭院毗邻，而在最佳的布置中，也能与邻里的偏僻小道衔接。在这些情况下，人与人见面十分容易且很自然。庭院内嬉戏

1.Connection to the ground and to neighbors.The strongest evidence comes from D.M.Fanning("Families in Flats," British Medical Journal,November 1967,pp.382-386). Fanning shows a direct correlation between incidence of mental disorder and high-rise living.These findings are presented in detail in FOUR-STORY LIMIT(21).High-rise living,it appears,has a terrible tendency to leave people alone,stranded,in their apartments.Home life is split away from casual street life by elevators,hallways,and long stairs.The decision to go out for some public life becomes formal and awkward;and unless there is some specific task which brings people out in the world,the tendency is to stay home,alone.

Fanning also found a striking lack of communication between families in the high-rise flats he studied.Women and children were especially isolated.The women felt they had little reason to take the trip from their apartment to the ground,except to go shopping.They and their children were effectively imprisoned in their apartments,cut off from the ground and from their neighbors.

It seems as if the ground,the common ground between houses,is the medium through which people are able to make contact with one another and with themselves.Living on the ground,the yards around houses join those of the neighbors,and,in the best arrangements,they also adjoin neighborhood byways.Under these conditions it is easy and natural to meet with people.Children playing in the yard,the flowers in the garden,or just the weather outside provide endless topics for conversations.This kind of contact is impossible to maintain in high-rise apartments.

的儿童，花园中盛开的鲜花，乃至户外的天气都可作为交谈的永恒话题。居住在高层公寓中是无法保持这种接触的。

接触是不可能的
Contact is impossible

2. 私人花园。J.F·迪莫斯在《丘状公园考察记》（J.F.Demors，"Park Hill Survey,"O.A.P.，February 1966，p.235）一文中说，在他采访过的高层公寓的居民中，约有1/3 的人不约而同地说，他们失去了在他们自己的花园里闲荡漫步的机会。

小花园或某种幽僻的户外空间是不可缺少的。人们的这种需要应当予以满足。其重要性，就家庭的规模而言，不亚于生物的生态需要，也即社会必须和它的乡村结合在一起——**指状城乡交错**（3）。在所有的传统建筑中，无论什么地方，房屋基本上由人掌管，这种需要或多或少地得到体现。如日式小巧玲珑的花园、室外工间、屋顶花园、庭院、后院的玫瑰园、公共野外炊事场、芳草园都是例子，真有成千上万。但在现代的公寓建筑中，这种空间简直难以寻觅了。

3. 每一单元各具特色。在伯克利环境结构中心举办的讨论会期间，肯尼思·雷丁做了如下的实验。他要求人们画出他们梦寐以求的公寓轮廓图，并把它贴在一张硬纸板上。接着，他要求他们把硬纸板放到一张地图的坐标方格内（每一方格表示一套大公寓住宅的正面），并要求他们到处"搬家"，直到他们满意地选中位置为止。不出所料，他们希望自己的公寓位在大楼的边缘或用无窗墙与其他单元隔开的地方。没有一个人希望他自己的住宅消失在公寓的方格内。

2.Private gardens.In the Park Hill survey(J.F.Demors,"Park Hill Survey," O.A.P.,February 1966,p.235),about one-third of the high-rise residents interviewed said they missed the chance to putter around in their garden.

The need for a small garden,or some kind of private outdoor space,is fundamental.It is equivalent,at the family scale,to the biological need that a society has to be integrated with its countryside—CITY COUNTRY FINGERS(3).In all traditional architectures,wherever building is essentially in the hands of the people,there is some expression of this need. The miniature gardens of Japan,outdoor workshops,roof gardens,courtyards,backyard rose gardens,communal cooking pits,herb gardens-there are thousands of examples.But in modern apartment structures this kind of space is simply not available.

3.Identity of each unit.During the course of a seminar held at the Center for Environmental Structure,Kenneth Radding made the following experiment.He asked people to draw their dream apartment,from the outside,and stuck the drawing on a small piece of cardboard.He then asked them to place the cardboard on a grid representing the facade of a huge apartment house,and asked them to move their "homes" around,until they liked the position they were in.Without fail,people wanted their apartments to be on the edge of the building,or set off from other units by blank walls.No one wanted his own apartment to be lost in a grid of apartments.

In another survey we visited a nineteen-story apartment building in San Francisco.The building contained 190 apartments each with a balcony.The management had set very rigid restrictions on the use of these balconies—no political posters,no painting,no

在另一次调查中，我们采访了旧金山的一幢19层高的公寓大楼。这座大楼含有190套公寓住宅，每套都有一个阳台。大楼在管理方面有着严格限制使用阳台的规定，诸如在阳台上不准张贴政治标语，不准贴画，不准晾晒衣服，不准放活动装置，不准举行野宴，不准悬吊挂毯。但是，即使有上述种种限制，一半以上的居民依然能以某种方式装饰阳台，使它个性化，使它具有鲜明的特色，如用种在花盆里的奇葩异草、名贵苗木装点阳台，或是铺上地毯，或是摆设家具。总而言之，居民们面对严格的管理制度，仍然想方设法使自己的公寓不落俗套，独具风格。

什么样的建筑造型才符合这三项基本要求呢？首先，为了同地面保持频繁而又直接的接触，楼房最高不得超过四层——**不高于四层楼**（21）。其次，我们认为，也许更为重要的是，每一"住宅"必须在若干跨步之内有相当宽度；直接从地面升起的平缓的楼梯，如果是室外的露明楼梯，有点凌乱，而且十分平缓，那么它将和街道以及街道上的生活连成一片，融为一体。再者，如果我们严肃认真地对待这一需要，楼梯必须同地面上的居民共有的一片公共用地连接起来。这片公共用地经过筹划将成为半私密的绿地。

关于私人花园。私人花园需要阳光和私密性，普通的阳台布置是难以满足这两种要求的。露台必须面南，面积大，且和住宅亲密相连。露台要很结实，能铺上土，且种以灌木和小树。这就使人联想起一种丘状住宅。这种住宅有一缓坡向南延伸，在"丘"之下有一停车库。

至于各具特色——解决这一问题的唯一有效办法就是让每个家庭在露台的上层结构上建设或重建自己的家园。如果这种结构的地面能支承住一幢住宅或一定数量的泥土，那么每一单元均可自由地呈现出自己的特色，并发展它自己的小花园。

虽然这些要求会使人想起一种类似于塞夫迪的居留地模式，但是应当承认，这种居留地未能解决这里论述的三

clothes drying,no mobiles,no barbecues,no tapestries.But even when confined by such restrictions,over half of the residents were still able,in some way,to personalize their balconies with plants in pots,carpets,and furniture.In short,in the face of the most extreme regimentation people try to give their apartments a unique face.

What building form is compatible with these three basic requirements? First of all,to maintain a strong and direct connection to the ground,the building must be no higher than four stories—FOUR-STORY LIMIT(21).Also,and perhaps more important,we believe that each "house" must be within a few steps of a rather wide and gradual stair that rises directly from the ground.If the stair is open,somewhat rambling,and very gradual,it will be continuous with the street and the life of the street.Furthermore,if we take this need seriously,the stair must be connected at the ground to a piece of land,owned in common by the residents-this land organized to form a semiprivate green.

Concerning the private gardens.They need sunlight and privacy—two requirements hard to satisfy in ordinary balcony arrangements.The terraces must be south-facing,large,and intimately connected to the houses,and solid enough for earth,and bushes,and small trees.This suggests a kind of housing hill,with agradual slope toward the south and a garage for parking below the "hill."

And for identity—the only genuine solution to the problem of identity is to let each family gradually build and rebuild its own home on a terraced superstructure.If the floors of this structure are capable of supporting a house and some earth,each unit is free to take its own character and develop its own tiny garden.

Although these requirements bring to mind a form similar to Safdie's Habitat,it is important to realize that Habitat fails

个问题中的两个。它有私人花园，但未能解决同地面的连接问题，各单元和随便的街道生活明显脱节了。还有，成批出现的住宅均无个性特征和特色。

下面这幅公寓楼草图原定为斯德哥尔摩附近马斯塔的瑞典社区而设计的，它包括了丘状住宅的全部基本特征。

斯德哥尔摩附近马斯塔的公寓楼草图
Apartment building for märsta, near Stockholm

因此：

为了在每一标准英亩内建造33幢以上的住宅，或为了建造3层或4层楼高的住宅，将这些住宅建成丘状住宅，并使它们形成阶梯形露台，呈缓坡面南，中央有一条大型室外楼梯，也面南，并向公园延伸……

下面停车场　　阶梯形露台

中央公共楼梯

❧❦☙

让居民们在露台上如同在平地上一样建造具有鲜明特色的住宅——**自己的家**（79）。因为每个露台重叠在它下面的露台之上，所以每幢住宅下面的住宅顶部都有自己的花园——**屋顶花园**（118）。将中央楼梯建成露明的，但有顶棚，也许是玻璃顶棚，以避雨雪天气——**室外楼梯**（158）；将公共用地安排在室外楼梯底部开口的方向，开辟游戏场，栽培花卉，种植疏菜——**公共用地**（67）、**相互沟通的游戏场所**（68）、**菜园**（177）……

to solve two of the three problem discussed here.It has private gardens;but it fails to solve the problem of connection to the ground—the units are strongly separated from the casual life of the street—and the mass-produced dwellings are anonymous,far from unique.

The following sketch for an apartment building—originally made for the Swedish community of *märsta*,near Stockholm—includes all the essential features of a housing hill.

Therefore:

To build more than 30 dwellings per net acre,or to build housing three or four stories high,build a hill of houses.Build them to form stepped terraces,sloping toward the south,served by a great central open stair which also faces south and leads toward a common garden...

༄༅

Let people lay out their own houses individually,upon the terraces,just as if they were land—YOUR OWN HOME(79). Since each terrace overlaps the one below it,each house has its garden on the house below—ROOF GARDENS(118).Leave the central stair open to the air,but give it a roof,in wet or snowy climates—perhaps a glass roof—OPEN STAIRS(158);and place the common land right at the bottom of the stair with playgrounds,flowers,and vegetables for everyone—COMMON LAND(67),CONNECTED PLAY(68),VEGETABLE GARDEN (177)...

模式40　老人天地

……当邻里合理地形成之后，就给居民展示出一幅年龄和发展阶段的横剖面图——**易识别的邻里**（14）、**生命的周期**（26）、**户型混合**（35）；但是，在现代社会中老人往往被遗忘，他们孤单寂寞。现在有必要建立一种特殊的模式来强调老人的需要。

❀❀❀

老人需要老人，但老人也需要年轻人；而年轻人也需要和老人保持接触。

老人聚集在住宅团组和社区里是一种自然倾向。但当这些老龄社区太孤立无援或太大时，它们对年轻人和老人的损害是一样的。在城镇另外一些地方的青年没有机会为老人效劳，而老人又太与世隔绝了。

40   OLD PEOPLE EVERYWHERE**

...when neighborhoods are properly formed they give the people there a cross section of ages and stages of development—IDENTIFIABLE NEIGHBORHOOD(14),LIFE CYCLE(26),HOUSEHOLD MIX(35);however,the old people are so often forgotten and left alone in modern society,that it is necessary to formulate a special pattern which underlines their needs.

🕸

Old people need old people,but they also need the young, and young people need contact with the old.

There is a natural tendency for old people to gather together in clusters or communities.But when these elderly communities are too isolated or too large,they damage young and old alike.The young in other parts of town,have no chance of the benefit of older company,and the old people themselves are far too isolated.

Treated like outsiders,the aged have increasingly clustered together for mutual support or simply to enjoy themselves.A now familiar but still amazing phenomenon has sprung up in the past decade:dozens of good-sized new towns that exclude people under 65.Built on cheap,outlying land,such communities offer two-bedroom houses starting at$18,000 plus a refuge from urban violence...and generational pressures.(*Time*,August 3,1970)

But the choice the old people have made by moving to these communities and the remarks above are a serious and painful reflection of a very sad state of affairs in our culture.The fact is that contemporary society shunts away old people;and the more shunted away they are,the deeper the rift between the old and young.The old people have no choice but to segregate

由于年迈体衰，老人被当作局外人看待。他们日益频繁地集合在一起，目的是相互支援，或是简单地自得其乐。最近十年内出现了一种熟悉而又令人迷惑不解的现象：几十个颇具规模的新城镇排斥不满 65 岁的人住进去。这样的社区因为建立在地价低廉且又偏僻的地区，提供两居室一套的住宅，每套的起始售价为 18000 美元，并附加防备城市暴力……和代沟压力的庇护所的费用。(*Time*，August 3，1970)

老人所做的选择——迁往这些社区以及上面引证的评论，反映出在我们文化领域内可悲的事态发展到了令人发指的地步。事实是：现代社会抛弃老人；越是抛弃老人，青老之间的鸿沟就越深。老人除了自我隔绝，别无选择余地。他们和其他任何人一样，都有自尊心，他们宁可不与不欣赏他们的年轻人在一起；他们为了证明自己确有社会地位而故意佯装心满意足。

老人的隔离状态在每个人的生活中都引起同样的裂痕：一旦老人进入老龄社区，他们和过去的关系就会被一笔勾销，就会被忘却，就会被破坏殆尽。他们的青春已经一去不复返了。世态炎凉，创巨痛深，他们蹉跎岁月，和过去判若两人了。

为了对照今天老人的境遇，请考虑一下人们在传统的文化中是如何尊重和需要老人的：

老人的一些特权威望在一切已知的社会中几乎是普遍存在的现象。事实上，这种现象太普遍了，它直接通过表现出的多种文化因素去决定与年龄有关的其他主题的潮流。(*The Role of Aged in Primitive Society*，Leo W.Simmons，New Haven：Yale University Press，1945，p.69)

下面一段叙述更为详尽：

……另一种众所周知的家庭关系对老人来说意义重大：稚童和老人亲密无间的关系。当体魄健壮的人外出挣钱养家时，他们"一老一小"经常一起待在家里。这些老人以自己的智慧和经验对年幼无知的小家伙备加爱护，并对他们进行启蒙教育，而孩子们又反过来成为他们虚弱的老朋友的"眼、耳、手、足"了。对小孩的照顾也让老人能做点有用处的事，让他们在漫长而沉闷的衰老期感到人生的乐趣。(出处同上，第 199 页)

themselves—they,like anyone else,have pride;they would rather not be with younger people who do not appreciate them,and they feign satisfaction to justify their position.

And the segregation of the old causes the same rift inside each individual life:as old people pass into old age communities their ties with their own past become unacknowledged,lost,and therefore broken.Their youth is no longer alive in their old age—the two become dissociated;their lives are cut in two.

In contrast to the situation today,consider how the aged were respected and needed in traditional cultures:

Some degree of prestige for the aged seems to have been practically universalin all known societies.This is so general,in fact,that it cuts across many cultural factors that have appeared to determine trends in other topics related to age.(*The Role of Agedin Primitive Society*,Leo W.Simmons,New Haven:Yale University Press,1945,p.69.)

More specifically:

...Another family relationship of great significance for the aged has been the commonly observed intimate association between the very young and the very old.Frequently they have been left together at home while the able-bodied have gone forth to earn the family living.These oldsters,in their wisdom and experience,have protected and instructed the little ones,while the children,in turn,have acted as the "eyes,ears,hands,and feet" of their feeble old friends. Care of the young has thus very generally provided the aged with a useful occupation and a vivid interest in life during the long dull days of senescence. (Ibid.p.199.)

Clearly,old people cannot be integrated socially as in traditional cultures unless they are first integrated physically— unless they share the same streets,shops,services,and cornrnon land with everyone else.But,at the same time,they obviously need other old people around them;and some old people who are infirm need special services.

十分明显的事实是：除非老人先同外界结合，除非他们和别人一起共同使用同一街道、商店、服务机构和公共用地，否则他们就不能在传统的文化中和社会结合。与此同时，他们也显然需要和其他老人促膝谈心，共叙友情；还有一些老人体弱多病，需要特殊的服务设施。

当然，年龄相仿的老人的需要和愿望因人而异。他们的体格越健全，生活上越能独立自理，他们就越不需要和其他老人待在一起，他们可能离特殊的医疗设施就越远。他们所需要的照顾在程度上也是因人而异的：从全部护理到半护理（包括一天一次或一周两次的家访）；直到代购物品、烹调菜肴、洗刷清扫和照料生活上完全能自理的老人。就目前而言，照料老人还没有分得这么细。屡见不鲜的是：一些老人只要在烹饪和清洗方面稍获帮助生活就能自理，但他们却住进了能得到全面护理的疗养院，为此，他们自己、他们的家庭和社区都要支付一笔巨大的费用。这种情况是心理上虚弱的表现，结果这些老人由于过分体贴入微的关照而变得意志薄弱，反而不能自助了。

由此可见，我们必须向老人提供一种全面的而不是顾此失彼的照顾：

1. 一定要允许老人居住在他们最熟悉的邻里内。因此，每一邻里都会有若干老人。

2. 一定要允许老人三五成群地聚集在一起，但又不要让他们和社区的小字辈断绝往来。

3. 一定要让那些能独立生活的老人独自料理自己的生活，同时社区也要关怀他们。

4. 一定要让那些需要护理和精制食品的老人称心如意，不要让他们到远离邻里的疗养院去取食品。

如果每一邻里都有一个小规模的老人居住点，不是统统集中在社区的一个地方，而是像一群蜜蜂似的，模糊一片地分布在各邻里的边界上。这既有助于维持青老之间的

And of course old people vary in their need or desire to be among their own age group.The more ablebodied and independent they are,the less they need to be among other old people,and the farther they can be from special medical services.The variation in the amount of care they need ranges from complete nursing care;to semi-nursing care involving house calls once a day or twice a week;to an old person getting some help with shopping,cooking,and cleaning;to an old person being completely independent.Right now,there is no such fine differentiation made in the care of old people-very often people who simply need a little help cooking and cleaning are put into rest homes which provide total nursing care,at huge expense to them,their families,and the community.It is a psychologically debilitating situation,and they turn frail and helpless because that is the way they are treated.

We therefore need a way of taking care of old people which provides for the full range of their needs:

1.It must allow them to stay in the neighborhood they know besthence some old people in every neighborhood.

2.It must allow old people to be together,yet in groups small enough not to isolate them from the younger people in the neighborhood.

3.It must allow those old people who are independent to live independently,without losing the benefits of communality.

4.It must allow those who need nursing care or prepared meals,to get it,without having to go to nursing homes far from the neighborhood.

All these requirements can be solved together,very simply,if every neighborhood contains a small pocket of old people,not concentrated all in one place,hut fuzzy at the edges like a swarm of bees.This will both preserve the symbiosis between young and old,and give the old people the mutual support they need within the pockets.Perhaps 20 might live in a central group house,another

共生现象，又能为居住点内的老人提供他们所需的相互支援。可能有 20 名老人住在一幢中心集体住宅内，另外的 10～15 名分别住在靠近这幢住宅的住所内，这些住所与其他住宅相互间杂交错，还有另外 10～15 名也分别住在离邻里中心住宅更远的住所内，但这些距离不超过 100yd 或 200yd，以使他们很容易步行到中心住宅去下下棋、吃点东西或请护士帮点忙。

这 50 名老人的数字是从孟福德的论据中得来的：

要作出决定的第一件事就是一个邻里单位内能容纳多少老人；我认为，对这个问题的回答是，从整体来说，社区应当保持正常的年龄分布。这意味着每 100 人中应有 5～8 名 65 岁以上的老人；因此，举例来说，一个 600 人的邻里单位应有老人 30～50 名。(Lewis Mumford, The Human Prospect, New York, 1968, p.49.)

至于谈到集体住宅的性质，它会因情况不同而起变化。在一些情况下，集体住宅可能就是群居村，老人们一起烧饭煮菜，并在一部分时间里能得到青年男女或职业护士的照料。然而，全国的老人中约有 5% 的人需要全时照料。这就是说每 50 名中就有 2～3 人将需要全面护理。一般一位护士能护理 6～8 名老人，所以，有人认为，每两幢或每三幢邻里集体住宅由一位护士负责全面护理是可行的。

因此：

在每一邻里内要为大约 50 名老人建造住宅，并将这些住宅布置在三个圈内……

1. 有一中心住宅提供膳食和护理。

2. 老人住所靠近中心住宅。

3. 离中心住宅较远的一些住所要和邻里中的其他住宅混杂相间，但离中心住宅的距离绝不超过 200yd。

……这样，50 幢住宅在一起形成一个组织严密的"蜂群"，它的中心清晰可见，而在它的边缘地区这些住宅和邻

10 or 15 in cottages close to this house,but interlaced with other houses,and another 10 to 15 also in cottages,still further from the core,in among the neighborhood,yet always within 100 or 200 yards of the core,so they can easily walk there to play chess,have a meal,or get help from the nurse.

The number 50 comes from Mumford's argument:

The first thing to be determined is the number of aged people to be accommodated in a neighborhood unit;and the answer to this,I submit,is that the normal age distribution in the community as a whole should be maintained.This means that there should be from five to eight people over sixty-five in every hundred people;so that in a neighborhood unit of,say,six hundred people,there would be between thirty and fifty old people.(Lewis Mumford,The Human Prospect,New York,1968,p.49.)

As for the character of the group house,it might vary from case to case.In some cases it might be no more than a commune,where people cook together and have part-time help from young girls and boys,or professional nurses.However,about 5 per cent of the nation's elderly need full-time care.This means that two or three people in every 50 will need complete nursing care.Since a nurse can typically work with six to eight people,this suggests that every second or third neighborhood group house might be equipped with complete nursing care.

Therefore:

Create dwellings for some 50 old people in every neighborhood.Place these dwellings in three rings...

1.A central core with cooking and nursing provided.

2.Cottages near the core.

3.Cottages further out from the core,mixed among the other houses of the neighborhood,but never more than 200yards from the core.

...in such a way that the 50 houses together form a single

里的其他普通住宅相互交错混杂。

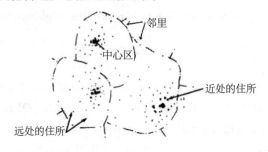

邻里
中心区
近处的住所
远处的住所

※

对待中心住宅如同对待任何集体住宅一样；所有的老人住所都要小型化，无论它们是靠近中心住宅还是离中心住宅较远——**老人住所**（155），其中一些老人住所也许和较大的家庭住宅相连接——**家庭**（75）；为每两幢或每三幢中心住宅提供适当的护理设施：在老人居住点边缘区的一些地方，给老人提供某种力所能及的工作，尤其是教育和照顾年幼无知的儿童——**学习网**（18）、**儿童之家**（86）、**固定工作点**（156）、**菜园**（177）。

在住宅团组之间，在中心周围，尤其在邻里之间的边界内，鼓励形成工作社区；

41. 工作社区

42. 工业带

43. 像市场一样开放的大学

44. 地方市政厅

45. 项链状的社区行业

46. 综合商场

47. 保健中心

48. 住宅与其他建筑间杂

coherent swarm,with its own clear center,but interlocked at its periphery with other ordinary houses of the neighborhood.

⊰⊱

Treat the core like any group house;make all the cottages,both those close to and those further away,small—OLD AGE COTTAGE(155),some of them perhaps connected to the larger family houses in the neighborhood—THE FAMILY(75);provide every second or third core with proper nursing facilities;somewhere in the orbit of the old age pocket,provide the kind of work which old people can manage best—especially teaching and looking after tiny children—NETWORK OF LEARNING(18),CHILDREN'S HOME(86),SETTLED WORK (156),VEGETABLE GARDEN(177)...

between the house clusters,around the centers,and especially in the boundaries between neighborhoods,encourage the formation of work communities;

41.WORK COMMUNITY

42.INDUSTRIAL RIBBON

43.UNIVERSITY AS A MARKETPLACE

44.LOCAL TOWN HALL

45.NECKLACE OF COMMUNITY PROJECTS

46.MARKET OF MANY SHOPS

47.HEALTH CENTER

48.HOUSING IN BETWEEN

模式41　工作社区**

41 WORK COMMUNITY**

...according to the pattern SCATTERED WORK(9),work is entirely decentralized and woven in and out of housing areas.The effect of SCATTERED WORK—can be increased piecemeal,by building individual work communities,one by one,in the boundaries between the neighborhoods;these work communities will then help to form the boundaries— SUBCULTURE BOUNDARY(13),NEIGHBORHOOD BOUNDARY (15)—and above all in the boundaries,they will help to form ACTIVITY NODES(30).

∞○☆

If you spend eight hours of your day at work,and eight hours at home,there is no reason why your workplace should be any less of a community than your home.

When someone tells you where he "lives", he is always talking about his house or the neighborhood his house is in.It sounds harmless enough.But think what it really means. Why should the people of our culture choose to use the word "live," which,on the face of it applies to every moment of our waking lives,and apply it only to a special portion of our lives—that part associated with our families and houses.The implication is straightforward.The people of our culture believe that they are less alive when they are working than when they are at home;and we make this distinction subtly clear,by choosing to keep the word "live" only for those places in our lives where we are not working.Anyone who uses the phrase "where do you live" in its everyday sense,accepts as his

……根据模式**分散的工作点**（9），工作完全是分散的，工作点交错分布于住宅区内外。因为在邻里边界内纷纷建立独立的工作社区，**分散的工作点**的影响将与日俱增。这些工作社区将有助于形成两种边界——**亚文化区边界**（13）和**邻里边界**（15）——最重要的是，在这些边界内工作社区将有助于形成**活动中心**（30）。

∞∞∞

如果你一天 8 小时工作，8 小时在家里，就没有理由说你的工作地点不是在家里而是在社区的任何一处。

当某人告诉你，他"生活"在何处时，他总会谈及他的住宅或他的住宅所在的邻里。这个词听起来是无害的，但仔细一想，确有语病。具有我们文化背景的人民竟会选用"生活"一词是值得我们深思的：顾名思义，此词本适用于我们醒着时的每一瞬间，但人们却只用于我们生活的特定部分，即与我们的家庭和住宅联系在一起的那部分时间。这种含意是容易理解的。他们认为，工作的乐趣少于家庭的乐趣。而我们对这种区别微妙地进行澄清，使它只适用于我们一生中不工作的那些地方。无论什么人从日常意义上使用"你生活在什么地方"这句话时，都可从普遍的文化背景上理解它，即没有人在他的工作地点真正地"生活"——那里没有歌声，没有音乐，没有爱情，没有食品——他工作时不像是在活着，不是在生活，只是在消磨时间，犹如一个死人。

一旦我们理解了这种底细，这一用语就会引起人们的愤慨。为什么我们会把一天 8 小时工作的地方视为"死人"待的地方呢？为什么我们创造不出一种崭新的工作环境呢？在新的工作环境里，我们干活如同在家里和亲朋好友一道

own the widespread cultural awareness of the fact that no one really "lives" at his place of work—there is no song or music there,no love,no food—that he is not alive while working,not living,only toiling away,and being dead.

As soon as we understand this situation it leads at once to outrage.Why should we accept a world in which eight hours of the day are "dead" ;why shall we not create a world in which our work is as much part of life,as much alive,as anything we do at home with our family and with our friends?

This problem is discussed in other patterns—SCATTERED WORK (9),SELF-GOVERNING WORKSHOPS AND OFFICES(80).Here we focus on the implications which this problem has for the physical and social nature of the area in which a workplace sits.If a person spends eight hours a day working in a certain area,and the nature of his work,its social character,and its location,are all chosen to make sure that he is living,not merely earning money,then it is certainly essential that the area immediately around his place of work be a community,just like a neighborhood but oriented to the pace and rhythms of work,instead of the rhythms of the family.

For workplaces to function as communities,five relationships are critical:

1.Workplaces must not be too scattered,nor too agglomerated,but clustered in groups of about 15.

We know from SCATTERED WORK(9)that workplaces should be decentralized,but they should not be so scattered that a single workplace is isolated from others.On the other hand,they should not be so agglomerated that a single workplace is lost in a sea of others.The workplaces should therefore be grouped to form strongly identifiable communities. The communities need to be small enough so that one can know most of the people working in them,at least by sight—

干活那样富有生气，而工作也就成为我们生活的一部分了。这一点难道做不到吗？

这个问题已在其他模式——**分散的工作点**（9）和**自治工作间和办公室**（80）中论述过了。这里我们要把注意力集中到如下两层含意上：设工作点的地区的物质环境性质和社会性质。如果某人在某一地区一天工作8小时，其工作性质、社会意义及工作地点都一一进行过适当的选择，确保他在工作就是在生活而不光是在挣钱，那么在他的工作点周围应当是一个社区，正如一个邻里一样，但这个社区应是与工作的速度和节奏相适应，而不是同家庭的节奏相适应。

发挥社区功能的工作点具有五种重要的关系：

1. 工作点既不应当太分散，又不应当太集中，呈团组分布，以15个左右为宜。

我们从**分散的工作点**（9）认识到，工作点应当是分散的，但不应当分散到一个工作点与其他的工作点相互隔绝。另外，工作点不应当太集中，以至于一个工作点淹没在其他工作点之中。因此，工作点应当呈团组分布，以便形成非常容易识别的社区。社区规模要小，小到足以使人们能认识在社区工作的大多数人——至少能看见他们；与此同时，社区规模又要大，大到足以能为工人们提供尽可能多的、令人愉快的服务设施：如便餐馆、地方运动场所、商店等。我们猜测，合适的规模可能是8～20个机构。

2. 工作社区包含混合工种：体力劳动、办公室工作、手工工艺、销售业务等。

今天，大多数人在专门化的领域工作，如从事医院建筑、汽车修理、广告设计、仓库管理、金融业务等。这种"隔

and big enough to support as many amenities for the workers as possible—lunch counters,local sports,shops,and so on.We guess the right size may be between 8 and 20 establishments.

2.The workplace community contains a mix of manual jobs,desk jobs,craft lobs,selling,and so forth.

Most people today work in areas which are specialized: medical buildings,car repair,advertising,warehousing,financ ial,etc.This kind of segregation leads to isolation from other types of work and other types of people,leading in turn to less concern,respect,and understanding of them.We believe that a world where people are socially responsible can only come about where there is a value intrinsic to every job,where there is dignity associated with all work.This can hardly come about when we are so segregated from people who do different kinds of work from us.

3.There is a common piece of land within the work community, which ties the individual workshops and offices together.

A shared street does a little to tie individual houses and places together;but a shared piece of common land does a great deal more.If the workplaces are grouped around a common courtyard where people can sit,play volleyball,eat lunches,it will help the contact and community among the workers.

4.The work community is interlaced with the larger community in which it is located.

A work community,though forming a core community by itself,cannot work well in complete isolation from the surrounding community.This is already discussed to some extent in SCATTERED WORK(9)and MEN AND WOMEN (27).In addition,both work community and residential community can gain by sharing facilities and services-restaurants,cafes,libraries.Thus it makes sense for the work community to be open to the larger community with shops and cafes at the seam between them.

行如隔山"的现象导致各行各业和就业人员相互"老死不相往来",导致相互漠不关心、不尊重和不理解。我们认为,各种职业都赋有内在价值和尊严,不分高低贵贱,只在此时,人们对社会负责的局面才会出现。如果我们因分工不同而处于隔离状态,上述局面是难以出现的。

3. 在工作社区内有一片公共用地,它是各工间和办公室的联系纽带。

一条公共街道在连接各住宅和地方方面所起的作用有限。但一片公共用地在这方面起的作用就大得多了。如果许多工作点坐落在一个公共庭院四周,工人就可以来这里坐坐,或是打排球,或是进午餐,这将有助于他们的相互接触和相互了解。

4. 工作社区与它所处的较大社区有着密切的联系。

虽然工作社区本身就是核心社区,如果它跟周围的社区断绝来往,就起不到什么作用。这一点早已在**分散的工作点**(9)和**男人和女人**(27)中有所论述了。此外,由于共用公共设施和服务机构,如餐厅、咖啡店、图书馆等的存在,工作社区和居住社区都可以得到方便。因此,工作社区向较大的社区开放那些位于它们边界内的商店和餐馆是有意义的。

5. 最后,工作社区的公共用地或公共庭院必须有不同的两级。

一方面,庭院用作乒乓球场和排球场,在其周围至多只需要 6 个工作组,假如工作组更多的话,就会使庭院无法承受。另一方面,便餐馆、洗衣店和理发店等需要 20 个或 30 个工作组才能维持经营。为此,工作社区必须有两级团组。

*5.Finally,it is necessary that the common land,or courtyards, exist at two distinct and separate levels.*On the one hand,the courtyards for common table tennis,volleyball,need half-a-dozen workgroups around them at the most-more would swamp them.On the other hand,the lunch counters and laundries and barbershops need more like 20 or 30 workgroups to survive.For this reason the work community needs two levels of clustering.

Therefore:

Build or encourage the formation of work communities-each one a collection of smaller clusters of workplaces which have their own courtyards,gathered round a larger common square or common courtyard which contains shops and lunch counters.The total work community should have no more than 10 or 20 workplaces in it.

🙟🙜

Make the square at the heart of the community a public square with public paths coming through it—SMALL PUBLIC SQUARES(61);either in this square,or in some attached space,place opportunities for sports—LOCAL SPORTS(72);make sure that the entire community is always within three minutes' walk of an ACCESSIBLE GREEN (60);lay out the individual smaller courtyards in such a way that people naturally gather there—COURTYARDS WHICH LIVE(115);keep the workshops small—SELFGOVERNING WORKSHOPS AND OFFICES(80);encourage communal cooking and eating over and beyond the lunch counters—STREET CAFE(88),FOOD STANDS(93),COMMUNAL EATING (147)...

因此：

建设或鼓励形成工作社区。每一工作社区都有若干较小的工作点，这些工作点有自己的庭院，它们集中在设有商店和便餐馆的较大的公共广场或公共庭院周围。全工作社区的工作点应不超过 20 个。

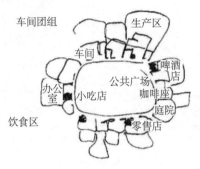

&ОСВ

使社区的中心广场成为公共广场，上面有公共小路穿越——**小广场（61）**；在该广场或在其某一附设场所处，提供运动的机会——**地方性运动场地（72）**；确保全社区离**近宅绿地（60）**只有 3min 步行的路程：安排单独的规模较小的庭院，以便人们十分自然地聚首聊天——**有生气的庭院（115）**；车间规模要小——**自治工作间和办公室（80）**；鼓励在便餐馆外用餐——**临街咖啡座（88）**；**饮食商亭（93）**、**共同进餐（147）**……

模式42　工业带*

　　……城市内的工作由于**分散的工作点**（9）而分散开来了。通常工业需要一定程度的集中，所以工业的布局具有特殊的重要性。像**工作社区**（41）一样，工业能十分容易地安排在亚文化区之间较大的边界内——**亚文化区边界**（13）。

❀❀❀

　　被夸大了作用的划区法令把工业和城市生活的其余部分完全分离开了，并促使有居民住宅的邻里变成非常不符合现实生活的了。

　　十分明显，工业造成污染：滚滚浓烟、奇腥恶臭、震耳欲聋的噪声和川流不息的重型卡车，这一切都是千真万确的。所以，有必要来防止工业尤其是重工业干扰当地居

42   INDUSTRIAL RIBBON*

...in a city where work is decentralized by SCATTERED WORK(9),the placing of industry is of particular importance since it usually needs a certain amount of concentration.Like WORK COMMUNITIES (41),the industry can easily be placed to help in the formation of the larger boundaries between subcultures—SUBCULTURE BOUNDARY(13).

❧❧❧

Exaggerated zoning laws separate industry from the rest of urban life completely,and contribute to the plastic unreality of sheltered residential neighborhoods.

It is true,obviously,that industry creates smoke, smells,noise,and heavy truck traffic;and it is therefore necessary to prevent the heaviest industry,especially,from interfering with the calm and safety of the places where people live.

But it is also true that in the modern city industry gets treated like a disease.The areas where it exists are assumed to be dirty and derelict.They are kept to the "other side of the tracks," swept under the rug.And people forget altogether that the things which surround them in their daily lives—bread,c hemicals,cars,oil,gaskets,radios,chairs—are all made in these forbidden industrial zones.Under these conditions it is not surprising that people treat life as an unreal charade,and forget the simplest realities and facts of their existence.

Since the 1930's various efforts have been made,on behalf of the workers,to make factories green and pleasant.

民的宁静和安全。

　　但是，同样千真万确的是，在现代的城市中，对待工业如同对待疾病一样。有工业项目的地区被认为是肮脏不堪的、被人抛弃的地方。这些地区被列为"另一类性质的问题"而置之不理。人们完全忘却他们周围一切的日常生活用品——面包、药品、汽车、油、垫圈、无线电和椅子等——都是在这些工业禁区制造出来的。在这些条件下，他们对待生活就像对待一个不现实的字谜似的，忘却他们自身的存在这一最简单的现实和事实是不足为奇的。

　　从20世纪30年代起，为了工人们的切身利益，在使厂区环境绿化和舒适愉快方面进行了各种努力，做了大量工作。为工业环境所采取的社会福利措施再一次被证明是不切实际的、事与愿违的。一个生产产品的工厂并不是一个花园或医院。环绕在新的工业"公园"四周的花园主要是为了炫耀，无论如何不是为了工人，因为几个小小的内部庭院或花园对工人们来说将更有用处。工业公园对周围的城市的社会生活和精神生活的贡献几乎等于零。

　　现在需要的是一种工业模式，它规模小，小到使它自己无须被如此尖锐地隔离开来，小到它自身不是看起来像个工厂，而确确实实是个工厂。工厂确定选址的条件是：工厂所引起的川流不息的卡车交通不危及附近的邻里，而且它建在邻里边缘区内，因而，邻里边界区就不会是一个危险的、被人遗忘了的地区，而是现实生活的一部分，和城市生活之网交织在一起，是住在附近的孩子们容易去玩耍的地方。这样，工厂就恰如其分地反映出它在各种事情方案布局中的巨大重要性。

This social wel-fare approach to the nature of industries is once again unreal,in the opposite direction.A workshop,where things are being made,is not a garden or a hospital.The gardens which surround the new industrial "parks" are more for show than for the workers anyway since a few small inner courts or gardens would be far more useful to the workers themselves. And the contribution of an industrial park to the social and emotional life of the surrounding city is almost nil.

What is needed is a form of industry which is small enough so that it does not need to be so sharply segregated; genuine, so that it seems like a workshop,because it is a workshop;placed in such a way that the truck traffic which it generates does not endanger near-by neighborhoods;and formed along the edge of neighborhoods so that it is not a dangerous,forgotten zone,but so that it is a real part of life,accessible to children from the surrounding houses, woven into the fabric of city life,in a way that properly reflects its huge importance in the scheme of things.

But many industries are not small.They need large areas to function properly.A survey of planned industrial districts shows that 71.2 per cent of the industries require 0 to 5.0 acres,13.6 per cent require 5 to 10 acres,and 9.9 per cent require 10 to 25 acres.(Robert E.Boley,Industrial Districts Restudied:An Analysis of Characteristics,Urban Land Institute,Technical Bulletin No.41,1961.)These industries can only fit into a NEIGHBORHOOD BOUNDARY (15)or SUBCULTURE BOUNDARY(13)if the boundary is wide enough.Ribbons whose width varies between 200 and 500 feet,with sites varying in length between 200 and 2000 feet,will be able to provide the necessary range of one to 25-acre sites in compact blocks, and

社会福利："绿化的"工业公园
The social welfare"green"industrid park

但是，许多工厂规模不小。它们需要大片土地才能发挥效能。对几个已规划好的工业区进行调查的结果表明：71.2%的工厂需要土地的面积为 0 ～ 5.0acre，13.6%的工厂需要 5 ～ 10acre，9.9%的工厂需要 10 ～ 25acre（Robert E.Boley，Industrial Districts Restudied：An Analysis of Characteristics，Urban Land Institute，Technical Bulletin No.41，1961.）。这些工厂只能适合于**邻里边界**（15）或**亚文化区边界**（13），如果这些边界有足够的宽度。工业带基地的宽度为 200 ～ 500ft，长度为 200 ～ 2000ft，就有可能在密集的街区内腾出一个必不可少的面积为 1 ～ 25acre 的场地，虽仍然是狭窄的，但足以在合理连接的工业带相对的两侧保持社区。

工业带必须具有卡车通道和某种铁路运输。卡车公路和铁路支线应当始终位于工业带的中央部分，所以工业带的边缘区仍然和社区保持畅通。

从一个工业区至附近一条高速公路去的卡车车流正毁坏着一个邻里
Truck trafficfrom an industrial area to a nearby freeway destrays a neighborhood

are still narrow enough to keep communities on opposite sides of the ribbon reasonably connected.

The industrial ribbons require truck access and some rail transport.Truck roads and rail spurs should always be located in the center of the ribbon,so that the edges of the ribbon remain open to the community.Even more important,the ribbons must be placed so that they do not generate a heavy concentration of dangerous and noisy truck traffic through neighborhoods.Since most truck traffic comes to and from the freeways,this means that the industrial ribbons must be placed fairly near to RING ROADS(17).

Therefore:

Place industry in ribbons,between 200 and 500 feet wide,which form the boundaries between communities. Break these ribbons into long blocks,varying in area between 1 and 25 acres;and treat the edge of every ribbon as a place where people from nearby communities can benefit from the offshoots of the industrial activity.

❧❧❧

Place the ribbons near enough to RING ROADS(17)so that trucks can pass directly from the ribbons to the ring road, without having to pass through any other intermediate areas. Develop the internal layout of the industrial ribbon like any other work community,though slightly more spread out— WORK COMMUNITY(41).Place the important buildings of each industry,the "heart" of the plant,toward the edge of the ribbon to form usable streets and outdoor spaces—POSITIVE OUTDOOR SPACE(106),BUILDING FRONTS(122).

更为重要的是：工业带的布局必须合理，避免高度密集的卡车车流通过邻里，酿成危险的车祸和噪声污染。大量的卡车在高速公路上往返奔驶，这意味着工业带的位置必须相当靠近**环路**（17）。

因此：

将工业分布在工业带内。工业带的宽度为200～500ft，它形成社区之间的边界。把这些工业带分散到长长的街区去，其面积变化的幅度为1～25acre：并且使每一工业带的边缘区成为从附近社区来的居民可以从支援工业的铁路支线的运输活动中获益匪浅的地方。

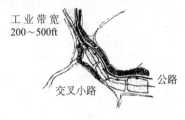

工业带宽
200～500ft

公路

交叉小路

❀✿❀

将工业带安排在十分靠近**环路**（17）的地方，以便卡车直接通过工业带进入环路而不必经过任何其他的中间区。发展工业带的内部的合理布局，如同任何其他的社区一样，即使稍有扩大——**工作社区**（41）。将每一工业的重要建筑，即工厂的"心脏"，安排在工业带的边缘区，以便形成可使用的街道和户外空间——**户外正空间**（106）和**建筑物正面**（122）。

模式43 像市场一样开放的大学

……**学习网**（18）业已使全社会认识到致力于提供分散的学习机会的学习过程是何等的重要。创办大学能极大地有益于学习网，而这种大学视学习过程为社会全体成员成人生活的正常的一部分。

⊰⊱

集中的、与世隔绝的大学扼杀各种学习机会。这些大学录取学生的政策是闭关自守的，聘任授课教师的手续是严格而死板的。

43　UNIVERSITY AS A MARKETPLACE

...the NETWORK OF LEARNING(18)has established the importance of a whole society devoted to the learning process with decentralized opportunities for learning.The network of learning can be greatly helped by building a university,which treats the learning process as a normal part of adult life,for all the people in society.

৪৩০৫৪

Concentrated,cloistered universities,with closed admission policies and rigid procedures which dictate who may teach a course,kill opportunities for learning.

The original universities in the middle ages were simply collections of teachers who attracted students because they had something to offer.They were marketplaces of ideas,located all over the town,where people could shop around for the kinds of ideas and learning which made sense to them.By contrast,the isolated and over-administered university of today kills the variety and intensity of the different ideas at the university and also limits the student's opportunity to shop for ideas.

To recreate this kind of academic freedom and the opportunity for exchange and growth of ideas two things are needed.

First,the social and physical environment must provide a setting which encourages rather than discourages individuality and freedom of thought.Second,the environment must provide a setting which encourages the student to see for himself which ideas make sense-a setting which gives him the maximum opportunity and exposure to a great variety of ideas, so that he

中世纪最早的大学是教师的荟萃之地。教师都能讲出个一二，所以大受学生欢迎。大学是思想的市场，遍布城镇各地，人们从那里能够极力寻求各种有意义的思想。与此相反，今天行政干预过多的孤立式的大学扼杀校内不同思想的多样性和鲜明性，并且也限制学生探求各种思想的机会。

为了重新创造这种学术自由以及交流和发展思想的机会，必须具备两个条件：

第一，社会环境和外部环境必须提供一种活动场所去鼓励而不是妨碍个性的发展和思想自由。第二，周围环境必须提供一种活动场所去鼓励学生理解、辨别哪些思想是有意义的；并向他们提供非常充分的机会去接触各种不同的思想，以便他们自己作出决断。

这种活动场所犹如传统的市场，对于这种市场有许多精彩的描写。那里聚集着成百上千个小商摊，展销各自的特制产品和风味小吃，由于货真价实，能招徕许多顾客。货摊排列有致，有购买力的购买者能在其间自由来回走动，商品可以反复挑选，中意后再买。

按照这一模式开办的别出心裁的大学意味着什么呢？

1. 来者不拒。首先，在市场式大学没有什么入学录取手续。不管是谁，也不管年龄大小，都可前往报名，编班听课。事实上，这种大学的"课程安排表"要通过报纸、电台宣传或在公共场所张贴，在全区域内进行广泛宣传，做到家喻户晓。

2. 能者为师。同样，在这种市场式大学中，无论什么人都能前往授课。教师和其余的市民阶层之间没有严格的区别。如果有人去授课，这门课就算开设了。当然，会有许多教师联合起来，讲授相关的课程；并且教师可以规定一些必要的条件，调整招生办法，他们认为怎么合适就可以怎么去调整。但是，像一个真正的市场一样，

can make up his mind for himself.

The image which most clearly describes this kind of setting is the image of the traditional marketplace,where hundreds of tiny stalls,each one developing some specialty and unique flavor which can attract people by its genuine quality,are so arranged that a potential buyer can circulate freely,and examine the wares before he buys.

What would it mean to fashion the university after this model?

1.*Anyone can take a eourse.*To begin with,in a university marketplace there are no admission procedures.Anyone,at any age,may come forward and seek to take a class.In effect,the "course catalog" of the university is published and circulated at large,in the newspapers and on radio,and posted in public places throughout the region.

2.*Anyone can give a course.*Similarly,in a university marketplace,anyone can come forward and offer a course.There is no hard and fast distinction between teachers and the rest of the cit izenry.If people come forward to take the course,then it is established.There will certainly be groups of teachers banding together and offering interrelated classes;and teachers may set prerequisities and regulate enrollment however they see fit.But,like a true marketplace,the students create the demand.If over a period of time no one comes forward to take a professor's course,then he must change his offering or find another way to make a living.

Many courses,once they are organized,can meet in homes and meeting rooms all across the town.But some will need more space or special equipment,and all the classes will need access to libraries and various other communal facilities. The university marketplaee,then,needs a physical structure to support its social structure.

学生也可提出自己的要求。如果经过一段时间后，无人再来听某位教授的课，那么他就必须改变讲课内容或离开这里另谋生计。

许多课程一旦开设，就可以在全城镇的家庭中或会议室里进行讲授。但有些课程需要更多的空间和专门设备，所有的班级都需要图书馆和其他各种公共设施。由此可见，市场式大学需要有一种实体结构来维持它的社会结构。

当然，市场绝不可能具有孤立的校园的形式。倒不如说市场倾向于敞开的和公开的，也许通过一两条集中着大学设施的街道与城市交织在一起。

我们在明确叙述尤金俄勒冈大学也即本模式最初方案时，详细描写了外界的物质环境。我们认为这种环境会使思想市场充实。我们提出的建议如下：

使大学成为许多小楼房的集合点，坐落在人行道两侧，每一座楼房包含一两项教育项目。在这些项目之间，在公共场所或一层楼内，一律要形成水平方向的流通。这意味着所有的项目都要面向人行道开放，而且楼房的上层通过楼梯和入口直接同地面连接。再把所有的人行道连接起来，最终就像市场一样，形成一个有许多入口的人行道主干系统。本模式总的结果是环境正在成为一个有几幢比较低矮的、向人行道主干系统敞开的楼房的汇集地。每一幢楼房都有一系列的入口和楼梯，相距约50ft。

我们现在仍然认为，这样的大学形象——像市场一样分散在整个城镇中——是正确无误的。它的大部分细节将由本书中的其他模式予以说明：**建筑群体**（95）、**步行街**（100）、**拱廊**（119）和**室外楼梯**（158）。

最后，应如何来管理市场式大学呢？我们一无所知。但可以肯定，担保人制度看来是切实可行的，人人都有平等的权利向担保人交付学费。为了使收支费与班级的规模达到平衡，某种管理办法是必不可少的。所以，向教师支

Certainly,a marketplace could never have the form of an isolated campus.Rather it would tend to be open and public, woven through the city,perhaps with one or two streets where university facilities are concentrated.

In an early version of this pattern,written expressly for the University of Oregon in Eugene,we described in detail the physical setting which we believe complements the marketplace of ideas.We advised:

Make the university a collection of small buildings,situated along pedestrian paths,each containing one or two educational projects.Make all the horizontal circulation among these projects,in the public domain,at ground floor.This means that all projects open directly to a pedestrian path,and that the upper floors of buildings are connected directly to the ground,by stairs and entrances.Connect all the pedestrian paths,so that,like a marketplace,they form one major pedestrian system,with many entrances and openings off it.The overall result of this pattern,is that the environment becomes a collection of relatively low buildings,opening off a major system of pedestrian paths,each building containing a series of entrances and staircases,at about 50 foot intervals.

We still believe that this image of the university,as a marketplace scattered through the town,is correct.Most of these details are given by other patterns,in this book:BUILDING COMPLEX(95),PEDESTRIAN STREET(100),ARCADES (119),and OPEN STAIRS(158).

Finally,how should a university marketplace be administered?We don't know.Certainly a voucher system where everyone has equal access to payment vouchers seems sensible. And some technique for balancing payment to class size is required,so teachers are not simply paid according to how many students they enroll.Furthermore,some kind of evaluation

付的酬金并不是简单地按他们所招收的学生人数的多少而定的。此外，还需要某种评定学生成绩的管理办法，这样，有关教师和课程质量的可靠信息将会传遍城镇的大街小巷。

目前，在高等教育方面正在进行着几项实验，这些实验可能有助于解决这些行政管理问题。英国的开放大学，各式各样的"自由"大学，如旧金山的赫里奥特罗普，遍及整个美国的无围墙大学的 20 个分校和完全适合劳动人民学习的大学补习班——这一切都是各种教育机构对思想市场的各个不同的方面进行实验的例证。

因此：

把大学办成高等教育的市场。作为一种社会观念来看，这意味着大学向各种年龄的人开放，他们可全时上课、半时上课或依次旁听各门课程。讲课——能者为师；听课——来者不拒。从物质环境来说，市场式大学有一处中心十字路口，大学的主要建筑和办公室就位于此处，会议室和实验室从该十字路口向外扩散——首先集中于沿步行街两侧的小楼房内，然后逐步分散并和全城镇融为一体。

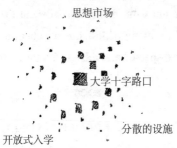

思想市场

大学十字路口

分散的设施

开放式入学

为大学的中心十字路口提供一**散步场所**（31）：在十字路口周围沿街道两侧集中一些楼房——**建筑群体**（95）和**步行街**（100）。为该中心区提供去安静的绿化区——**僻静区**（59）——的通路和正常的住宅分布——**住宅与其他建筑间杂**（48）：至于班级，按模式**师徒情谊**（83）的要求办理，什么地方可以，就在什么地方设置……

technique is needed,so that reliable information on courses and teachers filters out to the towns people.

There are several experiments going forward in higher education today which may help to solve these administrative questions.The Open University of England,the various "free" universities,such as Heliotrope in San Francisco, the 20 branches of the University Without Walls all over the United States,the university extension programs,which gear their courses entirely to working people—they are all examples of institutions experimenting with different aspects of the marketplace idea.

Therefore:

Establish the university as a marketplace of higher education.As a social conception this means that the university is open to people of all ages,on a full-time,part-time,or course by course basis.Anyone can offer a class.Anyone can take a class. Physically,the university marketplace has a central crossroads where its main buildings and offices are,and the meeting rooms and labs ripple out from this crossroads—at first concentrated in small buildings along pedestrian streets and then gradually becoming more dispersed and mixed with the town.

❧❧

Give the university a PROMENADE(31)at its central crossroads;and around the crossroads cluster the buildings along streets—BUILDING COMPLEX(95),PEDESTRIAN STREET(100).Give this central area access to quiet greens— QUIET BACKS(59);and a normal distribution of housing— HOUSING IN BETWEEN(48);as for the classes,wherever possible let them follow the model of MASTER AND APPRENTICES (83)...

模式44 地方市政厅*

……按照**7000人的社区**（12），城市的政治和经济生活划分为小规模的自治社区。在这种情况下，日理万机的地方政府需要一个有形的工作场所；它的设计和布置有助于形成和维持**7000人的社区**（12）——它是物质环境和社会的焦点。

છ

如果每一社区都有其有形的构成政治活动中心的市政厅，那么社区地方政府和居民行使地方管辖权才会相继成为现实。

我们在**亚文化区的镶嵌**（8）、**7000人的社区**（12）和**易识别的邻里**（14）中已经论证了，每一个城市都必须由自治的团体来构成，这些团体存在于两种不同的级别之中，即人口为5000～10000的社区和人口为200～1000的邻里。

44   LOCAL TOWN HALL*

...according to COMMUNITY OF 7000(12),the political and economic life of the city breaks down into small,self-governing communities.In this case,the process of local government needs a physical place of work;and the design and placing of this physical place of work can help to create and to sustain the COMMUNITY OF 7000 by acting as its physical and social focus.

⌘

Local government of communities and local control by the inhabitants,will only happen if each community has its own physical town hall which forms the nucleus of its political activity.

We have argued,in MOSAIC OF SUBCULTURES(8), COMMUNITY OF 7000(12),and IDENTIFIABLE NEIGHBORHOOD(14),that every city needs to be made of self-governing groups,which exist at two different levels,the communities with populations of 5000 to 10,000 and the neighborhoods with populations of 200 to 1000.

These groups will only have the political force to carry out their own,locally determined plans,if they have a share of the taxes which their inhabitants generate,and if the people in the groups have a genuine,daily possibility of access to the local government which represents them.Both require that each group has its own seat of government,no matter how modest,where the people of the neighborhood feel comfortable,and where they know that they can get results.

This calls up a physical image of city government which is quite the opposite of the huge city halls that have been built in the

如果这些团体共用居民们所缴纳的税款，如果这些团体的居民每天都真正能够代表他们的地方政府去走访，则这些团体才有政治力量去实施地方上决定的规划。团体和居民两者均要求每个团体在政府中占一席位，无论它是何等的低微，这样邻里的居民就会感到心情舒畅，并且他们是知道在政府中他们是能取得成绩的。

这会使人想起和过去 75 年内建立起来的庞大的城市市政厅的形象截然不同的一种有形的市政府形象。地方市政厅将含有两个基本特征：

1. 地方市政厅是它所服务的人群的社区领域。它采取的办法是吸引居民参加服务工作，让群众自发地辩论各项政策，并在它的楼房四周留出一片空地，以便居民前来集会或散步闲逛。

2. 地方市政厅位于地方社区的中心，它所服务的每一居民都能步行到达此地。

1. 地方市政厅作为社区领域。

社区政府的弱点一部分归因于城市市政厅官僚机构所制定和推行的各项政策。我们认为，城市市政厅实体形象的性质极大地助长着上述情况。换句话说，城市市政厅的有形存在破坏着地方社区政府，即使城市市政厅的全体职员同情"邻里参与"。

这一问题的解决途径就在于从社区一级的无权状态中汲取经验教训。当一居民上访城市市政厅并要求它采取行动解决邻里或社区的问题时，他马上就处于守势地位了。城市市政厅大厦和城市市政厅全体职员都是为全市服务的；他提出的问题显得无足轻重，除全市性的问题以外都排不上议事日程。此外，城市市政厅的每个工作人员还忙忙碌碌，不熟悉民情。他要填写表格，确定约见时间，即使他不甚明白这些表格和约见与他所提出的问题之间的联系。不久，邻里居民就会感到他们和城市市政厅越来越疏远了，他们

last 75 years.A local town hall would contain two basic features:

1.It is community territory for the group it serves;it is made in a way which invites people in for service,spontaneously,to debate policy,and the open space around the building is shaped to sustain people gathering and lingering.

2.It is located at the heart of the local community and is within walking distance of everyone it serves.

1.*The town hall as community territory.*

The weakness of community government is due in part to the kinds of policies created and maintained by the city hall bureaucracy.And we believe this situation is largely supported and bolstered by the physical nature of city hall.In other words,the physical existence of a city hall undermines local community government,even where the city hall staff is sympathetic to "neighborhood participation."

The key to the problem lies in the experience of powerlessness at the community level.When a man goes to city hall to take action on a neighborhood or community issue,he is at once on the defensive:the building and the staff of city hall serve the entire city;his problem is very small beside the problems of the city as a whole.And besides,everyone is busy-busy and unfamiliar.He is asked to fill out paper forms and make appointments,though perhaps the connection between these forms and appointments and his problem are not very clear.Soon the people in the neighborhoods feel more and more remote from city hall,from the center of decision-making and from the decisions themselves which influence their lives.Quickly the syndrome of powerlessness grows.

In an earlier publication,we presented a body of evidence to substantiate the growth of this syndrome(*A Pattern Language Which Generates Multi-Service Centers*,Center for Environmental

和决策中心越来越不同心同德了，他们对和他们的生活息息相关的各项决定越来越漠不关心了。无权的并发症迅速蔓延。

我们在较早发表的著作中提供了大量证据来证实这种无权并发症的发展趋势（*A Pattern Language Which Generates Multi-Service Centers*，Center for Environmental Structure，Berkeley，1968，pp.80～87）。我们在书中揭露了城市市政厅集中化的服务计划在它的管辖范围内几乎无人问津；这些服务中心的全体职员很快沾染上了盛气凌人的衙门作风，即使他们是专门被挑选出来支持邻里规划的；而最糟糕的是，服务中心本身被居民视为异己的地方。谁去使用它们都会叫苦连天的。大体情况就是如此。

这一并发症，如同所有的并发症一样，只有同时在几条战线上予以抨击，才能见效。这意味着，例如，将邻里和社区组织起来，控制和他们切身利益有关的各种公共机构；修改城市的宪章，以便授权给地方团体；在作为家庭基地的社区和邻里内，为这种权力机构——地方市政厅——的巩固而选定地址。

如果地方市政厅在粉碎无权并发症方面十分奏效，那么将是个什么样子呢？

有证据表明，如果人们具备适当的环境和手段，他们能够并且愿意把他们的各种需要连接起来。创造这种环境要和社区组织合作共事。如果地方市政厅逐步成为现实邻里的权力机构，它必定有助于社区组织的发展过程。这实质上就是说，建筑物必须兴建在社区组织不断发展的地区的四周，这种地方作为社区领域应当是非常容易识别的。

当我们把社区组织和社区领域的概念转变为有形的实体条件时，它们产生两个建筑要素：公共会场和社区行业区。

Structure,Berkeley,1968,pp.80-87).There we discovered that centralized service programs reached very few of the people in their target areas;the staff of these centers quickly took on the red tape mentality,even where they were chosen specifically to support neighborhood programs;and,most damaging of all,the centers themselves were seen as alien places,and the experience of using them was,on the whole,debilitating to the people.

Like all syndromes,this one can only be broken if it is attacked on its several fronts simultaneously.This means,for example,organizing neighborhoods and communities to take control of the functions that concern them;revising city charters to grant power to local groups;*and making places,in communities and neighborhoods,that act as home bases for the consolidation of this power—the local town halls.*

What might these local town halls be like if they are to be effective in breaking down the syndrome of powerlessness?

The evidence shows that people can and will articulate their needs if given the proper setting and means.Creating this setting goes hand in hand with community organization. If the local town hall is gradually to become a source of real neighborhood power,it must help the process of community organization.This means,essentially,that the building be built around the *process* of community organization,and that the place be clearly recognizable *as community territory*.

When we translate the idea of community organization and community territory into physical terms,they yield two components:an arena and a zone of community projects.

The community needs a public forum,equipped with sound system,benches,walls to put up notices,where people are free to gather;a place which belongs to the community where people

社区需要一个装有扩音系统的论坛、长凳和张贴通知的墙壁。在那里人们可以自由集合：这块属于社区所有的土地，人们自然可以什么时候想去做点什么事，就什么时候去。我们称这种公共论坛为公共会场。

社区还需要一块地方，从那里人们可以进入店面、工作点、会议室和办公室。一旦某一团体准备搬迁，它要取走打字机、复印机、电话等，以便坚持完成一项任务并发展基础广泛的社区支援——而这种情况又转过来需要价格低廉的、容易到达的办公空间。我们称这种空间为社区行业区——参阅**项链状的社区行业**（45），细节俱详。

2. 地方市政厅的位置。

如果这些地方市政厅打算在吸引居民方面取得成功，选址问题就必须加以认真考虑。根据早期著作对多服务中心选址问题的论述，我们深信不疑，如果地方市政厅选址不当，则会名存实亡，形同虚设：当地方市政厅设在主要交叉路口附近时，到社区中心去访问的人数为当其淹没于居住街区中心时的20倍。

例如，下面一份表格表明一个地方市政厅因选址的不同而引起访问服务中心的客流量的变化，并对两种客流量的情况作了对比。第一种客流量是当一个地方市政厅位于居住街时形成的，第二种客流量是当它搬迁到靠近一处主要人行道交叉口的主要商业街后形成的。

	每日客流人数	每日约见人数
搬迁前	1～2	15～20
搬迁后2个月	15～20	约50
搬迁后6个月	约50	约50

关于这份调查报告的细节可参阅《形成多服务中心的一种模式语言》（*A Pattern Language Which Generates MultiService Centers* pp.70-75）。本书得出的结论是：社区

would naturally come whenever they think something should be done about something.*We call this public forum the arena.*

And the community needs a place where people can have access to storefronts,work space,meeting rooms,office equipment.Once a group is ready to move,it takes typewriters, duplicating machines,telephones,etc.,to carry through with a project and develop broad based community support—and this in turn needs cheap and readily accessible office space.We call this space the community projects zone-see NECKLACE OF COMMUNITY PROJECTS (45)for details.

2.*The location of local town halls.*

If these local town halls are to be successful in drawing people in,the question of their location must be taken seriously. From earlier work on the location of multi-service centers,we are convinced that town halls can die if they are badly located:*twenty times as many people drop into community centers when they are lo-cated near maior intersections* as when they are buried in the middle of residential blocks.

Here,for example,is a table which shows the number of people who dropped in at a service center while it was located on a residential street,versus the number of people who dropped in after it was relocated on a maior commercial street,close to a main pedestrian intersection.

	Number of people dropping in, per day	Number of people with appointments, per day
Before the move	1~2	15~20
Two months after the move	15~20	about 50
Six months after the move	about 40	about 50

The details of this investigation are given in A *Pattern*

中心位于主要人行道交叉口的一个街区内能吸引大客流量，若是离上述地点较远，作为地方服务中心则名存而实亡了。

这种结论性的看法必须另作解释后才适用于不同规模的邻里和社区。我们设想，在 500 人的邻里中，邻里市政厅应是小规模的，非官方的；也许根本不是一幢独立的楼房，但在邻里的一个重要拐角处有一个房间及和它相连的室外空间。在 7000 人的社区，需要的东西就多一些：例如一幢规模有大住宅那么大的楼房，户外区发展成为一个论坛和会场，楼址就选在社区的主要散步场所。

因此：

为了使地方机构行使政治控制成为现实，在每一个 7000 人的社区内，甚至在每一个邻里内建立一个小规模的地方市政厅；将它设在社区最繁华的交叉路口附近。使建筑物分为三个部分：一个供公众发表议论的公共会场、公共会场周围的公共服务机构和特定的供社区行业所租赁的一块地方。

要使公共会场成为社区交叉路的中心；公共会场要小，以便群众在那里集合——**活动中心**（30）、**小广场**（61）和**行人密度**（123）。这一小广场四周的公共服务机构的规模要尽可能小——**小服务行业**（81）。在地方市政厅建筑周围，为社区行业提供充足的呈圈状分布的用地，以便社区行业形成地方市政厅的外观——**项链状的社区行业**（45）……

Language Which Generates Multi-Service Centers(pp.70-73).The conclusion reached there,is that community centers can afford to be within a block of the major pedestrian intersections,but if they are farther away,they are virtually dead as centers of local service.

This information must be interpreted to suit the different scales of neighborhood and community.We imagine,in a neighborhood of 500,the neighborhood town hall would be quite small and informal;perhaps not even a separate building at all,but a room with an adjoining outdoor room,on an important corner of the neighborhood. In a community of 7000,something more is required:a building the size of a large house,with an outdoor area developed as a forum and meeting place,located on the community's main promenade.

Therefore:

To make the political control of local functions real, establish a small town hall for each community of 7000, and even for each neighborhood;locate it near the busiest intersection in the community.Give the building three parts:an arena for public discussion,public services around the arena,and space to rent out to ad hoc community projects.

❧⌘☙

Arrange the arena so that it forms the heart of a community crossroads;and make it small,so that a crowd can easily gather there—ACTIVITY NODES(30),SMALL PUBLIC SQUARES(61),PEDESTRIAN DENSITY(123).Keep all the public services around this square as small as possible— SMALL SERVICES WITHOUT RED TAPE(81);and provide ample space for the community projects,in a ring around the building,so that they form the outer face of the town hall— NECKLACE OF COMMUNITY PROJECTS(45)...

模式45 项链状的社区行业

……**地方市政厅**（44）在每一社区中心都需要许多地方政府的小中心。本模式充实了地方市政厅和其他类似的公共机构——**像市场一样开放的大学**（43）和**保健中心**（47），以及供社区活动的一块用地。

❦

如果地方市政厅的周围没有居民们为他们自身创办起来的各种小的社区活动中心和行业，那么它将不会是该社区的一个履行职责的部分了。

社区自治政府的蓬勃发展取决于为数众多的、特定的政治团体和服务团体，这些团体可以自由发挥功能，并有合适的机会在市民们面前检验自己的思想。这一思想的空

45   NECKLACE OF COMMUNITY PROJECTS

...LOCAL TOWN HALL(44)calls for small centers of local government at the heart of every community.This pattern embellishes the local town hall and other public institutions like it—UNIVERSITY AS A MARKETPLACE(43)and HEALTH CENTER(47)—with a ground for community action.

ঙু৩৫৪৬

The local town hall will not be an honest part of the community which lives around it,unless it is itself surrounded by all kinds of small community activities and projects,generated by the people for themselves.

A lively process of community self-government depends on an endless series of ad hoc political and service groups,functioning freely,each with a proper chance to test its ideas before the towns-people.The spatial component of this idea is crucial:this process will be stymied if people cannot get started in an office on a shoestring.

We derive the geometry of this pattern from five requirements:

1.Small,grass roots movements,unpopular at their inception,play a vital role in society.They provide a critical opposition to established ideas;their presence is a direct correlate of the right to free speech;a basic part of the self-regulation of a successful society,which will generate counter movements whenever things get off the track.Such movements need a place to manifest themselves,in a way which puts their

间因素是很关键的：如果居民们不能用很少的资本开始进行投资，社区的发展就会受到影响。

我们从下列 5 种要求得出本模式的几何形状：

1. 一开始，不受欢迎的小型民众运动在社会中起着性命攸关的作用。它们和传统观念格格不入，针锋相对；它们的存在是和居民享有言论自由权直接有关的，同时也是卓有成效的社会进行自我调整的一个基本组成部分。但无论什么时候，事情偏出了自己的轨道，就会事与愿违，适得其反。这样的民众运动需要有显示自己并将其思想直接公之于众的场所。在撰写此文时，有关东部海湾地区进行的快速调查表明，那里有 30 个或 40 个自力更生的团体苦于缺少这种活动场所，如阿尔卡特拉兹印地安人协会、孟加拉共和国赈济会、团结影片社、居住者行动计划组、"11·7"运动、同性恋者合法保护团、不要退回中世纪协会和人民翻译服务社等……

2. 但是一般来说，这些团体规模小，资金严重不足。为了蓬勃开展这类活动，社区必须向它们分别提供一块不收租金的小小的活动用地，但对于租借期有某种限制。这块活动用地必须像个小店面，备有打字机、复印机和电话，并能通往会议室。

3. 为了鼓励推心置腹、畅所欲言的辩论气氛，这些店面应设在地方市政厅附近，公共生活的主要交叉路口。如果它们分散在整个城镇，远离主要的地方市政厅，那么它们就不能应付各种势力了。

4. 这些店面应当十分醒目。它们建造得妙趣横生，街上的行人能一眼看明白各团体的思想和宗旨。从外部环境方面说，它们也必须巧妙安排，以便抵制地方政府一旦有权就必须用围墙把自己围起来或与社区隔离开来的这种习以为常的倾向。

5. 最后，为了使这些团体同社区进行十分自然的接触

ideas directly into the public domain.At this writing,a quick survey of the East Bay shows about 30 or 40 bootstrap groups that are suffering for lack of such a place:for example,Alcatraz Indians,Bangla Desh Relief,Solidarity Films,Tenant Action Project,November 7th Movement,Gay Legal Defense,No on M,People's Translation Service...

2.But as a rule these groups are small and have very little money.To nourish this kind of activity,the community must provide minimal space to any group of this sort,rent free,with some limit on the duration of the lease.The space must be like a small storefront and have typewriters,duplicating machines,and telephones;and access to a meeting room.

3.To encourage the atmosphere of honest debate,these store-front spaces must be near the town hall,the main crossroads of public life.If they are scattered across the town,away from the main town hall,they cannot seriously contend with the powers that be.

4.The space must be highly visible.It must be built in a way which lets the group get their ideas across,to people on the street.And it must be physically organized to undermine the natural tendency town governments have to wall themselves in and isolate themselves from the community once they are in power.

5.Finally,to bring these groups into natural contact with the community,the fabric of storefronts should be built to include some of the stable shops and services that the community needs-barber-shop,cafe,laundromat.

These five requirements suggest a necklace of rather open store-front spaces around the local town hall.This necklace of spaces is a physical embodiment of the political process in an

交往，店面结构应包括社区所需要的一些稳定的商店和服务行业，如理发店、饮食店、自动洗衣店等。

上述 5 种要求提出，在地方市政厅四周，需要有一个呈项链状的、相当开阔的店面空间。这种项链状空间是一个开放社会政治过程的具体体现：人人都能利用设备、空间去发动一种运动，人人都有机会让自己的思想公之于公共会场。

因此：

允许在地方市政厅四周发展小店铺商业区和其他合适的社区建筑。使这些店铺坐落在繁华的人行道旁，并以极低的租金租借给各具特色的社区团体，以便它们开展政治活动、试销业务、研究和律师事务等。思想意识上不受限制。

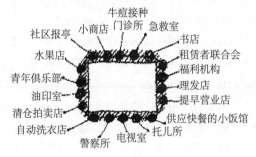

每个商店的规模要小而紧凑，且便于人们到达，就像**个体商店**（87）那样；在商店之间建立供人们散心休憩的小公共空间——**户外亭榭**（69）。利用商店形成建筑物边缘——**建筑物正面**（122）和**建筑物边缘**（160），并使它们向街道敞开——**向街道的开敞**（165）……

open society:everyone has access to equipment,space to mount a campaign,and the chance to get their ideas into the public arena.

Therefore:

Allow the growth of shop-size spaces around the local town hall,and any other appropriate community building. Front these shops on a busy path,and lease them for a minimum rent to ad hoc community groups for political work,trial services,research,and advocate groups.No ideological restrictions.

❧❧❧

Make each shop small,compact,and easily accessible like INDIVIDUALLY OWNED SHOPS(87);build small public spaces for loitering amongst them—PUBLIC OUTDOOR ROOM(69).Use them to form the building edge—BUILDING FRONTS(122),BUILDING EDGE(160),and keep them open to the street—OPENING TO THE STREET(165)...

模式46 综合商场**

……我们已经建议，商店应当广泛地分散并开设在最易进入商店所服务的社区的那些地方——**商业网**（19）。最大的商店群排列起来就形成步行街或**商业街**（32），它们几乎总是需要市场才能生存下去。本模式描写市场的形式和经济性质。

✦✦✦

人们需要提供方便的市场是极其自然的。在那里，在一个屋顶下，你能购买到你所需要的各种不同的食品和家庭用品。但当市场像超级市场那样独家经营时，食品平淡乏味，去那里采购东西就没有什么愉快可言了。

确实，大超级市场拥有各色俱全、名目繁多的食品。但这些"各色俱全"仍然是集中进货，集中存放在仓库，

46   MARKET OF MANY SHOPS**

...we have proposed that shops be widely decentralized
and placed in such a way that they are most accessible to the
communities which use them-WEB OF SHOPPING(19).
The largest groups of shops are arranged to form pedestrian
streets or SHOPPING STREETS(32) which will almost always
need a market to survive.This pattern de-scribes the form and
economic character of markets.

❧❧❧

**It is natural and convenient to want a market where
all the different foods and household goods you need can be
bought under a single roof.But when the market has a single
management,like a supennarket,the foods are bland,and
there is no joy in going there.**

It is true that the large supennarkets do have a great variety
of foods.But this "variety" is still centrally purchased,centrally
warehoused,and still has the staleness of mass merchandise.
In addition,there is no human contact left,only rows of shelves
and then a harried encounter with the check-out man who takes
your money.

The only way to get the human contact back,and the
variety of food,and all the love and care and wisdom about
individual foods which shopkeepers who know what they are
selling can bring to it,is to create those markets once again
in which individual owners sell different goods,from tiny
stalls,under a common roof.

往往是不新鲜的。此外，在商场内没有富于人情味的交往，只有一排排的货架，以及与收款员短暂而乏味的接触。

为了使富于人情味的接触交往的情景再现，使食品的色、香、味、形应有尽有，让知道自己在销售什么的店主发挥自己的聪明才智，施展各种绝招，唯一的办法就是重新开放综合商场。在那里，在同一个屋顶下，个体商贩可以在小摊上出售各种不同风味的食品。

就目前而论，超级市场可能变得越来越大，和其他的工业联合在一起，并在各个方面使人感到市场的非人性化。举例来说，霍恩和哈达特一直反复推敲这一方案：

……顾客驱车前来或者徒步走上自动扶梯，她就被彬彬有礼地传送通过整个商店，凭观看发光的橱窗内显示出来的样品（也可以是用一把特殊的钥匙或她的信用卡启开一些盒子），通过闭路电视选购杂货、食用肉和农产品。然后她驱车到一个分离的仓库领取她选购的物品，利用一种通用的信用卡来支付货款……大多数的人是隐而不见的了……（*Jennifer Cross*，*The Supermarket Trap*，New York：Berkeley Medallion，1971）

现将上述情况与下面描写旧金山老式市场的情况进行对照比较：

如果你定时光顾这一市场，你就会看中心里喜欢的小商摊，如经销从沃森维拉运来的翠玉苹果和霍尔苹果的商铺。当果农为你挑选苹果时，总是逐个精挑细拣，轻轻放入袋中，并提醒你要把苹果存放在阴凉处，以便它们保持脆甜爽口的滋味。如果你表现出兴趣，他就会满怀夸耀之情地告诉你盛产良种苹果的果园，苹果是如何开花结果的，以及如何照料护理它们等，此时此刻，他的一双蓝眼睛留意着你的眼神。他操着稍带意大利口音的英语，告诉你他出生在意大利北部的一个地方，于是你对他那纯蓝的眼睛、浅棕色的头发和瘦削的身材就不会感到惊奇了。

在一排小商摊的尽头，有一个英俊的黑人在卖瓜，他的瓜堆成一座座小山似的。当你告诉他你不是挑瓜的行家，想挑个后天吃的上等瓜。他不仅会挑出包你称心如意的优质瓜（结果证明果真如此），而且会给你讲解下次如何挑选，不管它是"客来尝"瓜、蜜瓜还是西瓜，也无

As it stands,supermarkets are likely to get bigger and bigger,to conglomerate with other industries,and to go to all lengths to dehumanize the experience of the marketplace. Horn and Hardart,for example,have been contemplating this scheme:

...the customer either drives her car or walks onto a moving ramp,is conveyeddecorously through the whole store,selects her groceries by viewing samples displayed in lighted wall panels(or unlocking the cases with a special key or her credit card),and chooses her meat and produce via closed circuit TV.She then drives around to a separate warehouse area to collect her order,paid for by a universal credit card system...Most of the people would be invisible...(Jennifer Cross,*The Supermarket Trap*,New York:Berkeley Medallion, 1971).

Now contrast this with the following description of an oldfashioned market place in San Francisco:

If you visit the Market regularly you come to have favorite stalls,like the one with the pippin and Hauer apples from Watsonville.The farmer looks at each apple as he chooses it and places it in the bag,reminding you to keep them in a cool place so they will remain crisp and sweet.If you display interest,he tells you with pride about the orchard they come from and how they were grown and cared for,his blue eyes meeting yours.His English is spoken with a slight Italian accent so you wonder about the clear blue eyes,light brown hair and long-boned body until he tdls you about the part of northern Italy where he was born.

There is a handsome black man offering small mountains of melons where the stalls end.Tell him you are not enough of an expert to choose one you would like to have perfect for the day after tomorrow,and he will not only pick one out that he assures you will be just right(as it tums out to be),but gives you a lesson in choosing your next melon,whether cranshaw,honeydew or watermelon,wherever you may happen to buy it.He cares that you will

论你在什么地方买瓜。他所关心的是你总会买到好瓜并享用它。("The Farmers Go to Market", *California Living*, San Francisco Chronicle Sunday Magazine, February 6, 1972.)

毫无疑问，与超级市场的传送带相比，上述情况更富于人情味，更有生气。问题的重点在于经济管理方面。综合商场是否有一个合理的经济基础？抑或市场会被超级市场的效率排挤掉？

看来"万事开头难"，创业维艰。经济管理方面的主要问题是协调一致——协调个体商店以便形成一个连贯的市场；协调若干市场内许多相似的商店以便形成大宗进货的综合商场。

如果个体商店的位置良好，就能够开展竞争，利润率可高达销售额的 5 ％（"Expenses in Retail Business", National Cash Register, Dayton, Ohio, p.15）。根据国家现金收入记录机的数据，对所有的方便食品店来说，这一利润率保持不变，不管规模大小如何。小商店经常因超级市场减价而被抢走生意。因为小商店的店址是店主自己选定的，所以不能为购买者一次性提供和超级市场一样的花色品种。但是，如果许多小商店联合起来，集中选址，共同提供可与超级市场相媲美的花色品种，那么它们就能够有效地与一系列的超级市场进行竞争。

许许多多小商店的综合效率就在于大量进货的效率。如果遍布城镇的许多相似的小商店群协调它们的需要，并建立大宗进货的综合商场，那么小店效率不高这一点就可得到弥补。例如，在海湾地区有几名花商，他们在大街小巷推着轻便送货车出售鲜花。虽然各人独立营业，但他们却一齐去采购鲜花。他们通过大批量购买鲜花而获得巨大利益，并且以"三分本一分利"的价钱廉价售给那些固定的花卉爱好者。

always get a good one and enjoy it.("The Farmers Go to Market," *California Living*,San Francisco Chronicle Sunday Magazine,February 6,1972.)

There is no doubt that this is far more human and enlivening than the supermarket conveyor belt.The critical question lies with the economics of the operation.Is there a reasonable economic basis for a marketplace of many shops? Or are markets ruled out by the efficiencies of the supermarket?

There do not seem to be any economic obstacles more serious than those which accompany the start of any business. The major problem is one of coordination-coordination of individual shops to form one coherent market and coordination of many similar shops,from several markets,to make bulk purchase arrangements.

If individual shops are well located,they can operate competitively,at profit margins of up to 5 per cent of sales("Expenses in Retail Business," National Cash Register, Dayton,Ohio,p.15).According to National Cash Register figures,this profit margin stays the same,regardless of size,for all convenience food stores.The small stores are often undercut by supermarkets because they are located by themselves,and therefore cannot offer shoppers the same variety at one stop,as the supermarket.However,if many of these small shops are clustered and centrally located,and together they offer a variety comparable to the supermarket,then they can compete effectively with the chain supermarkets.

The one efficiency that chain stores do maintain is the efficiency of bulk purchase.But even this can be offset if groups of similar shops,all over the town,coordinate their needs and set

秘鲁的一座市场
A market in Peru

当然，一个综合商场开始创建时会遇到重重困难。它很难找到一块立锥之地，也很难筹措资金。我们建议在开始阶段它具有一个非常粗糙而又非常简单的结构，往后可以逐步充实和完善。上图中秘鲁利马的市场，就是从几根主柱和几条走廊开始发展起来的。柱间的商店都是逐步建设起来的，其中大多数商店的面积为 6ft×9ft。

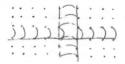

……起初仅是几根立柱
…began with nothing more than columns

一个场面壮观的实例就是华盛顿州西雅图市的派克市场。它开始时只不过是一个简陋的木结构，通过过去几年的不断改建和扩建，才呈现出今天这样欣欣向荣的新貌。

派克市场——西雅图一综合商场
The Pike Place Market—a market of many shops in Seattle

因此：

建立名目繁多的市场来取代现代的超级市场。每一市场均由许多小商店构成，而这些小商店都是自治的、专门化的（如专售奶酪、肉类、粮食、水果等）。市场建筑结构

up bulk purchase arrangements.For example,in the Bay Area there are a number of flower vendors running their business from small carts on the street.Although each vendor manages his own affairs independently,all the vendors go in together to buy their flowers.They gain enormously by purchasing their flowers in bulk and undersell the established florists three to one.

Of cotrse,it is difficult for a market of many shops to get started-it is hard to find a place and hard to finance it.We propose a very rough and simple structure in the beginning, that can be filled in and improved over time.The market in the photo,in Lima,Peru,began with nothing more than freestanding columns and aisles.The shops—most of them no more than six feet by nine—were built up gradually between the columns.

A spectacular example of a simple wood structure that has been modified and enlarged over the years is the Pike Place Market in Seattle,Washington.

Therefore:

Instead of modern supermarkets,establish frequent marketplaces,each one made up of many smaller shops which are autonomous and specialized(cheese,meat,grain, fruit,and so on).Build the structure of the market as a minimum,which provides no more than a roof,columns which define aisles,and basic services.Within this structure allow the different shops to create their own environment,according to their individual taste and needs.

要尽量简易，它只提供屋顶、限定走廊走向的主柱和基本的服务设施。在这一结构之内，允许各种不同的商店根据各自的特长和需要，创造出他们自己的环境。

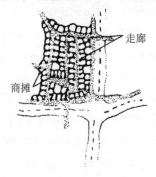

务必使走廊具有足够的宽度，以便小型送货车和密集的人群自由通过——也许要 6 ～ 12ft 宽——**有顶街道**（101）；商摊要非常小，这样可降低租金——也许不超过 6ft×9ft——需要更大空间的商店可占两个商摊的面积——**个体商店**（87）；只在拐角处规定商摊数量——**角柱**（212）；也许甚至让经营者自己搭顶篷——**帆布顶篷**（244）；把走廊同外部连接起来，以便使市场与环绕着它的市内人行小道直接沟通——**步行街**（100）……

Make the aisles wide enough for small delivery carts and for a dense throng of pedestrians—perhaps 6 to 12 feet wide—BUILDING THOROUGHFARE(101);keep the stalls extremely small so that the rent is low—perhaps no more than six feet by nine feet-shops which need more space can occupy two—INDIVDUUALLY OWNED SHOPS(87);define the stalls with columns at the comers only—COLUMNS AT THE CORNERS(212);perhaps even let the owners make roofs for themselves—CANVAS ROOFS(244);connect the aisles with the outside so that the market is a direct continuation of the pedestrian paths in the city just around it—PEDESTRIAN STREET(100)...

模式47　保健中心*

……毫不含糊地承认生命的周期是每个人的生活基础，这对社区人民的健康将有极大裨益——**生命的周期**（26）。本模式描述有助于人民去关心他们自己的生活和健康的更加专门化的机构。

છાછ

在一个普通的邻里内，根据简单的生物学标准来进行评定，90%以上的来往行人是不健康的。这种健康欠佳的状况靠医院和药品是无法治愈的。

现在的医院强调治病，医疗费十分昂贵；而且由于过分集中，造成不便；医院不是在医治疾病，而是倾向于制造疾病，因为医生只在人们有病时才能得到酬金。

相反，据传统的中医称，病人只在康复时才付给医生酬金；人们患病时，医生免费为他们治病，责无旁贷；医生的动机就是使人民身心健康。

保健医疗制确实能够起到保护人民身心健康的作用，它侧重保健而不是侧重治病。因此，它必须在空间上是分散的，尽可能和人们的日常活动接近，鼓励大家平时开展有益于健康的体育锻炼活动。解决这一问题的关键，就我们所知，在于保健医疗制必须提供许多规模小、分布广的保健中心，鼓励大家发展体育运动、增强体质，如游泳、跳舞、运动会和新鲜空气浴等，提供仅作为应付某些意外情况所需的医疗护理。

47   HEALTH CENTER*

...the explicit recognition of the life cycle as the basis for every individual life will do a great deal to help people's health in the community—LIFE CYCLE(26);this pattern describes the more specific institutions which help people to care for themselves and their health.

ဆာလ

More than 90 per cent of the people walking about in an ordinary neighborhood are unhealthy,judged by simple biological criteria.This ill health cannot be cured by hospitals or medicine.

Hospitals put the emphasis on sickness.They are enormously expensive;they are inconvenient because they are too centralized;and they tend to create sickness,rather than cure it,because doctors get paid when people are sick.

By contrast,in traditional Chinese medicine,people pay the doctor only when they are healthy;when they are sick,he is obliged to treat them,without payment.The doctors have incentives to keep people well.

A system of health care which is actually capable of keeping people healthy,in both mind and body,must put its emphasis on health,not sickness.It must therefore be physically decentralized so that it is as close as possible to people's everyday activities.And it must be able to encourage people in daily practices that lead to health.The core of the solution,as far as we can see,must be a system of small,widely distributed,health centers,which encourage physical activities—

在有关保健医疗的文献中，大量的证据和见解表明，按照保健原理而组织起来的、具有上述特征的保健中心是十分重要的。(See, for example : William H.Glazier, "The Task of Medicine", *Scientific American*, Vol.228, No.4, April 1973, pp.13 ~ 17 ; and Milton Roemer, "Nationalized Medicine for America", *Transaction*, September 1971, p.31)

我们了解到，存在着与本建议并行不悖的发展保健医疗规划的多种尝试。可是，在大多数情况下，这些规划成功的希望极为渺茫。尽管这些规划者有良好的愿望，但他们的规划仍然是迎合病人的，而不是符合保健需求的。试举所谓"社区精神保健中心"为例，这些中心是美国全国精神保健协会于20世纪60年代后期倡导而建立的。它们想提倡保健而非治病，但只不过是纸上谈兵而已。

实际上它们搞的是另外一套。我们在加利福尼亚圣昂赛尔莫采访了最先进的中心之一。病人整天围坐在一起；他们的目光呆滞无神；他们半疯狂地在接受"涂泥疗法"和"涂油彩疗法"。一个患者走近我们跟前说道："大夫，"他的眼睛流露出欣喜之情，"这是一所非常出色的精神保健中心：这是我曾经待过的中心里最好的一所了。"总而言之，病人被当作病人治，病人也清楚自己是病人。在某些情况下，他们居然因自己是病人而显得洋洋得意。他们没有有益于身心健康的职业，没有工作，一天结束之时，他们感到无所事事，百无聊赖。以人道为宗旨的中心事实上加重了病人认为自己有病的思想负担，甚至当中心在大力提倡保健时，病人的病态行为却有增无减。

swimming,dancing,sports,and fresh air—and provide medical treatment only as an incidental side of these activities.

There is converging evidence and speculation in the health care literature that health centers with these characteristics, organized according to the philosophy of health maintenance,are critical.(See,for example:William H.Glazier, "The Task of Medicine," *Scientific American*,Vol.228,No.4,April 1973,pp.13-17;and Milton Roemer, "Nationalized Medicine for America," *Transaction*, September 1971,p.31.)

We know of several attempts to develop health care programs which are in line with this proposal.In most of the cases,though,the programs fall short in their hopes because,despite their good intentions,they still tend to cater to the sick,they do not work to maintain health.Take, for example,the so-called "community mental health centers" encouraged by the United States National Institute of Mental Health during the late 1960's.On paper,these centers are intended to encourage health,not cure sickness.

In practice it is a very different story.We visited one of the most advanced,in San Anselmo,California.The patients sit around all day long;their eyes are glazed;they are half-enthusiastically doing "clay therapy" or "paint therapy." One patient came up to us and said, "Doctor," his eyes shining with happiness, "this is a wonderful mental health center;it is the very best one I have ever been in." In short,the patients are kept as patients;they understand themselves to be patients;in certain cases they even revel in their role as patients.They have no useful occupation,no work,nothing useful they can show at the end of a day,nothing to be proud of.The center,for all its intentions to be human,in fact reinforces the patients' idea of their own sickness and encourages the behavior of

加利福尼亚的凯西—珀曼内特规划和上述的如出一辙。在最近一篇论文中，凯西医院一直作为"把重点从治病转向保健的医院"而备受赞扬（William H.Glazier，"The Task of Medicine"）。凯西医院的病人有权每年进行全面体检，以使了解各自的总的健康状况。可是这种全面体检的方案所形成的保健概念仍然是"解除病痛"的老一套。它基本上是消极的。凯西医院在培养积极的创造性和保持真正的精力旺盛的健康方面，没有作出切实的努力。除此之外，凯西中心依然仅仅是一所大医院而已。病人被作为一批商品来对待，该中心是如此巨大和如此集中，致使医生无法将他们视为自然社区的居民，只好将他们看成病人了。

我们知道的真正致力于保健而非治病的唯一保健中心就是英国著名的佩卡姆保健中心。佩卡姆保健中心是一个俱乐部，由两名医生负责管理。医生将注意力集中于一个游泳池，一个舞厅和一个露天餐厅。除此之外，还有几间医生办公室。众所周知，接受该中心定期体检的绝不是单独的个人，而是全家人。这种体检是一个个家庭开展游泳和跳舞等活动所不可缺少的一个组成部分。在此条件下，无论在白天还是黑夜，人们都可以有条不紊地合理使用这一中心。他们的保健措施和社区的日常生活交织在一起，从而开创了保健医疗的一个非常特殊的时期。

举例说吧，看来，在英国战前的工人阶级中间，许多母亲对她们自己的身体感到羞愧。这种羞愧感有时竟然达到非常严重的程度，她们羞色满面，不敢抱着婴儿给他们哺乳，结果在许多情况下她们索性就不要孩子了。佩卡姆中心一贯强调保健，能有办法消除这种综合征。游泳和跳舞的方案，再配合家庭体检，就能改变妇女的偏见，使她

sickness,even while it is preaching and advocating health.

The same is true for the Kaiser-Permanente program in California.The Kaiser hospitals have been hailed in a recent article as "ones which shift the emphasis away from treatment of illness and toward the maintenance of health(William H.Glazier, "The Task of Medicine").Members of Kaiser are entitled to a multi-phasic examination yearly,intended to give every member a complete picture of the state of his health.But the conception of health which is created by this multi-phasic program is still "freedom from sickness." It is essentially negative.There is no effort made toward the positive creation and maintenance of actual,blooming,health.And besides,the Kaiser Center is still nothing but a giant hospital.People are treated as numbers;the center is so large and concentrated that the doctors cannot possibly see their patients as people in their natural communities.They see them as patients.

The only health center we know which actually devoted itself to health instead of sickness was the famous Peckham Health Center in England.The Peckham Center was a club,run by two doctors,focused on a swimming pool,a dance floor,and a cafe. In addition,there were doctors'offices,and it was understood that families—never individuals—would receive periodic check-ups as part of their activities around the swimming and dancing.Under these conditions,people used the center regularly,during the day and at night.The question of their health became fused with the ordinary life of the community,and this set the stage for a most extraordinary kind of health care.

For example,it seems that many of the mothers in working-class pre-war England,were ashamed of their own bodies.This shame reached such proportions that they were ashamed of

们为自己的身体感到骄傲而不再是羞愧，使她们为自己的新生儿感到欣喜若狂而不再是担惊受怕。近几年来，在佩卡姆保健中心，儿童患情绪波动症和童年精神变态症的人数，与其初创期相比急剧减少。

这种具有深远意义的、在身体保健、家庭生活和精神愉快三者之间的生物学联系确实是人性生物学新纪元的开端。这一点已由该中心的两位医生作了非常精彩而又详尽的描述（Innes Pearse and Lucy Crocker, *The Peckham Experiment*, *A study in the Living Structure of Society*, New Haven : Yale University Press, 1946）。只有当具有这种深度和力度的生物学观念被认真研究时，才有可能创建名副其实的保健中心，而不是治疗中心。

因此：

逐步发展小型保健中心网，遍布全市，比如一个 7000 人的社区设一个保健中心。每个中心的医疗设备都要配备齐全，医治常见疾病包括医治儿童和成人的精神方面和身体方面的疾病，但要进行组织，基本上设施的重点放在娱乐和教育活动上，如游泳和跳舞，以有益于人们的身心健康。

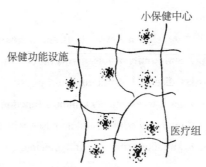

小保健中心

保健功能设施

医疗组

suckling and holding their own babies,and in many cases they actually did not want their babies as a result.The Peckham Center was able to dismantle this syndrome entirely by its emphasis on health.The program of swimming and dancing,coupled with the family check-ups,allowed women to become proud of their own bodies;they no longer felt afraid of their own newborn babies,no longer felt shame about their bodies;the babies felt wanted;and the incidence of emotional disturbance and childhood psychosis among the children in later years was drastically reduced within the Peckham population,starting exactly from the year when the health center began its operation.

This kind of profound biological connection between physical health,family life,and emotional welfare was truly the beginning of a new era in human biology.It is described, beautifully,and at length,by two doctors from Peckham Center (Innes Pearse and Lucy Crocker,*The Peckham Experiment, A Study in the Living Structure of Society*,New Haven:Yale University Press,1946).Only when biological ideas of this depth and power are taken seriously will it be possible to have real health centers,instead of sickness centers.

Therefore:

Gradually develop a network of small health centers, perhaps one per community of 7000,across the city;each equipped to treat everyday disease-both mental and physical,in children and adults-but organized essentially around a functional emphasis on those recreational and educational activities which help keep people in good health,like swimming and dancing.

医疗组要小而又独立——**小服务行业**（81），但它们彼此之间以及和其他门诊所都要协调一致，如**分娩场所**（65）——遍布整个城镇。给每个中心提供一些设施，以便与地方工作的一般进程和娱乐活动交融在一起，如游泳池、车间、桑拿浴室、体育馆、菜园和温室。但是不要强迫这些设施形成连续的"保健公园"——使它们和城镇的其他地方松散地连接起来——**住宅与其他建筑间杂**（48）、**地方性运动场地**（72）、**冒险性的游戏场地**（73）、**家庭工作间**（157）、**菜园**（177）。也许，有助于人们保持健康的最重要的辅助模式就是提供游泳的机会；十分理想的是在每个街区试设一个游泳池——**池塘**（71）……

Keep the medical teams small and independent—SMALL SERVICES WITHOUT RED TAPE(81),but coordinated with each other and other clinics,like BIRTH PLACES(65)— throughout the town.Give each center some functions that fuse with the ordinary course of local work and recreation:swimming pool,workshops,sauna,gym,vegetable garden,greenhouse. But don't force these facilities to form a continuous "health park"—knit them together loosdy with other parts of the town— HOUSING IN BETWEEN(48),LOCAL SPORTS (72),ADVENTURE PLAYGROUND(73),HOME VORKSHOP (157),VEGETABLE GARDEN(177).Perhaps the most important subsidiary pattern for helping people to keep healthy is the opportunity for swimming;ideally,try and put a swimming pool on every block—STILL WATER(71)...

模式48　住宅与其他建筑间杂**

　　……大部分住宅位于邻里住宅区内和邻里住宅团组内——**易识别的邻里**（14）和**住宅团组**（37）；根据我们的模式，这些住宅区需要由包含公用土地和工作社区的各种边界分隔开来——**亚文化区边界**（13）、**邻里边界**（15）、**工作社区**（41）。但是，就连这些工作社区、边界和商业街也必须包含居民住宅。

❧❀❦

　　无论什么地方，城镇的住宅区和非住宅区都截然分开，非住宅区将迅速变成贫民窟。

48   HOUSING IN BETWEEN**

...most housing is in residential neighborhoods,and in the clusters within neighborhoods—IDENTIFIABLE NEIGHBORHOOD(14),HOUSE CLUSTER(37);and according to our patterns these housing areas need to be separated by boundaries which contain public land and work communities—SUBCULTURE BOUNDARY(13),NEIGHBORHOOD BOUNDARY(15),WORK COMMUNITY(41).But even these work communities,and boundaries,and shopping streets,must contain houses which have people living in them.

 ∞⋙⋘

Wherever there is a sharp separation between residential and nonresidential parts of town, the nonresidential areas will quickly turn to slums.

The personal rhythms of maintenance and repair are central to the well being of any part of a community,because it is only these rhythms which keep up a steady sequence of adaptations and corrections in the organization of the whole. Slums happen when these rhythms break down.

Now in a town,the processes of maintenance and repair hinge on the fact of user ownership.In other words,the places where people are user-owners are kept up nicely;the places where they are not,tend to run down.When people have their own homes among shops,workplaces,schools,services ,the university,these places are enhanced by the vitality that is natural to their homes.They extend themselves to make it

个人进行房屋维护修整的节奏对社区各部分的环境质量十分重要。因为只有这种节奏才能保持在整体环境组织中不断进行适应和修正。这种节奏一旦遭到破坏，贫民窟就出现了。

目前，在城镇内，维修的过程取决于使用者的所有权。换句话说，在使用者就是所有者的那些地方，情况良好；在使用者不是所有者的那些地方，呈现出日趋衰落的景象。当人们在商店、车间、中小学、服务机构和大学之间拥有住宅时，这些地方的环境质量就会被住宅拥有者家庭的内在活力所提高。他们施展才能，改造环境，使它具有个性，使人们感到舒适。一个人在精心布置家庭方面总比在其他地方花费的时间多，要他在感情上平分秋色是不可能的。由此我们得出结论，周围环境的许多地方之所以毫无生气，无人照料关心，其简单的原因就是确实无人居住在那里。

只有在住宅与其他建筑间杂、三三两两地联成排或团组时，住户的个性和建宅活动才会给车间、办公室和服务机构带来生气。

因此：

商店、小工业、中小学、公共机构、大学等市内白天吸引人的地方正在成为"无人居住区"。所以，要把住宅间杂地建在上述各种建筑的结构之中。住宅可以建成联排式住宅或下面设有商店的"丘状"住宅，抑或是独立式住宅，只要它们与其他建筑间杂，并使整个地区适于"居住"。

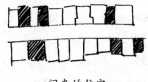

间杂的住宅
Occasional houses

personal and comfortable.A person will put more of himself into his home than into any of the other places where he spends his time.And it is unlikely that a person can put this kind of feeling into two places,two parts of his life.We conclude that many parts of the environment have the arid quality of not being cared for personally,for the simple reason that indeed nobody lives there.

It is only where houses are mixed in between the other functions,in twos and threes,in rows and tiny clusters,that the personal quality of the households and house-building activities gives energy to the workshops and offices and services.

Therefore:

Build houses into the fabric of shops,small industry, schools,public services,universities—all those parts of cities which draw people in during the day,but which tend to be "non-residential." The houses may be in rows or "hills" with shops be-neath,or they may be free-standing,so long as they mix with the other functions,and make the entire area "lived-in."

⊰⊱

Make sure that,in spite of its position in a public area,each house still has enough private territory for people to feel at home in it—YOUR OWN HOME(79).If there are several houses in one area,treat them as a cluster or as a row—HOUSE CLUSTER(37),ROW HOUSES(38)...

between the house clusters and work communities,allow the local road and path network to grow in formally, piecemeal:

确保每幢住宅,尽管它的位置是在公共地区内,都要有一方足够大的私人土地,以便使人感到像在自己家里一样——**自己的家**(79)。如果有好几幢住宅在一个区域内,就把它们当作住宅团组或联排式住宅处理——**住宅团组**(37)和**联排式住宅**(38)……

在住宅团组和工作社区之间,要允许地方公路和小路网络不拘一格地逐渐发展起来:

49. 区内弯曲的道路

50. 丁字形交点

51. 绿茵街道

52. 小路网络和汽车

53. 主门道

54. 人行横道

55. 高出路面的便道

56. 自行车道和车架

57. 市区内的儿童

模式49 区内弯曲的道路**

……我们认为我们对邻里、住宅团组、工作社区和主干公路多少作了说明，参阅**地方交通区（11）**、**易识别的邻里（14）**、**平行路（23）**、**住宅团组（37）**、**工作社区（41）**。现在我们来谈谈地方公路的布局。

☙❧

谁都不希望快速直达交通经过自己的家门。

直达交通速度快，噪声大，且非常危险。同时，汽车却很重要，在居住区，汽车不能完全被排除掉。地方公路必须给住宅提供通路，但要避免直达交通经过家门。

如果区内一切附近有住宅的公路都设计成拐弯抹角的弯曲道路，那么这个问题就迎刃而解了。我们把区内弯曲

49   LOOPED LOCAL ROADS**

...assume that neighborhoods,house clusters,work communities,and major roads are more or less defined— LOCAL TRANSPORT AREAS(11),IDENTIFIABLE NEIGHBORHOOD(14),PARALLEL ROADS(23),HOUSE CLUSTER(37),WORK COMMUNITY(41).Now,for the layout of the local roads.

෴

Nobody wants fast through traffic going by their homes.

Through traffic is fast,noisy,and dangerous.At the same time cars are important,and cannot be excluded altogether from the areas where people live.Local roads must provide access to houses but prevent traffic from coming through.

This problem can only be solved if all roads which have houses on them are laid out to be "loops." We define a looped road as any road in a road network so placed that no path along other roads in the road network can be shortened by travel along the "loop."

The loops themselves must be designed to discourage high volumes or high speeds:this depends on the total number of houses served by the loop,the road surface,the road width,and the number of curves and corners.Our observations suggest that a loop can be made safe so long as it serves less than 50 cars.At

的道路定义为公路网中的任何一条公路，但它应当这样来布置：没有任何一条沿着该公路网内其他公路的道路能以"弯道"为捷径。

弯道必须被设计成能够防止大量汽车涌入或高速通过的模式。这取决于弯道所服务的住宅总数、公路表面、路面宽度、弯曲数和拐角数。据我们观察所知，只要弯道服务的汽车数量低于 50 辆，交通安全就有保障。以每一住宅拥有 1.5 辆汽车计算，这样一条弯道就能为 30 幢住宅服务；以每一住宅拥有 0.5 辆汽车计算，它可为 50 幢住宅服务；以每一住宅拥有 0.5 辆汽车计算，它可为 100 幢住宅服务。

下面是一个区内弯曲道路完整体系的实例，它是为秘鲁的一个拥有 1500 幢住宅的社区而设计的。

利马市区内弯曲的道路
Looped local roads in Lima

甚至在一个简单的地图坐标方格内道路也可以改变成弯道。

形成区内弯曲的道路的方法之一：堵死街道
A way of closing streets to form looped local roads

根据所下的定义，有死端的街道也是弯道。可是，从社会的观点来看，只有一端与外界相通的街道是十分糟糕的。这些死端相互影响，会使人感到一种幽闭的恐怖感，因为仅有一个入口。当形成汽车交通的死端时，要确保步行道成为唯一的直通小路，它从一个方向通入死巷，从另一个方向通出死巷。

one and one-half cars per house,such a loop serves 30 houses;at one car per house,50 houses;at one-half car per house,100 houses.

Here is an example of an entire system of looped local roads designed for a community of 1500 houses in Peru.

Even a simple grid can be changed to have looped local roads.

Dead-end streets are also loops,according to the definition.However,cul-de-sacs are very bad from a social standpoint—they force interaction and they feel claustrophobic,because there is only one entrance.When auto traffic forms a dead end,make sure that the pedestrian path is a through path,leading into the cul-de-sac from one direction,and out of it in another direction.

Recognize also that many roads which appear looped are actually not.This map looks as though it has looped roads. Actually,only one or two of these roads are looped in the functional sense defined.

Therefore:

Lay out local roads so that they form loops.A loop is defined as any stretch of road which makes it impossible for cars that don't have destinations on it to use it as a shortcut.Do not allow any one loop to serve more than 50 cars,and keep the road really narrow—17 to 20 feet is quite enough.

穿越死端的步行道
Pedestrian paths which go beyond a dead end

同样要承认，有许多道路看起来似乎是弯道，但实际上却不是。下面这张地图看起来仿佛有许多弯道，实际却不然。按其功能来说，其中只有一两条公路才是弯曲的道路。

这些不是区内弯曲的道路
These are not looped local roads

因此：

将区内的公路设计成弯曲的道路。区内弯曲的道路被定义为公路的任何一段路程，使过路汽车在此路上没有目的地，从而无法利用它作为捷径。不允许区内任何一条弯曲的道路为 50 辆以上的汽车服务，务使道路真正成为狭窄的：17 ~ 20ft 宽就完全足够了。

    ❧❧❦❧

使地方公路之间的所有交叉点成为三向的 T 形交点，而绝不要成为四向交点——**丁字形交点**（50）：在那些有可能有人在临街的房屋内生活的地方，要使街道的路面变得十分粗糙，上面长着杂草，铺上碎石和供汽车轮子滚动的条状石路面——**绿茵街道**（51）；在有车道的公路旁建立停车场——**小停车场**（103）、**与车位的联系**（113）；除了要使公路保持非常安静之外，还要使步行道和公路成直角相交，而不是沿着它们并行，并使住宅和偏离的小路保持畅通，但不和偏离的公路保持畅通——**小路网络和汽车**（52）……

❦

Make all the junctions between local roads three-way T junctions,never four-way intersections—T JUNCTIONS(50); wherever there is any possibility of life from buildings being oriented toward the road,give the road a very rough surface of grass and gravel,with paving stones for wheels of cars—GREEN STREETS(51);keep parking off the road in driveways—SMALL PARKING LOTS(103)and CAR CONNECTION(113);except where the roads are very quiet,run pedestrian paths at right angles to them,not along them,and make buildings open off these paths,not off the roads— NETWORK OF PATHS AND CARS(52)...

模式50 丁字形交点*

……如果主干公路布局适当——**平行路**（23），而你又正在处理如何限定地方公路的问题，本模式会给你提供有关交点性质的说明。这将极大地影响地方公路的设计，并将有助于地方公路建成弯曲的道路——**区内弯曲的道路**（49）。

❧❧❧

和在丁字形交点上相比，在两条公路的十字交叉点上发生交通事故更加频繁。

本模式是根据下列几何图形得出的。在那些有两条双向公路相交的地方，有 16 个主要碰撞点，对比丁字形交点则只有 3 个（John Callendar，*Time Saver Standards*，Fourth Edition，New York，1966，p.1230）。

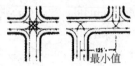

16个碰撞点……3个碰撞点

根据我们自己的调查研究，即对 5 年来在各种不同的街道模式内所发生的事故数量进行的比较分析而得出的结果，已画成地图列举如下。它们清楚地表明丁字形交点上发生的交通事故比在十字形交点上要少得多（from *Planning for Man and Motor*，by Paul Ritter，p.307）。

充分的证据进一步表明，如果丁字形交点上呈垂直相交，它最安全。当角度偏离直角时，司机就很难看清拐角的地方，交通事故就会增加（Swedish National Board of Urban Planning，"Principles for Urban Planning with Respect to Road Safety," The Scafi Guide-lines 1968,Publication

50   T JUNCTIONS*

...if major roads are in position—PARALLEL ROADS (23),and you are in the process of defining the local roads,this pattern gives the nature of the intersections.It will also greatly influence the layout of the local roads,and will help to generate their looplike character—LOOPED LOCAL ROADS(49).

❧❦

Traffic accidents are far more frequent where two roads cross than at T junctions.

This follows from the geometry.Where two two-way roads cross,there are 16 major collision points,compared with three for a T junction(John Callendar,*Time Saver Standards*,Fourth Edition,New York,1966,p.1230).

Maps from an empirical study which compares the number of accidents over a period of five years for different street patterns are shown below.They show clearly that T junctions have many fewer accidents than four-way intersections(from *Planning for Man and Motor*,by Paul Ritter,p.307).

Further evidence shows that the T junction is safest if it is a right-angled junction.When the angle deviates from the right angle,it is hard for drivers to see round the corner, and accidents increase(Swedish National Board of Urban Planning, "Principles for Urban Planning with Respect to Road Safety," The Scaft Guide-lines 1968,Publication No.5,Stockholm,Sweden,p.11).

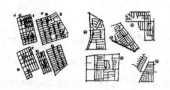

不同交点上的交通事故率
Accidents at different intersections

No.5,Stockholm,Sweden.p.11）。

因此：

设计公路系统时，务必使任何两条公路在同一平面上相交，都应做成丁字形相交，并尽可能以 **90°** 直角相交。避免十字交叉点和十字交叉运动。

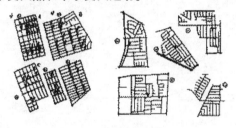

๛ఔ

在那些交通繁忙、步行道都汇集在一起的交叉点上，要做成一种特殊的高出路面的人行横道，它与常见的有所不同——**人行横道**（54）……

Therefore:

Lay out the road system so that any two roads which meet at grade,meet in three-way T junctions as near 90 degrees as possible.Avoid four-way intersections and crossing movements.

୫୦ଔଓ

At busy junctions,where pedestrian paths converge,make a special raised crossing for pedestrians,something more than the usual crosswalk—ROAD CROSSING(54)...

模式51 绿茵街道**

51   GREEN STREETS**

...this pattern helps to give the character of local roads. Even though it only defines the surface of the road,and the position of parking,the gradual emergence of this pattern in an area,can be used,piecemeal,to create LOOPED LOCAL ROADS(49),T JUNCTIONS (50),and COMMON LAND (67). This pattern was inspired by a beautiful road in the north of Denmark,built by Anne-Marie Rubin,and illustrated here.

❧☙

There is too much hot hard asphalt in the world.A local road,which only gives access to buildings,needs a few stones for the wheels of the cars;nothing more.Most of it can still be green.

In a typical low density American suburb,more than 50 percent of the land is covered with concrete or asphalt paving.In some areas,like downtown Los Angeles,it is more than 65 percent.

This concrete and asphalt have a terrible effect on the local environment.They destroy the microclimate;they do nothing useful with the solar energy that falls on them;they are unpleasant to walk on;there is nowhere to sit;nowhere for children to play;the natural drainage of the ground is devastated;animals and plants can hardly survive.

The fact is that asphalt and concrete are only suitable for use on high speed roads.They are unsuitable,and quite unnecessary,on local roads,where a few cars are moving in and out.When local roads are paved,wide and smooth,like major roads,drivers are encouraged to travel past our houses

……本模式有助于显示地方公路的特征。虽然本模式只对公路路面和停车场的位置作出若干限制,但在一个区内逐渐出现的绿茵街道却可以被逐步用来建成**区内弯曲的道路**(49)、**丁字形交点**(50)和**公共用地**(67)。本模式是我们受到丹麦北部一条风光绮丽的公路的启示而构思出来的,那条公路是由安妮-玛丽·鲁宾所建造,并引为实例在此加以说明。

⁂

在这个世界上,热的硬沥青使用过多了。一条通往住宅的道路铺上寥寥几块石头供汽车轮子滚动就足够了;别的则可一概不用。这样,路面的大部分可能依然绿茵一片。

在典型的人口密度低的美国城市郊区,铺上混凝土和沥青的地面达 50% 以上。在某些地区如洛杉矶的商业区,则超过 65%。

这种混凝土和沥青地面对地方环境具有严重的不良影响。它们破坏微气候;它们无法配合照射在其上的太阳能而发挥任何有益的作用;人们步行其上,心中感到不悦;真是行人无可坐之地,孩子无可玩之处;地表的天然排水遭到破坏;动植物几乎无法继续生存。

事实上,沥青和混凝土只适用于铺筑高速公路,而不适用于铺筑地方公路,而且也完全是不必要的。因为来往于地方公路的汽车寥寥无几。一旦地方公路的路面铺筑得又宽又平,像主干公路一样,就会鼓励汽车司机以每小时 35mi 或 40mi 的速度驱车通过我们的住宅区。然而,地方公路所需要的是路面绿草丛生,以便符合住宅间公共土地的主要用途,只在为数不多的汽车往返行驶的路面铺砌硬实就行了。

最佳的解决方案就是在一片草地里稀稀拉拉地嵌入石块。这种布置为孩子们和动物提供了活动场所,并使道路成为邻里的集合地点了。在炎热的盛夏,在绿草如茵的路

at 35 or 40 miles per hour.What is needed,instead,on local roads is a grassy surface that is adapted to the primary uses of the common land between the buildings,with just enough hard paving to cope with the few cars that do go on it.

The best solution is a field of grass,with paving stones set into it.This arrangement——provides for animals and children and makes the street a focal point for the neighborhood.On hot summer days the air over the grass surface is 10 to 14 degrees cooler than the air over an asphalt road.And the cars are woven into this scheme,but they do not dominate it.

Of course,such a scheme raises immediately the question of parking.How shall it be organized? It is possible to arrange for parking on green streets,so long as it is parking for residents and their guests,only.When overflow parking from shopping streets and work communities sprawls onto streets that were intended to be quiet neighborhoods,the character of the neighborhood is drastically altered.The residents generally resent this situation. Often it means they cannot park in front of their own homes. The neighborhood becomes a parking lot for strangers who care nothing about it,who simply store their cars there.

The green street will only work if it is based on the principle that the street need not,and should not,provide for more parking than its people need.Parking for visitors can be in small parking lots at the ends of the street;parking for people in the individual houses and workshops can either be in the same parking lots or in the driveways of the buildings.

This does not imply that commercial activities,shops,and businesses should be excluded from residential areas.In fact,as we have said in SCATTERED WORK (9),it is extremely important to build such functions into neighborhoods.The point is,however,that

面之上的空气与柏油路面之上的空气相比，前者的气温比后者的低10～14℃。汽车交织在这似画一般美丽的图案中，但并不会凌驾于其上。

铺砌石块的路面
Paving stones

当然，这样一个方案会提出停车场的问题。停车场应当怎样来组织安排呢？在绿茵街道上建立停车场是可能的，只要这种停车场仅仅是为居民及其客人而设。一旦从商业街和工作社区蜂拥而至的汽车多得泛滥成灾时，停车场势必要向本应安静的邻里扩展，邻里的性质就会急剧改变。对此居民通常极为忿恨。这往往意味着居民不能在自己的家门口停放汽车。邻里却成为陌生人的停车场了。他们仅仅是停放汽车而已，对该邻里是漠不关心的。

如果绿茵街道是依照下列原理建造的，即它不必而且也不应当提供超过其居民所需的停车场，才能发挥效益。来宾小停车场可以设在街道的尽头。为私人住宅的居民和工厂的工人提供的停车场，既可以是街道尽头的同一来宾停车场，又可以是住宅区的车道。

这并不意味着商业活动、商店以及其他营业应当从居民区排除出去。事实上，正如我们在**分散的工作点**（9）中所说的那样，在邻里内建设这样的公共设施是极为重要的。

businesses cannot assume when they move into a neighborhood that they have the right to a huge amount of free parking.They must pay for their parking;and they must pay for it in a way which is consistent with the environmental needs of the neighborhood.

Therefore:

On local roads,closed to through traffic,plant grass all over the road and set occasional paving stones into the grass to form a surface for the wheels of those cars that need access to the street.Make no distinction between street and sidewalk.Where houses open off the street, put in more paving stones or gravel to let cars turn onto their own land.

❧❧❧

When a road is a green street,it is so pleasant that it naturally tends to attract activity to it.In this case,the paths and the green street are one—COMMON LAND(67).However,even when the street is green,it may be pleasant to put in occasional very small lanes,a few feet wide,at right angles to the green streets,according to NETWORK OF PATHS AND CARS(52). In order to preserve the greenness of the street,it will be essential,too,to keep parked cars in drive-ways on the individual lots,or in tiny parking lots,at the ends of the street,reserved for the house owners and their visitors—SMALL PARKING LOTS(103).Fruit trees and flowers will make the street more beautiful—FRUIT TREES(170),RAISED FLOWERS(245)— and the paving stones which form the beds for cars to drive on,can themselves be laid with cracks between them and with grass and moss and flowers in the cracks between the stones-PAVING WITH CRACKS BETWEEN THE STONES(247)...

问题在于一些商店进入邻里进行贸易时不能逞强，认为它们有权获得大量的自由停车点；它们必须支付停车费；它们必须支付符合邻里环境保护需要的费用。

因此：

在临近直达交通的地方公路的全部路面上栽种绿草，使之成为绿草如茵的街道，在它上面铺砌少量石头，形成供需要通过这条绿茵街道的汽车车轮滚动的路面。在绿茵街道和人行道之间不要形成明显的区别。在住宅通往街道的地方，多铺砌些石块和鹅卵石，使汽车能行驶在它们自己的路面上。

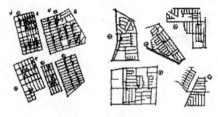

ഗ‍ോ‍ഗ

当公路是绿茵街道时，这是多么令人高兴啊！它自然而然地会吸引人们前来活动。在此情况下，小路和绿茵街道融为一体——**公共用地**（67）。可是，即使在长满绿草的绿茵街道上，偶尔穿插几条数英尺宽的小巷也是会使人感到惬意的。这些小巷按照**小路网络和汽车**（52），与绿茵街道成直角相交。为了使绿茵街道保持一片葱绿，同样重要的是使汽车停放在私人停车场的车道内或停放在专供宅主及其亲朋好友使用的街道尽头的小停车场内——**小停车场**（103）。果树和鲜花会使绿茵街道增添美色——**果树林**（170）和**高花台**（245）——并使铺砌石头的路面成为汽车的过道；在石块之间的缝隙中芳草萋萋，鲜花怒放，青苔满地——**留缝的石铺地**（247）……

模式52　小路网络和汽车**

　　……公路可以由**平行路**（23）、**区内弯曲的道路**（49）和**绿茵街道**（51）来加以控制；主要的小路可以由**活动中心**（30）、**散步场所**（31）以及**小路和标志物**（120）来加以控制。本模式控制上述两类模式之间的相互作用。

☙

　　汽车对于行人来说是危险的；可是各种活动恰恰发生在汽车和行人会合的地方。

　　使行人和汽车分开而各行其道——这是普通的规划实践。这使得步行区更富有人情味和安全感。但是，这种实践活动没有考虑到下列情况：汽车和行人相互需要，而且，实际上大量的城市生活活动恰恰发生在这两种系统的汇合点上。在城市内许多面积最大的地方，如皮卡迪利广场、时代广场和爱丽舍宫广场都生机勃勃，因为它们地处行人和车辆的汇合点。新的城镇如苏格兰的坎伯诺尔特，处于

52   NETWORK OF PATHS AND CARS**

...roads may be governed by PARALLEL ROADS(23),
LOOPED LOCAL ROADS(49),GREEN STREETS(51);major
paths by ACTIVITY NODES (30),PROMENADE(31),and
PATHS AND GOALS(120).This pattern governs the interaction
between the two.

※※※

**Cars are dangerous to pedestrians;yet activities occur
just where cars and pedestrians meet.**

It is common planning practice to separate pedestrians
and cars.This makes pedestrian areas more human and
safer.However,this practice fails to take account of the
fact that cars and pedestrians also need each other:and
that,in fact,a great deal of urban life occurs at just the
point where these two systems meet.Many of the greatest
places in cities,Piccadilly Circus,Times Square,the
Champs Elysées,are alive because they are at places
where pedestrians and vehicles meet.New towns like
Cumbernauld,in Scotland,where there is total separation
between the two,seldom have the same sort of liveliness.

The same thing is true at the local residental scale.
A great deal of everyday social life occurs where cars and
pedestrians meet.In Lima,for example,the car is used as an
extension of the house:men,especially,often sit in parked
cars,near their houses,drinking beer and talking.And in one
way or another,something like this happens everywhere.

人和车辆完全隔离的地方，很少有那种生机盎然的情景。

在地方居民区情况也是如此。大量的日常社会生活活动发生在汽车和行人的汇合处。例如在利马，汽车被当作住宅的延伸物：尤其是男人们坐在离他们家门不远的停放着的汽车里，痛饮啤酒，扎堆聊天。像这种情况到处都有，屡见不鲜。在人们擦洗汽车的停车场周围，居民间的交谈和讨论十分自然。许多商贩在行人和汽车的汇合处摆摊设点，他们需要一切来往的行人和车辆。孩子们在停车场玩耍——也许因为他们感到这是主要的集散点；当然，还因为他们喜欢汽车。可是，与此同时，重要的是务必使行人和车辆分开，各行其道，以便保护孩子和老人，保障行人的生命安全及安宁。

孩子们喜欢汽车
Children like cars

为了解决这一矛盾，很有必要寻找出一种步行小路和车行公路合理分布的方案。这两种道路是分开的，但在公认的集中点频频相交。一般来说，这需要两种正交网络，一种是公路的正交网络，另一种是人行道的正交网络。每一网络连接起来，并不断延伸。两种网络每隔一定距离成直角相交（据我们的观察表明，小路网络上的大多数交点应以距最近的公路不超过150ft为宜）。

两种正交网络
Two orthogonal networks

实际上，形成两种道路之间的这种关系的可能途径不止一种。

在单向高速公路系统中，两条公路相距约300ft，上述

Conversation and discussion grow naturally around the lots where people wash their cars.Vendors set themselves up where cars and pedestrians meet;they need all the traffic they can get.Children play in parking lots-perhaps because they sense that this is the main point of arrival and departure;and of course because they like the cars.Yet,at the same time,it is essential to keep pedestrians separate from vehicles:to protect children and old people;to preserve the tranquility of pedestrian life.

To resolve the conflict,it is necessary to find an arrangement of pedestrian paths and roads,so that the two are separate,but meet frequently,with the points where they meet recognized as focal points.In general,this requires two orthogonal networks,one for roads,one for paths,each connected and continuous,crossing at frequent intervals(our observations suggest that most points on the path network should be within 150 feet of the nearest road),meeting,when they meet,at right angles.

In practice,there are several possible ways of forming this relationship between the roads and paths.

It can be done within the system of fast one-way roads about 300 feet apart described in PARALLEL ROADS(23). Between the roads there are pedestrian paths running at right angles to the roads,with buildings opening off the pedestrian paths.Where the paths intersect the roads there are small parking lots with space for kiosks and shops.

It can be applied to an existing neighborhood—as it is in the following sequence of plans drawn by the People's Architects,Berkeley,California.This shows a beautiful and simple way of creating a path network in an existing grid of

关系就能实现，参阅**平行路**（23）。在公路之间有人行道与它们成直角相交,沿着这些人行道常有住宅,交通畅通无阻。在小路和公路相交的地方，设有停车场、商亭和商店。

这一点可应用于一个现存的邻里中，正如应用于下列的平面图顺序中一样。这些平面图是由加利福尼亚伯克利的人民建筑师协会所绘制的。这表明了一种出色而又简单的方法，该方法在现存的街道地图坐标方格中创造出在各个方向上关闭交替街道的小路网络。正如下图所示，小路网络可以逐步实现。

平行路之间的小路
Path between parallel roads

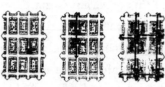

小路网络在一条街道的地图坐标方格中发展的情况
The growth of a path network in a street grid

还有与上述不同的方案，就是我们为利马设计的住宅建筑工程。两种正交系统列举如下:

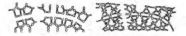 

公路　　　　步行小路　　　公路和步行小路连接成网络

在所有这些情况中，我们发现一个全球性模式。在此模式中，公路和小路大致同时形成，因此使两者具有恰如其分的关系。可是，重要的是要承认，这一模式在大多数的实际应用中表明，没有必要同时将公路和小路一起选定位置。最有代表性的一点是现存的公路系统:在此系统中，小路可以一条条地逐步布置，并与现存的公路成直角相交。一种内部连贯的小路网络将十分缓慢地逐渐发展而成。

最后，需注意汽车和行人的这种分离只在交通密度是中等或中上等的地方才是适用的。在交通密度低的地方（如

streets,by closing off alternate streets,in each direction.As the drawings show,it can be done gradually.

Different again,is our project for housing in Lima.Here the two orthogonal systems are laid out as follows:

In all these cases,we see a global pattern,in which roads and paths are created more or less at the same time—and therefore brought into the proper relationship.However,it is essential to recognize that in most practical applications of this pattern,it is not necessary to locate the roads and paths together.Most typically of all,there is an existing road system:and the paths can be put in one by one,piecemeal,at right angles to the existing roads.Slowly,very slowly,a coherent path network will be created by the accumulation of these piecemeal acts.

Finally,note that this kind of separation of cars from pedestrians is only appropriate where traffic densities are medium or medium high.At low densities(for instance,a cul-de-sac gravel road serving hall-a-dozen houses),the paths and roads can obviously be combined.There is no reason even to have sidewalks—GREEN STREETS (51). At very high densities,like the Champs Elysées,or Piccadilly Circus,a great deal of the excitement is actually created by the fact that pedestrian paths are running along the roads. In these cases the problem is best solved by extra wide sidewalks—RAISED WALKS (55)—which actually contain the resolution of the conflict in their width.The edge away from the road is safe—the edge near the road is the place where the activities happen.

一条为 6 幢住宅服务的死巷），小路和公路显然可以合二为一，甚至没有任何理由再辟人行道——**绿茵街道**（51）。在最高交通密度区，像爱丽舍宫广场或皮卡迪利广场，许多激动人心的活动场面的出现是由于如下这一事实：小路与公路并行。所以，在这些情况下，解决问题的最好办法就是特别宽的人行道——**高出路面的便道**（55）。便道的宽度包含着解决这一矛盾的实际可能性。离开公路的一侧是安全的，靠近公路的一侧是进行各种活动的场所。

因此：

除了交通密度很高或很低的地方外，把人行道设计成与公路成直角相交，而不是与公路并行，小路就会开始逐步形成一个与公路系统截然不同并与之成正交的第二网络。这一点可以缓慢地逐渐实现——即使你们一次布置一条小路，但总是把它们布置在"街区"中央，以便它们与公路成十字相交。

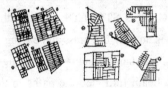

❧⚜❧

在小路不得不与主干公路相平行的地方——这种情况偶有发生——在公路的一侧，将小路建得高出公路路面 18in，比通常的宽度大一倍——**高出路面的便道**（55）；在**绿茵街道**（51）上，小路可能就是公路，因为那里仅仅是一片绿草和石铺的路面。即使如此，也偶有狭窄的小路与绿茵街道成直角相交，这些小道却显得格外美丽。按照**小路和标志物**（120）精心安排这些小路；根据模式**小路的形状**（121）来设计它们的形状。最后，将重要的街道十字路口作为人行横道来处理，这种人行横道的高度和高出路面的便道相等——这样，当汽车穿越人行横道时就不得不减速——**人行横道**（54）……

Therefore:

Except where traffic densities are very high or very low,lay out pedestrian paths at right angles to roads,not along them,so that the paths gradually begin to form a second network,distinct from the road system,and orthogonal to it.This can be done quite gradually—even if you put in one path at a time,but always put them in the middle of the "block," so that they run across the roads.

❧❦❧

Where paths have to run along major roads—as they do occasionally—build them 18 inches higher than the road,on one side of the road only,and twice the usual width—RAISED WALK(55);on GREEN STREETS(51)the paths can be in the road since there is nothing but grass and paving stones there;but even then,occasional narrow paths at right angles to the green streets are very beautiful.Place the paths in detail according to PATHS AND GOALS(120);shape them according to PATH SHAPE(121).Finally,treat the important street crossings as crosswalks,raised to the level of the pedestrian path— so cars have to slow down as they go over them—ROAD CROSSING(54)...

模式53 主门道**

53 MAIN GATEWAYS**

...at various levels in the structure of the town,there are identifiable units.There are neighborhoods—IDENTIFIABLE NEIGHBORHOOD (14),clusters—HOUSE CLUSTER(37),communities of work—WORK COMMUNITY(41);and there are many smaller building complexes ringed around some realms of circulation—BUILDING COMPLEX(95),CIRCULATION REALMS(98). All of them get their identity most clearly from the fact that you pass through a definite gateway to enter them—it is this gateway acting as a threshold which creates the unit.

❧❦

Any part of a town—large or small—which is to be identified by its inhabitants as a precinct of some kind,will be reinforced,helped in its distinctness,marked,and made more vivid,if the paths which enter it are marked by gateways where they cross the boundary.

Many parts of a town have boundaries drawn around them. These boundaries are usually in people's minds.They mark the end of one kind of activity,one kind of place,and the beginning of another.In many cases,the activities themselves are made more sharp,more vivid,more alive,if the boundary which exists in people's minds is also present physically in the world.

A boundary around an important precinct,whether a neighborhood,a building complex,or some other area,is most critical at those points where paths cross the boundary.If the point where the path crosses the boundary is invisible,then to all intents and purposes the

……在城镇结构的不同层次上有许多易识别的单元。其中有邻里——**易识别的邻里**（14）、团组——**住宅团组**（37）、进行工作的社区——**工作社区**（41）；还有许多较小的建筑群体呈环状排列在一些畅通区的四周——**建筑群体**（95）、**内部交通领域**（98）。它们都具有十分明显的可识别性：你通过一个轮廓鲜明的门道进入它们的内部——正是这种门道起到形成这种单元的门槛的作用。

∞⋘

城镇的任何一个地方——无论大小如何——都是它的居民可以识别的某种地区。如果进入这一地区的小路都由跨越边界的门道一一标志出来，那么该地区的独特性就会被加强、被突出，从而就会变得更加生动活泼。

城镇许多地方的四周都有边界。这些边界通常浮现在人们的脑海中。边界标志着一个地方、一种活动的结束，另一个地方、另一种活动的开始。在许多情况下，如果萦绕于人们脑际的边界在外界也确实存在，这些活动本身就会变得更加光彩夺目，更加生动活泼。

一个重要地区四周的边界，无论是邻里的、建筑群体的，还是某一其他地区的，小路穿过边界的那些点是最重要的。如果小路穿越边界的点是隐而不见的，那么实际上，边界就不在那里。唯有在十字路口建立标志物，那里才会有边界，才会被人们所感知。通过边界的小路基本上只能用门道来标志。这就是为什么各种形状的门道在周围环境中起到如此重要的作用。

门道有许多种形式：如名副其实的一扇门、一座桥、在分离的建筑物间一条狭窄的通道、一条树木郁闭的林荫道、一条穿过楼房的门道。所有这一切都具有同样的功能：它们标志出小路穿越边界的各个点，并有助于保持边界的鲜明性。所有这一切都是"物"——不仅是洞和空隙，而且是完整的实体。

boundary is not there.It will be there,it will be felt,only if the crossing is marked.And essentially,the crossing of a boundary by a path can only be marked by a gateway.That is why ali forms of gateway play such an important role in the environment.

A gateway can have many forms:a literal gate,a bridge,a passage between narrowly separated buildings,an avenue of trees,a gateway through a building.All of these have the same function:they mark the point where a path crosses a boundary and help maintain the boundary.All of them are "things" — not merely holes or gaps,but solid entities.

In every case,the crucial feeling which this solid thing must create is the feeling of transition.

Therefore:

Mark every boundary in the city which has important human meaning—the boundary of a building cluster,a neighborhood,a precinct—by great gateways where the major entering paths cross the boundary.

༺✿༻

Make the gateways solid elements,visible from every line of approach,enclosing the paths,punching a hole through a building,creating a bridge or a sharp change of level—but above all make them "things," in just the same way specified for MAIN ENTRANCE (110),but make them larger.Whenever possible,emphasize the feeling of transition for the person passing through the gateway,by allowing change of light,or surface,view,crossing water,a change of level—ENTRANCE TRANSITION(112).In every case,treat the main gateway as the starting point of the pedestrian circulation inside the precinct—CIRCULATION REALMS(98)...

门道标志过界点
Gateways mark the point of transition

在任何情况下，这种完整的实体必然造成一种强烈的气氛；即使人感到这是过界点。

因此：

将市内每一条富有人情味的边界——住宅团组的、邻里的、一个区的——都用大门道来标志，主要的小路就在那些有门道的地方穿越边界。

门道

边界

৪০০৪

使门道成为完整的建筑构件。人们从各条道路走近门道时，都能看得见它。将小路围堵住，再打洞穿过一幢建筑，形成一座桥形物，或是明显改变标高——但是首先要使门道成为"物"，正如在**主入口**（110）中所描述的那样，而且规模要大一些。无论何时，只要可能，就要强调创造一种强烈氛围，使穿过门道的人感觉到这是过界点。为了获得这一效果，可以利用光线、表面、景物的变化，跨过流水或改变标高等办法——**入口的过渡空间**（112）。在任何情况下，将主门道视为一个地区内行人畅通区的起点——**内部交通领域**（98）……

模式54　人行横道

54   ROAD CROSSING

...under the impetus of PARALLEL ROADS(23) and NETWORK OF PATHS AND CARS(52),paths will gradually grow at right angles to major roads—not along them as they do now.This is an entirely new kind of situation,and requires an entirely new physical treatment to make it work.

❧☙

Where paths cross roads,the cars have power to frighten and subdue the people walking,even when the people walking have the legal right-of-way.

This will happen whenever the path and the road are at the same level.No amount of painted white lines,crosswalks,traffic lights,button operated signals,ever quite manage to change the fact that a car weighs a ton or more,and will run over any pedestrian,unless the driver brakes.Most often the driver does brake.But everyone knows of enough occasions when brakes have failed,or drivers gone to sleep,to be perpetually wary and afraid.

The people who cross a road will only feel comfortable and safe if the road crossing is a physical obstruction,which physically guarantees that the cars must slow down and give way to pedestrians.In many places it is recognized by law that pedestrians have the right-of-way over automobiles.Yet at the crucial points where paths cross roads,the physical arrangement gives priority to cars.The road is continuous,smooth,and

……在**平行路**（23）、**小路网络和汽车**（52）的推动下，小路将逐步与主干公路发展成直角相交——不是像现在似的与主干公路平行。这是一种全新的情况，需要全新的实际处理，才能使小路系统发挥效用。

❧

在小路和公路相交的地方，即使行人具有合法的先行权时，汽车也会威胁到他们的安全，把他们吓得晕头转向。

无论何时，小路和公路处于同一水平面内，这种情况是总会发生的。无数漆成白色的分界线、人行横道、用按钮操作的交通信号灯都无济于事：载有 1t 或超过 1t 重物的汽车，除非司机及时急刹车，不然就会从任何一个行人身上开过去。司机只好频频刹车。但是，人人都知道，许多场合刹车会失灵或司机在开车时睡着，使行人总是小心翼翼和提心吊胆的。

如果人行横道是一种保证汽车减速并给行人提供通路的外界环境的有形障碍物，那么横穿公路的行人就会有舒适感和安全感。经法律认可，在许多地方，行人有优先于汽车的先行权。但是在小路和公路相交的重要的各点上，外界环境的安排是优先考虑汽车。公路是连续的、平整的高速公路，它在交叉点上中断人行横道。这种连续的公路路面实际表明，汽车有先行权。

为了适应行人的需要，避免车祸，人行横道究竟应当是什么样子的呢？

当行人比汽车高出 18in 时，他们就较少地感到汽车会肇事、伤害他们。这一事实将在下一个模式**高出路面的便道**（55）中详加论述。在行人不得不穿过公路的地方，可以更强有力地运用上述同一原理。横穿公路的行人从公路上看去应是十分清楚的。当汽车行经人行横道时，也应当被迫减速。如果人行横道比公路高出 6 ～ 12in，并且有一

fast,interrupting the pedestrian walk-way at the junctions.This continuous road surface actually implies that the car has the right-of-way.

What should crossings be like to accommodate the needs of the pedestrians?

The fact that pedestrians feel less vulnerable to cars when they are about 18 inches above them,is discussed in the next pattern RAISED WALK(55).The same principle applies,even more powerfully,where pedestrians have to cross a road.The pedestrians who cross must be extremely visible from the road.Cars should also be forced to slow down when they approach the crossing.If the pedestrian way crosses 6 to 12 inches above the roadway,and the road-way slopes up to it,this satisfies both requirements.A slope of 1 in 6,or less,is safe for cars and solid enough to slow them down.To make the crossing even easier to see from a distance and to give weight to the pedestrian's right to be there,the pedestrian path could be marked by a canopy at the edge of the road-CANVAS ROOFS (244).

We know that this pattern is rather extraordinary.For this reason,we consider it quite essential that readers do not try to use it on every road,for forrnalistic reasons,but only on those roads where it is badly needed.We therefore complete the problem state-ment by defining a simple experiment which you can do to decide whether or not a given crossing needs this treatment.

Go to the road in question several times,at different times of day.Each time you go,count the number of seconds you have to wait before you can cross the road.If the average of these waiting times is more than two seconds,then we recommend you use the pattern.(On the basis of Buchanan's

斜坡与公路衔接，那么这会满足双方的要求。坡度为1:6或小于1:6，这对汽车来说是安全的，并且斜坡坚实得足以使汽车放慢速度。为了使人行横道从远处更容易看清楚，以及强调行人在人行横道上的权利，可以在公路边上设一顶篷作为人行横道的标志——**帆布顶篷**（244）。

差不多是条人行横道……但路面没有隆起部分
Almost a road crossing...but no bump

我们知道，本模式是相当特殊的。为此，我们认为十分重要的是，读者不要试图在每一条公路上形式主义地乱用它，而只在那些非常急需的公路上才使用它。因此，为了充分说明这一点，我们现在设定一个你们都能做的简单实验，以便决定某一人行横道是否需要这样来处理。

请你们在白天的不同时间里，到所考察的公路去几次。每次都要记下在你们能穿过公路之前所等候的时间。如果等候时间的平均值超过2s，那么我们建议你们采用本模式。根据布坎南原理，当公路交通量造成对行人平均2s或2s以上的延误时，公路就成为试图步行穿越者的威胁了（See the extensive discussion，Colin Buchanan et al.，*Traffic in Towns*，HMSO，London，1963，pp.203~213）。如果你们无法做此实验，或公路尚未建成，你可以利用下列图表进行推测。该图表表明，公路交通量和公路宽度如何结合才会典型地造成超过2s的平均时延。

statement that roads become threatening to pedestrians when the volume of traffic on them creates an average delay of two seconds or more,for people trying to cross on foot.See the extensive discussion,Colin Buchanan et al.,Traffic in Towns,HMSO,London,1963,pp.203-213.)If you cannot do this experiment,or the road is not yet built,you may be able to guess,by using the chart below.It shows which combinations of volume and width will typically create more than a two-second average delay.

One final note.This pattern may be impossible to implement,in places where traffic engineers are still in control. Nevertheless the functional issue is vital,and must not be ignored. A big wide road,with several lanes of heavy traffic can form an almost impenetrable barrier.In this case,you can solve the problem,at least partially,by creating islands—certainly one in the middle,and perhaps extra islands,between adjacent lanes.This has a huge effect on a person's capacity to cross the road,for a very simple reason.If you are trying to cross a wide road,you have to wait for a gap to occur simultaneously in each of the lanes.It is the waiting for this coincidence of gaps that creates the problem.But if you can hop,from island to island,each time a gap occurs in any one lane,one lane at a time,you can get across in no time at all— because the gaps which occur in individual lanes are many many times more frequent,than the big gaps in all lanes at the same time. So,if you can't raise the crossing,at least use islands,like stepping stones.

Therefore:

At any point where a pedestrian path crosses a road that has enough traffic to create more than a two second delay to people crossing,make a "knuckle" at the crossing:narrow

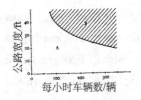

在阴影区的公路需要特殊的人行横道

Roads that fall in the shaded region require special crossings

最后，还要说明一点。本模式也许在交通工程师仍然负责的那些地方难以实行。然而，公路的功能问题是至关重要的，是不应被忽略的。一条宽阔的、拥有几股有大交通量的车道的大公路能够造成几乎难以跨越的障碍。在此情况下，你们设置安全岛，至少可以部分地解决问题——毫无疑问，安全岛设在街心，也许在相邻的车道之间设有特殊的安全岛。这对于穿过公路的人的能力会有极大的影响，理由是十分简单的。如果你们正在试图穿过一条宽阔的公路，就不得不等待同时发生在每股车道中的空档。这是因等待空档的重合而引起的问题。但是，如果你能跳跃，从一个安全岛跳跃到另一个安全岛，每当一个空档发生在任何一股车道内也即一股车道有一次时，你能很快穿过公路——因为发生在单股车道内的空档比在所有的车道内同时发生的大空档频繁不知多少倍。所以，如果你们不能加高人行横道，至少要利用像台阶石那样的安全岛。

因此：

在有人行横道与公路相交的任何交叉点上，公路交通量对于横穿公路的人足以形成 2s 以上的时延时，就要在交叉路口造一个"关节"：将公路宽度缩小到仅有直通车道的宽度；并使整条人行横道高出公路路面约 1ft；在车道间设置安全岛；人行横道与公路衔接的斜坡的坡度为 1 ： 6（最大值）；将人行横道用遮篷或顶棚作标记，以使它清晰可见。

the road to the width of the through lanes only;continue the pedestrian path through the crossing about a foot above the roadway;put in islands between lanes;slope the road up toward the crossing(1 in 6 maximum);mark the path with a canopy or shelter to make it visible.

8003

On one side or the other of the road make the pedestrian path swell out to form a tiny square,where food stands cluster round a bus stop—SMALL PUBLIC SQUARES(61),BUS STOP(92),FOOD STANDS (93);provide one or two bays for standing space for buses and cars—SMALL PARKING LOTS(103),and when a path must run from the road crossing along the side of the road,keep it to one side only,make it as wide as possible,and raised above the roadway—RAISED WALK (55).Perhaps build the canopy as a trellis or canvas roof—TRELLISED WALK(174),CANVAS ROOFS(244)...

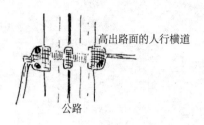

高出路面的人行横道

公路

❧❧❧

将公路的一侧或另一侧的人行道扩展，以便形成小
广场，此处有饮食商亭三三两两地开设在公共汽车站四
周——**小广场**（61）、**公共汽车站**（92）和**饮食商亭**（93）；
为公共汽车和轿车提供一两块凹入的停车场——**小停车场**
（103）；当一条小路必须从人行横道起与公路的一侧平行
时，使这一小路只在公路的一侧，并使它变得尽可能的宽
阔而又高出公路路面——**高出路面的便道**（55）。也许要建
造棚架式的遮篷或帆布顶棚——**棚下小径**（174）和**帆布顶
棚**（244）……

模式55　高出路面的便道*

……本模式有助于完善**小路网络和汽车**（52）及**人行横道**（54）。确实，在大多数情况下，遵循小路网络原理的人行道将穿越公路而不是挨着公路。但是，在一条人行横道和下一条人行横道之间，尤其沿主要**平行路**（23），仍然不时地需要有与公路平行的小路。本模式说明这些特殊的小路应有哪些特征。

৪০৫৪

快速行驶的汽车和行人在市内相遇的那些地方，汽车对行人占压倒优势。汽车是王，而行人却感到自己十分渺小。

55   RAISED WALK*

...this pattern helps complete the NETWORK OF PATHS AND CARS (52)and ROAD CROSSINGS(54). It is true that in most cases,pedestrian paths which follow the path network will be running across roads,not next to them.But still,from time to time,especially along major PARALLEL ROADS(23),between one road crossing and the next,there is a need for paths along the road.This pattern gives these special paths their character.

৪৩৫৩

Where fast moving cars and pedestrians meet in cities,the cars overwhelm the pedestrians.The car is king,and people are made to feel small.

This cannot be solved by keeping pedestrians separate from cars;it is in their nature that they have to meet,at least occasionally—NETWORK OF PATHS AND CARS(52).What can be done at those points where cars and pedestrians do meet?

On an ordinary street,cars make pedestrians feel small and vulnerable because the sidewalks are too narrow and too low. When the sidewalk is too narrow,you feel you are going to fall off,or get pushed off—and there is always a chance that you will step off just in front of a passing car.When the sidewalk is too low,you feel that cars can easily mount the sidewalk,if they go out of control,and crush you.It is clear,then,that pedestrians will feel comfortable,powerful,safe,and free in their movements

这种情况单靠把行人和汽车分开是无法解决通行问题的。就其性质而言，两者不得不相遇，少则偶然相遇——**小路网络和汽车**（52）。在行人和汽车相遇的那些点上应当采取什么措施呢？

　　在一条普通的街道上，汽车使行人感到渺小和易受伤害，因为人行道太窄太低。当人行道太窄时，你往往会感到要跌出去或被撞出去，而且常常会有这种情况，你刚走过去就来到正在疾驶而过的汽车面前。当人行道太低时会令人觉得汽车一旦失控，就很容易闯入人行道，把你压得粉身碎骨。当行人所处的人行道不但宽得足以使他们很好地避开汽车，而且也高得完全不可能让任何汽车偶然撞上他们时，他们就会感到舒适、安全、精神抖擞和行动自由了。这是不言而喻的。

墨西哥毕柯开利斯传统的高出路面的便道
Traditional raised walk in Pichucalis, Mexico

　　我们首先来考虑宽度。一条高出路面的便道究竟该多宽才算合适呢？当然，有名的例子就是爱丽舍宫广场，那里的人行道宽 30ft 有余，十分舒展。就我们的经验而言，与一条有汽车交通的商业街平行的人行道有它一半宽，仍然是舒展的。但是，人行道宽 12ft 或不足 12ft，行人就会感到狭窄，意识到汽车的威胁。一条常规的人行道的宽度往往不超过 6ft，行人会实实在在地感到汽车近在身旁。我们怎样才能提供使行人感到舒适安全的有特殊

when the walks they walk on are both wide enough to keep the people well away from the cars,and high enough to make it quite impossible for any car to drive up on them by accident.

We first consider the width.What is the appropriate width for a raised walk?The famous example,of course,is the Champs Elysées,where the sidewalk is more than 30 feet wide,and very comfortable.In our own experience,a walk of half this width,along a typical shopping street with traffic,is still comfortable;but 12 feet or less,and a pedestrian begins to feel cramped and threat-ened by cars.A conventional sidewalk is often no more than 6 feet wide;and people really feel the presence of the cars.How can we afford the extra width which people need in order to be comfortable?One way:instead of putting sidewalks along both sides of a road,we can put a double width raised walk along one side of the road only,with road crossings at intervals of 200 to 300 feet.This means,of course,that there can only be shops along one side of the road.

What is the right height for a raised walk?Our experiments suggest that pedestrians begin to feel secure when they are about 18 inches above the cars.There are a number of possible reasons for this finding.

One possible reason.When the car is down low and the pedestrin world physically higher,pedestrians feel, symbolically,that they are more important than the cars and therefore feel secure.

Another possible reason.It may be that the car overwhelms the pedestrian because of a constant,unspoken possibility that a runaway car might at any moment mount the curb and run him down.A car can climb an ordinary six-inch curb easily.For the pedestrian to feel certain that a car could not climb the curb,the

宽度的便道呢？一种途径是：不在公路的两侧皆设置人行道，而是只在其一侧设有人行道。我们可以使这种高出路面的便道的宽度加大一倍，并且在每隔 200 ～ 300ft 的路段上有人行横道。这意味着，只在公路的一侧才能开设商店。

一条高出路面的便道究竟高多少才算合理呢？我们的实验表明，当这种便道高出汽车行驶的路面约 18in 时，行人就开始有安全感。对此发现存在着几种可能的解释。

一种可能的解释：当人行道在空间上高出汽车时，行人会象征性地感到他们比汽车重要，因此他们会感到安全有了保障。

另一种可能的解释：或许是由于失控的汽车随时都会撞上路边的镶边石，并把行人撞倒这种可能性持续存在，汽车就对行人占绝对优势。汽车爬上一块 6in 高的镶边石十分容易，所以镶边石的高度必须大于汽车轮胎的半径（10 ～ 15in），行人才会真正感觉到汽车爬不上镶边石。

还有一种可能的解释：大多数人的水平视线的高度介于 51in 和 63in 之间。一辆典型的汽车的总高度为 55in。虽然身材高的人能俯视看清汽车，但对他们来说，汽车充塞着他们的视野，因为一个站着的人的正常视线低于水平面 10°（Henry Dreyfus, *The Measure of Man*, New York, 1958, sheet F）。若把一辆汽车停放在离行人 12in 的地方，要使汽车完全在他的视线以下，则该汽车所在的路面应低于行人 18 ～ 30in。

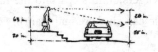

保持汽车处于行人的视线之下
Keep the cars below a person's line of sight

curb height would have to be greater than the radius of a car tire (10 to 15 inches).

Another possible reason.Most people's eye level is between 51 and 63 inches.A typical car has an overall height of 55 inches.Although tall people can see over cars,even for them,the cars fill the landscape since a standing person's normal line of sight is 10 degrees below the horizontal(Henry Dreyfus,*The Measure of Man*,New York,1958,sheet F).To put a car 12 feet away completely below a pedestrian's line of sight,it would have to be on a road some 18 to 30 inches below the pedestrian.

Therefore:

We conclude that any pedestrian path along a road carrying fast-moving cars should be about 18 inches above the road,with a low wall or railing,or balustrade along the edge,to mark the edge.Put the raised walk on only one side of the road—make it as wide as possible.

୫୬୯ଓ

Protect the raised walk from the road,by means of a low wall—SITTING WALL(243).An arcade built over the walk,will,with its columns,give an even greater sense of comfort and protection—ARCADES(119).At the end of blocks and at special points where a car might pull in to pick up or drop passengers,build steps into the raised walk,large enough so people can sit there and wait in comfort—STAIR SEATS(125)....

因此：

我们得出如下结论：**任何一条人行道在和高速汽车行驶的公路并行时，都要高出公路路面 18in，并在公路的沿边筑一低矮护墙或设置围栏或栏杆，以标记公路的边缘。只在公路一侧安排人行道，并使它成为尽可能宽阔的便道。**

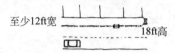

利用低矮护墙——**可坐矮墙**（243）来保护高出路面的便道。在便道上建筑带有立柱的拱廊，使行人有更多的安全感和舒适感——**拱廊**（119）。在街区的尽头或在过往旅客上下车的那些特殊的点上，修筑通往高出路面的便道的台阶，台阶要大，足以使行人能坐在那里舒舒服服地候车——**能坐的台阶**（125）……

模式56 自行车道和车架*

……在**地方交通区**（11）内，小型车辆如自行车、电动车等，也许还有马，它们高度集中，需要一种自行车道系统。自行车道将在促进建立地方交通区方面起到极其巨大的作用，也将有助于修改**区内弯曲的道路**（49）及**小路网络和汽车**（52）。

&&&&

自行车价廉物美，有益于人们的健康和环境的改善。可是环境的布局并不是为自行车而设计的。在公路上骑自行车受到汽车的威胁；在小路上骑自行车危及行人的生命安全。

56   BIKE PATHS AND RACKS*

...within a LOCAL TRANSPORT AREA(11)there is a heavy concentration of small vehicles like bikes,electric carts,perhaps even horses,which need a system of bike paths.The bike paths will play a very large role in helping to create the local transport areas,and may also help to modify LOOPED LOCAL ROADS(49) and NETWORK OF PATHS AND CARS(52).

৪০৫৪

Bikes are cheap,healthy,and good for the environment; but the environment is not designed for them.Bikes on roads are threatened by cars;bikes on paths threaten pedestrians.

In making the environment safe for bikes,the following problems must be solved:

1.Bikes are threatened where they meet or cross heavy automobile traffic.

2.They are also threatened by parked cars.Parked cars make it difficult for the bike rider to see other people,and they make it difficult for other people to see him.In addition,since the bike rider usually has to ride close to parked cars,he is always in danger of someone opening a car door in front of him.

3.Bikes endanger pedestrians along pedestrian paths;yet people often tend to ride bikes along pedestrian paths,not roads,because they are the shortest routes.

4.Where bikes are in heavy use,for instance around schools and universities,they can lay a pedestrian precinct to waste in their own way,just as cars can.

为了给自行车提供一种安全的环境，必须解决以下问题：

1. 自行车在与交通负荷较大的机动车流相遇或相交的地方会遭到危险。

2. 自行车同样遭到停放的汽车的危害。停放的汽车使骑自行车的人难以看清其他人，也使其他人难以看清骑自行车的人。此外，因为骑自行车的人往往不得不在停放的汽车中间贴近汽车而过，所以他总是受到正在他前面开车门的人的妨害。

3. 自行车危害人行道上行人的安全，可是骑自行车的人还总是乐于在人行道上而不是在公路上骑车，因为人行道是路程最短的路线。

4. 在大量使用自行车的地方，如在中小学和大学周围，正如汽车一样，自行车也要占去一块地方，虽然方式不同。

很明显，要解决这些问题就要创造一种完全独立的自行车道系统。可是，人们对这种解决办法是否可行，是否令人满意尚存疑虑。《车轮子上的学生》(Jany，Putney，and Ritter，Department of Landscape Architecture，University of Oregon，Eugene，Oregon，1972) 一文表明骑自行车的人和不骑自行车的人都希望有一种混合的系统，只要这种系统保证安全。

我们也认为，在街道上沿着人行道开辟自行车道是极为重要的。如果自行车被迫驶入一个分离的系统，那么这一系统几乎肯定会被那些抄近路通过其他网络的人所破坏。而且，那些明文规定自行车完全离开公路和小路系统的法律会使一直议论纷纷的骑自行车的人灰心丧气。无论什么地方，只要有可能，自行车道就应当与公路和主要人行道重合。

在自行车道与主干公路重合的地方，应和机动车道分隔开来。如果自行车道高出公路路面几英寸或由一排树木

An obvious solution to these problems is to create a completely independent system of bike paths.However,it is doubtful that this is a viable or desirable solution.The study *"Students on Wheels"* (Jany,Putney,and Ritter, Department of Landscape Architecture,University of Oregon,Eugene,Oregon,1972)shows that bike riders and nonbike riders want a mixed system,so long as it is reasonably safe.

We also think that it is essential for bike paths to run in streets and along pedestrian paths:if bikes are forced onto a separate system,it will almost certainly be violated by people taking shortcuts across the other networks.And laws which would keep bikes completely off the road and path systems would be discouraging to the already hasseled bike riders. Wherever possible,then,bike paths should coincide with roads and major pedestrian paths.

Where bike paths coincide with major roads,they must be separated from the roadway.It helps put the bike rider in a safer position with respect to the cars if the bike path is raised a few inches from the road;or separated by a row of trees.

Where bike paths run alongside local roads,parking should be removed from that side of the road;the bike surface may simply be part of the road and level with it.An article by Bascome in the Oregon Daily Emerald(October 1971)suggested that bike lanes along streets should always be on the sunny side of the street.

Where bike paths coincide with major pedestrian paths,they should be separate from the paths,perhaps a few inches below them.Here,the change in level gives the pedestrian a sense of safety from the bikes.

Quiet paths and certain pedestrian precincts should be completely protected from bikes for the same reasons that they

隔离，就人和汽车的关系来说，这一点会使骑自行车的人处于更加安全的位置。

在自行车道与地方公路平行的地方，停车场应当从公路的这一侧迁出。自行车道的路面简直就是公路路面的一部分，并处于同一水平面之内。贝斯科姆1971年10月在《俄勒冈绿宝石日报》上撰文说，沿街通行自行车的小巷始终应当位于街道的向阳一侧。

在自行车道与主要人行道重合的地方，应和小路分隔开来，或许要比小路低几英寸。这里路面水平位置的变化可避免自行车肇事，给行人以安全感。

安静的小路和一些行人区应当完全不受自行车干扰，同理也不应受汽车干扰。使自行车道绕过上述地区，或是用台阶和低矮护墙将这些地方围起来，从而迫使骑自行车的人推车而行，这样才能妥善处理这一问题。

因此：

建造一种专供自行车道用的小路系统，它具有下列特点：利用一种特殊的易识别的路面（例如红色沥青路面）醒目地标记出自行车道。自行车道要尽量与地方公路或主要人行道平行。自行车道与地方公路平行的地方，两者的路面处于同一水平面内——如果可能，自行车道应位于向阳的一侧。自行车道与人行道平行的地方，两者要分隔开，并且自行车道要低于人行道几英寸。将这种自行车道系统引入每幢楼房的 100ft 以内的范围，并在每幢楼房的主入口处附近放置一个自行车车架。

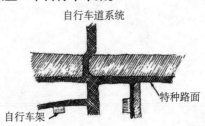

自行车道系统

特种路面

自行车架

need to be protected from cars.This can be handled by making the bike path system bypass these places,or by enclosing these places with steps or low walls which force bike riders to dismount and walk their bikes.

Therefore:

Build a system of paths designated as bike paths,with the following properties:the bike paths are marked clearly with a special,easily recognizable surface(for example,a red asphalt surface).As far as possible they run along local roads,or major pedestrian paths.Where a bike path runs along a local road,its surface may be level with the road—if possible,on the sunny side;where a bike path runs along a pedestrian path,keep it separate from that path and a few inches below it.Bring the system of bike paths to within 100 feet of every building,and give every building a bike rack near its main entrance.

ജ‍ൽ

Build the racks for bikes to one side of the main entrance,so that the bikes don't interfere with people's natural movement in and out—MAIN ENTRANCE(110),and give it some shelter,with the path from the racks to the entrance also under shelter—ARCADES(119);keep the bikes out of quiet walks and quiet gardens—QUIET BACKS (59),GARDEN WALL(173)....

    在主入口的一侧搁置自行车车架，以便自行车不妨碍行人出入的自然流动——**主入口**（110），给车架搭棚，并为从车架到入口的这一段小路也架设顶棚——**拱廊**（119）；不让自行车停放在安静的人行道和幽静的花园之内——**僻静区**（59）和**花园围墙**（173）……

模式57　市区内的儿童

57 CHILDREN IN THE CITY

...roads,bike paths,and main pedestrian paths
are given their position by PARALLEL ROADS(23),
PROMENADE(31),LOOPED LOCAL ROADS(49),GREEN
STREETS(51),NETWORK OF PATHS AND CARS(52),BIKE
PATHS AND RACKS(56).Some of them are safe for
children,others are less safe.Now,finally,to complete the paths
and roads,it is cssential to define at least one place,right in the
very heart of cities,where children can be completely free and
safe.If handled properly,this pattern can play a great role in
helping to create the NETWORK OF LEARNING(18).

❧❧❧

**If children are not able to explore the whole of the
adult world round about them,they cannot become adults.
But modern cities are so dangerous that children cannot be
allowed to explore them freely.**

The need for children to have access to the world of
adults is so obvious that it goes without saying.The adults
transmit their ethos and their way of life to children through
their actions,not through statements.Children learn by
doing and by copying.If the child's education is limited to
school and home,and all the vast undertakings of a modern
city are mysterious and inaccessible,it is impossible for
the child to find out what it really means to be an adult and
impossible,certainly,for him to copy it by doing.

This separation between the child's world and the adult
world is unknown among animals and unknown in traditional

……公路、自行车道和主人行道都要各得其所，但它们均由**平行路**（23）、**散步场所**（31）、**区内弯曲的道路**（49）、**绿茵街道**（51）、**小路网络和汽车**（52），以及**自行车道和车架**（56）等模式所制约。对儿童们来说，其中一些模式是安全的，另一些是欠安全的。最后，为了完善小路和公路，现在十分重要的是至少在市中心规定一个地方供儿童们安全地、自由自在地游玩。如果处理得当，本模式能在促成**学习网**（18）方面起到重要作用。

⊰⊱

如果儿童们不能去探索他们周围的整个成人世界，他们就不会成长为名副其实的成年人。但是，现代城市险象丛生，儿童们不会被允许去自由探索。

儿童们需要了解成人世界的心情，不言而喻，是非常明显的。成年人把他们的社会精神气质和生活方式通过行动而不是口口声声的说教潜移默化地传给他们。儿童们通过动手做和模仿而学习知识。如果儿童教育仅局限于学校和家庭，如果儿童们对于现代城市缺少广泛的理解，觉得它神秘莫测和难以接近，那么，他们就无法弄明白成为成年人的真正含义，当然，也无法通过动手做而去仿效他们了。

未成年和成年之间的这种脱节在动物界和传统的人类社会中是难以发现的。在简陋的村庄里，儿童们成天和农民在田野里待在一起，成天和建房的人凑热闹，事实上，和他们周围终日忙碌的男男女女形影不离：制作陶器、算账、医治病人、向上帝祈祷、磨面、争论本村未来的前景等。

但是，在城市里，生活是如此纷繁复杂和充满危险，所以没有人会把儿童独自留下，让他们到处游逛。快速

societies.In simple villages,children spend their days side by side with farmers in the fields,side by side with people who are building houses,side by side,in fact,with all the daily actions of the men and women round about them:making pottery,counting money,curing the sick,praying to God,grinding corn,arguing about the future of the village.

But in the city,life is so enormous and so dangerous,that children can't be left alone to roam around.There is constant danger from fast-moving cars and trucks,and dangerous machinery.There is a small but ominous danger of kidnap,or rape,or assault.And,for the smallest children,there is the simple danger of getting lost.A small child just doesn't know enough to find his way around a city.

The problem seems nearly insoluble.But we believe it can be at least partly solved by enlarging those parts of cities where small children can be left to roam,alone,and by trying to make sure that these protected children's belts are so widespread and so far—reaching that they touch the full variety of adult activities and ways of life.

We imagine a carefully developed childrens'bicycle path,with in the larger network of bike paths.The path goes past and through interesting parts of the city;and it is relatively safe. It is part of the overall system and therefore used by everyone.It is not a special children's "ride" —which would immediately be shunned by the adventurous young—but it does have a special name,and perhaps it is specially colored.

The path is always a bike path;it never runs beside cars.Where it crosses traffic there are lights or bridges.There are many homes and shops along the path—adults are nearby, especially the old enjoy spending an hour a day sitting along this path,themselves riding along the loop,watching the kids out of the corner of one eye.

奔驰的轿车和卡车常常酿成车祸，危险的机器也频频肇事。此外，还有绑架、强奸或暴力袭击等不祥之灾。年幼的稚童十分容易迷路，他们还认不清楚城里的大街小巷。

看来，这一问题非常棘手，难以解决。但我们认为，这一问题至少可以部分地得到解决，办法如下：扩大市内可以让儿童们单独留下游逛的地方，并竭力保证这些受到保护的儿童活动地带展宽和延伸，以便儿童们能接触到成年人丰富多彩的活动和生活方式。

我们构思出一种经过审慎研究的儿童自行车道，它设置在较大的自行车道网络之内。这种儿童自行车道可以穿过市内非常有趣的地方，并且是相当安全的。这是总系统的一部分，人人都可使用。这并不是一种儿童专用"车道"——爱冒险的青年人应当毫不迟疑地回避它——但它要有一个特殊的名称，也许还要涂上特别的颜色。

你准在开玩笑
You must be kidding

这种儿童自行车道始终是自行车道；它绝不在汽车旁边通过。在它穿越公路的地方，设置信号灯或桥形物。道旁林立着许多住宅和商店——成年人就在附近，尤其是老人们会乐意一天一小时坐在路边，或是自己沿着弯曲的路骑自行车，眯着眼看着儿童们。

而最重要的是，这种儿童自行车道的妙处在于：沿着它可穿过城镇的一些公共设施，一直延伸到通常难以到达的一些地方——报纸印刷所、牛奶瓶装厂、码头、制作汽车门窗的汽车间、餐厅后面的小巷和公墓等。

And most important,the great beauty of this path is that it passes along and even through those functions and parts of a town which are normally out of reach:the place where newspapers are printed,the place where milk arrives from the countryside and is bottled,the pier,the garage where people make doors and windows,the alley behind restaurant row,the cemetery.

Therefore:

As part of the network of bike paths,develop one system of paths that is extra safe——entirely separate from automobiles, with lights and bridges at the crossings,with homes and shops along it,so that there are always many eyes on the path.Let this path go through every neighborhood,so that children can get onto it without crossing a main road. And run the path all through the city,down pedestrian streets,through workshops,assembly plants,warehouses,interchanges,print houses,bakeries,all the interesting "invisible" life of a town——so that the children can roam freely on their bikes and trikes.

ଓଓଔ

Line the children's path with windows,especially from rooms that are in frequent use,so that the eyes upon the street make it safe for the children—STREET WINDOWS(164);make it touch the children's places all along the path—CONNECTED PLAY(68),ADVENTURE PLAYGROUND(73),SHOPFRONT SCHOOLS(85),CHILDREN'S HOME (86),but also make it touch other phases of the life cycle—OLD PEOPLE EVERYWHERE(40),WORK COMMUNITY (41),UNIVERSITY AS A MARKETPLACE (43),GRAVE SITES(70),LOCAL SPORTS(72), ANIMALS (74), TEENAGE SOCIETY(84)....

因此：

把一种小路系统作为自行车道网络的一部分来加以发展，它是非常安全的，完全和汽车分隔的，并在十字路口设置交通信号灯和桥形物，沿途住宅和商店林立，所以，始终有许多人注视这条小路——儿童自行车道。让它通过每一邻里，以便儿童们不用横穿主干公路就能到达邻里。让它通过全市，通过人行街、工厂、装配厂、货栈、换乘站、印刷所、面包烘房、通过一切有趣的"看不见的"城镇生活——最终儿童们能骑着自行车和三轮脚踏车自由自在地漫游了。

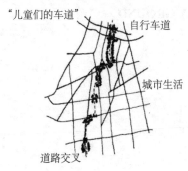

"儿童们的车道"　　自行车道

城市生活

道路交叉

☙❧

将儿童自行车道与住宅的窗户保持一致，尤其与那些频繁使用的房间的窗户保持一致，以便人们注视街道而使儿童们感到安全——**临街窗户**（164）；使儿童自行车道通往沿途儿童们喜欢的地方——**相互沟通的游戏场所**（68）、**冒险性的游戏场地**（73）、**店面小学**（85）、**儿童之家**（86）；而且也要使它通往生活圈的其他地方——**老人天地**（40）、**工作社区**（41）、**像市场一样开放的大学**（43）、**墓地**（70）、**地方性运动场地**（72）、**动物**（74）和**青少年协会**（84）……

社区和邻里要有宽阔的公共土地，以便居民能在那里休息，相互交际，消除疲劳，恢复精力和体力：

in the communities and neighborhoods provide public open land where people can relax,rub shoulders and renew themselves;

58.CARNIVAL

59.QUIET BACKS

60.ACCESSIBLE GREEN

61.SMALL PUBLIC SQUARES

62.HIGH PLACES

63.DANCING IN THE STREET

64.POOLS AND STREAMS

65.BIRTH PLACES

66.HOLY GROUND

58. 狂欢节

59. 僻静区

60. 近宅绿地

61. 小广场

62. 眺远高地

63. 街头舞会

64. 水池和小溪

65. 分娩场所

66. 圣地

模式58　狂欢节

　　……偶尔，在频繁举行狂欢节活动的亚文化区内，一个散步场所可能迸发出一种比较狂热的生活节奏——**散步场所**（31）和**夜生活**（33）。也许每一散步场所都具有这一特征。

〰〰〰

　　正如一个人想入非非才能迸发出在正常情况下迸发不出来的内在力量一样，**城市也需要异想天开。**

　　在今天的世界上，正常情况下，人们可以进行的娱乐活动要么是健康的和无害的——观看电影、收看电视、骑车、打网球、乘坐直升机、散步溜达，看足球赛，要么完全是病态的和伤风败俗的——吸毒、横冲直撞开车、结伙打架斗殴等。

58   CARNIVAL

...once in a while,in a subculture which is particularly open to it,a promenade may break into a wilder rhythm—PROMENADE(31),NIGHT LIFE(33)and perhaps every promenade may have a touch of this.

❧❧❧

Just as an individual person dreams fantastic happenings to release the inner forces which cannot be encompassed by ordinary events,so too a city needs its dreams.

Under normal circumstances,in today's world the entertainments which are available are either healthy and harmless—going to the movies,watching TV,cycling,playing tennis,taking helicopter rides,going for walks,watching football—or downright sick and socially destructive—shooting heroin,driving recklessly,group violence.

But man has a great need for mad,subconscious processes tocome into play,without unleashing them to such an extent that they become socially destructive.There is,in short,a need for socially sanctioned activities which are the social,outward equivalents of dreaming.

In primitive societies this kind of process was provided by the rites,witch doctors,shamans.In Western civilization during the last three or four hundred years,the closest available source of this outward acknowledgment of underground life has been the circus,fairs,and carnivals.In the middle ages,the market place itself had a good deal of

但是，人极其需要疯狂的、下意识的活动过程，不过也不能让他随心所欲，闹到破坏社会治安的严重地步。总而言之，需要有社会认可的娱乐活动，这些活动具有社会意义，是人们梦寐以求的感情外露的具体表现。

　　在原始社会中这类活动就是宗教仪式，巫医、巫师主持的各种宗教活动。在最近三四百年间的西方文明中，个人内心感情允许外露的最近的地方就是马戏团竞技场、集市和狂欢节了。在中世纪，市场本身就充满着这种感情外露的浓厚气氛。

　　今天，总体看来，这种经历正在成为过去。马戏团和狂欢节日趋衰落，但人们并未放弃这种需要。在海湾地区，一年一度的文艺复兴博览会在满足这种需要方面迈出了一小步。但它太平淡了，差强人意。我们根据下面一系列相关联的人和事物，想象出一些新名堂、新花样：在街头巷尾、在广场或室内开展各种娱乐活动，有露天剧场，有小丑粉墨登场，有使人着迷的游戏；人们可以一连几周欢度狂欢节；简陋的住所和粗茶淡饭一律免费供应；男女老少不分白天黑夜地厮混在一起；演员们混杂在人群里，无论你愿意不愿意，都会把你卷入人流；这种活动过程的结局如何难以预见：或会打架——两个汉子，手执口袋，踩在一根滑溜溜的圆木上，在大庭广众之前大打出手；或变成菲利尼（意大利著名电影导演——译注）的影中人——小丑、死神和一群疯子乱作一团。

　　请回忆一下影片《愚人之船》中的一个驼背的矮子吧，他是这条船上唯一有理智的人，他说："人人都有一本难念的经，但我交了好运气，我把它背在背上，好让大家都看见它。"

　　因此：

　　城镇要留出一部分空地供狂欢节用。狂欢节上有惹人拍手叫绝的杂耍、马上比武、精彩演出、商品展销、体育竞赛、音乐舞蹈、露天剧场，还有形形色色的小丑和异装癖者，一连串的表达内心欣喜若狂的任性举动。让宽阔的步行街迂回曲折地通过此地；沿大街小巷搭起有帐篷的货摊；在散步场所的一端

this kind of atmosphere.

Today,on the whole,this kind of experience is gone.The circuses and the carnivals are drying up.But the need persists.In the Bay Area,the annual Renaissance Fair goes a little way to meet the need—but it is much too bland.We imagine something more along the following lines:street theater,clowns,mad games in the streets and squares and houses;during certain weeks,people may live in the carnival;simple food and shelter are free;day and night people mixing;actors who mingle with the crowd and involve you,willy hilly,in processes whose end cannot be foreseen;fighting—two men with bags on a slippery log,in front of hundreds;Fellini—clowns,death,crazy people,brought into mesh.

Remember the hunchbacked dwarf in Ship of Fools,the only reasonable person on the ship,who says "Everyone has a problem;but I have the good fortune to wear mine on my back,where everyone can see it."

Therefore:

Set aside some part of the town as a carnival—mad sideshows,tournaments,acts,displays,competitions,dancing, music,street theater,clowns,transvestites,freak events, which allow people to reveal their madness;weave a wide pedestrian street through this area;run booths along the street,narrow alleys;at one end an outdoor theater; perhaps connect the theater stage directly to the carnival street,so the two spill into and feed one another.

设一露天剧场；也许把露天剧场的舞台直接和举行狂欢节的街道连接沟通，这样就可使两股人流涌入涌出，川流不息。

露天剧场　　有篷货摊
疯狂的竞技　　跳舞

🙰🙲

　　街头舞会、饮食商亭、一两个户外亭榭、一个有露天剧场的广场，以及帐篷和帆布——所有这一切都将有助于增加狂欢节的无比生动活泼和欢快的节日气氛——**小广场**（61）、**街头舞会**（63）、**户外亭榭**（69）、**饮食商亭**（93）、**步行街**（100）、**帆布顶棚**（244）……

Dancing in the street,food stands,an outdoor room or two,a square where the theater is,and tents and canvas will all help to make it even livelier—SMALL PUBLIC SQUARES(61),DANCING IN THE STREET(63),PUBLIC OUTDOOR ROOM(69),FOOD STANDS(93),PEDESTRIAN STREET(100),CANVAS ROOFS(244)...

模式59 僻静区*

……分散的工作点（9）决定工作点的一般地位，而工作社区（41）却决定工作点的详细组织和分布情况。可是，重要的是：某种僻静区支持并补充工作点。本模式和下面接连几个模式都将从不同角度说明僻静区的结构。

⊱⊰

凡是不得不在噪声中和人很多的办公室内工作的人，都需要有时暂停工作，在天然成趣的幽静环境里恢复精力。

59   QUIET BACKS*

...the work places are given their general position by
SCATTERED WORK(9)and their detailed organization and
distribution by WORK COMMUNITIES(41).It is essential
though,that they be supported by some kind of quiet,which
is complementary to the work.This pattern,and the next few
patterns,gives the structure of that quiet.

⅏

**Any one who has to work in noise,in offices with people
all around,needs to be able to pause and refresh himself
with quiet in a more natural situation.**

The walk along the Seine,through the middle of Paris,is a
classic "quiet back" in the middle of a fast city.People drop
down from the streets and the traffic and the commerce to stroll
along the river,where the mood is slow and reflective.

The need for such places has often been recognized in
universities,where there are quiet walks where people go to
think,or pause,or have a private talk.A beautiful case is the
University of Cambridge:each college has its "backs" —
quiet courts stretching down to the River Cam.But the need
for quiet backs goes far beyond the university.It exists
everywhere where people work in densely populated,noisy
areas.

To meet this need,we may conceive all buildings as having
a front and a back.If the front is given over to the street life—
cars,shopping paths,delivery—then the back can be reserved

一条沿着塞纳河畔通往巴黎市中心的人行道，就是这个高速度快节奏的城市中央的古典式的"僻静区"了。人们从街道、公路和商业区络绎不绝地来到塞纳河畔漫步徜徉，心旷神怡，浮想联翩。

现在，人们已经认识到大学应当有这样僻静的地方，寂静的小径是人们思考问题、休息或私下谈话的好去处。剑桥大学就是一例，校园风光秀丽：每个学院都有自己的"僻静区"——静悄悄的庭院一直延伸到剑桥河边。但需要僻静区的远不止于大学。凡在人声嘈杂的稠密区工作的人都有此需要。

为了满足这种需要，我们可以把所有的房屋都设想成有前院和后院。如果前院是服务于街道生活的，可停放汽车，开辟购物小路，投递货物；则后院就可用作僻静区了。

如果后院十分幽静，你就只能听见自然之音——风声、鸟鸣声、流水声——要珍惜并妥善保护。同时，僻静区必须离开它所服务的房屋一段距离。这就是说，在房屋后面一定范围之内有一条小道或私人花园，四周有坚固的围墙和茂密的植物防护。

就我们所知，一条通过吉彻斯特附近大教堂的人行道就是一个实例。在这条道路的两侧都有高大的砖墙，沿途长着各种花木。它以大教堂为起点，和城镇的主要公路平行，但相隔一段距离。在这条道上，距城镇主要十字路口不到一个街区的地方，你竟能听到蜜蜂嗡嗡的叫声。

如果这样一些人行道相互连接起来，那么就会在嘈杂混乱的街道的背后逐步形成一个带状的小僻静区系统，一个令人愉快的小胡同系统。在建造这种不可缺少的僻静区时，潺潺的流水声作用甚大，所以，这些小路始终应当与地方的**水池和小溪**（64）相连。僻静区越长越好。

因此：

在城镇的繁华地区，在房屋背后留块空地作为隔离噪

for quiet.

If the back is to be quiet,a place where you can hear only natural sounds—winds,birds,water—it is critical that it be protected.At the same time,it must be some way from the buildings which it serves.This suggests a walk,some distance behind the buildings,perhaps separated from them by their private small gardens,completely protected by substantial walls and dense planting along its length.

An example we know is the walk through the cathedral close in Chichester.There is a high brick wall on each side of this walk and flowers planted all along it.It leads away from the cathedral,parallel but set back from the town's major road. On this path,less than a block from the major crossroads of the town,you can hear the bees buzzing.

If a number of these walks are connected,one to another,then slowly,there emerges a ribbon—like system of tiny backs,pleasant alleyways behind the commotion of the street.Since the sound of water plays such a powerful role in establishing the kind of quiet that is required,these paths should always connect up with the local POOLS AND STREAMS(64). And the longer it can be,the better.

Therefore:

Give the buildings in the busy parts of town a quiet "back" behind them and away from the noise.Build a walk along this quiet back,far enough from the building so that it gets full sunlight,but protected from noise by walls and distance and buildings.Make certain that the path is not a natural shortcut for busy foot traffic,and connect it up with other walks,to form a long ribbon of quiet alleyways which converge on the local pools and streams and the local greens.

声的"僻静区"。在僻静区修一条人行道，与房屋保持相当距离，以便有充足的阳光照射，僻静区的四周利用围墙、建筑物和间距来防止噪声。确保小路不要成为频繁的步行交通的天然捷径，并将小路和其他人行道连接起来，以便形成长长的带状的安静的小胡同网，这些小胡同都簇集在地方的水池和河流边，以及地方绿化区内。

如果可能的话，把僻静区设在有水域的地方——**水池和小溪**（64）、**池塘**（71），也可以设在未遭交通破坏的大片树林里——**林荫空间**（171）；将僻静区和**近宅绿地**（60）连成一片；利用围墙或建筑物阻隔噪声——**花园围墙**（173）……

∞∞∞

If possible,place the backs where there is water—
POOLS AND STREAMS(64),STILL WATER(71),and
where there are still great trees unharmed by traffic—
TREE PLACES(171);connect them to ACCESSIBLE
GREENS(60);and protect them from noise with walls or
buildings—GARDEN WALL(173)...

模式60　近宅绿地**

……在邻里中心及靠近所有工作社区的地方，都需要有小块绿地——**易识别的邻里**（14）和**工作社区**（41）。当然，最重要的是：小块绿地要选址合理，有助于形成边界、邻里和僻静区——**亚文化区边界**（13）、**邻里边界**（15）和**僻静区**（59）……

❀❀❀

人们需要到开阔的绿地里去，他们居住在绿地附近时就会常去。但如果他们离绿地有 3min 以上的路程，就不会去了。这就是距离压倒需要。

60 ACCESSIBLE GREEN**

...at the heart of neighborhoods,and near all work communities,there need to be small greens— IDENTIFIABLE NEIGHBORHOOD(14),WORK COMMUNITY(41). Of course it makes the most sense to locate these greens in such a way that they help form the boundaries and neighborhoods and backs— SUBCULTURE BOUNDARY(13), NEIGHBORHOOD BOUNDARY(15), QUIET BACKS(59).

୫୦୪

People need green open places to go to;when they are close they use them.But if the greens are more than three minutes away,the distance overwhelms the need.

Parks are meant to satisfy this need.But parks,as they are usually understood,are rather large and widely spread through the city.Very few people live within three minutes of a park.

Our research suggests that even though the need for parks is very important,and even though it is vital for people to be able to nourish themselves by going to walk,and run,and play on open greens,this need is very delicate.The only people who make full,daily use of parks are those who live less than three minutes from them.The other people in a city who live more than 3 minutes away,don't need parks any less;but distance discourages use and so they are unable to nourish themselves,as they need to do.

公园本来是满足这种需要的。但是，按照一般的理解，公园是相当大的，并且疏散地分布于城市各处。很少有人住在离公园 3min 的路程之内。

我们的研究表明，即使人们迫切需要公园，即使公园在丰富他们的生活方面——如散步、跑步和在开阔的绿地里做游戏等——举足轻重，但这种需要是难以满足的。只有住在公园附近、走两三分钟就能到达的人才能天天充分利用公园。市内其他居民，同样强烈需要公园，但因他们的住处离公园超过 3min 的路程，也就不能如愿以偿了。

只有数量充足的、数以百计的小公园——即小块绿地——疏散地分布于全市，使每家每户和每一工作地点离最近的小公园只有 3min 的路程，这一问题才能解决。

说得再详细一点：市内需要公园是一致公认的。这一认识的典型例子是伯克利城市规划局根据 1971 年对市民进行有关开阔空间的调查所得出的结果而提出的。调查结果表明，住在公寓楼内的绝大多数居民最迫切希望有两类户外空间：(a) 惬意舒适的、可充分利用的私人阳台和 (b) 可以步行到达的安静的公园。

但是，人们不是很清楚也不是很理解公园的利用率受距离远近的影响十分严重。为了彻底搞明白这一问题，我们曾采访了伯克利的一个小公园，并征求了 22 名游客的意见，问及他们是否常来，要走多少路。我们向每一位游客还专门提出了下面 3 个问题：

a. 你是驾驶汽车来的还是步行来的？

b. 你到这里要经过多少个街区？

c. 你上次来公园是哪一天？

就第一个问题的回答来看，有 5 人是开汽车或骑自行车来的。第三个问题是测算每个人一周来公园的次数。例

This problem can only be solved if hundreds of small parks—or greens—are scattered so widely,and so profusely,that every house and every workplace in the city is within three minutes walk of the nearest one.

In more detail:The need for parks within a city is well recognized.A typical example of this awareness is given by the results of a 1971 citizen survey on open space conducted by the Berkeley City Planning Department.The survey showed that the great majority of people living in apartments want two kinds of outdoor spaces above all others:(a)a pleasant,usable private balcony and(b)a quiet public park within walking distance.

But the critical effect of distance on the usefulness of such parks is less well known and understood.In order to study this problem,we visited a small park in Berkeley,and asked 22 people who were in the park how often they came there,and how far they had walked to the park.Specifically,we asked each person three questions:

a.Did you walk or drive?

b.How many blocks have you come?

c.How many days ago did you last visit the park?

On the basis of the first question we rejected five subjects who had come by car or bike.The third question gave for each person a measure of the number of times per week that person comes to the park.For example,if he last came three days ago,we may estimate that he typically comes once per week. This is more reliable than asking the frequency directly,since it relies on a fact which the person is sure of,not on his judgment of a rather intangible frequency.

We now construct a table showing the results.In the first column,we write the number of blocks people walked to get to the

如，如果他上次是 3 天前来的，我们就可以估计出，他就是一周来公园一次的典型。这比直接询问一周几次更为可信，因为这是依赖于他所确认的事实，而不是依赖于他想当然说出的次数。

我们现已制出一份表格，说明调查的结果。在第一栏中，我们列出步行去公园的人所经过的街区数。在第二栏中，我们列出在该距离内的环形区面积的测算值。该环形区面积和两数的平方之差成正比。例如，3 个街区的环形面积测算值为 $3^2-2^2=5$。

表格列举如下：

采访局部绿地模式的分析

半径R 街区数	半径为R的环形 面积测算值	每周游园 总次数	P（每人相对游园概率）	Log P
1	1	19.5	19.5	1.29
2	3	26	8.7	.94
3	5	11	2.2	.34
4	7	6	0.9	T.95
5	9	0	—	—
6	11	0	—	—
7	13	0	—	—
8	15	6	0.4	T.60
9	17	0	—	—
10	19	3	0.2	T.30
11	21	0	—	—
12	23	2.5	0.1	T.0

在第三栏中我们列出经过那段距离去游园的人数，再乘上每人每周去公园的次数。这就为我们提供了那个环形区内每周游园的总数测算值。

在第四栏中我们列出每周游园数除以环形区的面积。如果我们假设居民以大致相等的密度分布于这一整个环状区内，这就为我们提供了任何一个人在某一环形区内，在某一周之内将去游园的概率测算值。

park.In the second column we write a measure of the area of the ring-shaped zone which lies at that distance.The area of this ring-shaped zone is proportional to the difference of two squares.For example,the measure of area of the ring at three blocks,is $3^2-2^2=5$.

Analysis of visiting pattern to a local green

Radius R Blocks	Measure of area of the ring at Radius R	Trips/week	P(Relative probability of trips,for any one person)	Log P
1	1	19.5	19.5	1.29
2	3	26	8.7	.94
3	5	11	2.2	.34
4	7	6	0.9	T.95
5	9	0	—	—
6	11	0	—	—
7	13	0	—	—
8	15	6	0.4	T.60
9	17	0	—	—
10	19	3	0.2	T.30
11	21	0	—	—
12	23	2.5	0.1	T.0

In the third column,we write the number of people who have come from that distance,each person multiplied by the number of trips per week he makes to the park.This gives us a measure of the total number of trips per week,which originate in that ring.

In the fourth column we write the number of trips per week divided by the area of the ring.If we assume that people are distributed throughout the entire area at approximately even density,this gives us a measure of the probability that any one person,in a given ring,will make a trip to the park in a given week.

In the fifth column we write the logarithm(base 10)of this probability measure P.

在第五栏中我们列出这一概率测算值 P 的对数（底数为 10）。

经简单核查这些数据表明，当概率测算值 P 在一个街区和两个街区之间下降一半时，在两个街区和 3 个街区之间就下降到 1/4。它的递减率从此往后就逐渐缩小。这表明，任何人只要居住在 3 个街区外的地方，他个人使用公园的性质就会发生根本变化。

为了更加精确起见，让我们来检验距离和 Log$P$ 之间的依赖关系。在正常情况下，去某一中心的次数将随某一距离衰减函数而变化，如 $P=Ae^{-Br}$，式中 A 和 B 为常数，r 是半径。这意味着，如果行为和动机相对于距离是常数，我们就能画出一幅 Log$P$ 和半径的关系图，我们本应得到一条直线。该直线的任何偏移都将向我们表明，一种行为和动机的阈值会改变成另一种行为和动机的阈值。这一关系图表示如下：

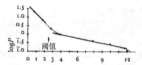

离绿化区的距离，街区数在两个或三个街区以外的住户对绿地的使用率急剧下降

Beyond two or three blocks use of the green drops off drastically

我们看到，所得出的曲线呈 S 形。它开始以一定的角度向下斜，然后更陡，再往后渐渐又趋平缓。很明显，在两个和 3 个街区之间的某一处存在着一个阈值，在这阈值内居民的行为和动机变化剧烈。

紧靠绿地居住的那些人遵循极频繁使用的原则——该曲线呈现出一个陡的梯度，并对增大距离十分敏感。但是，居住在远离绿地的那些人看来遵循很少使用的原则（该线段呈现出一个比较平缓的梯度），而他们的行为对距离并不那样敏感。好像那些准备去绿地的人对距离远近满不在乎。

Simple inspection of these data shows that while the probability measure,P,drops in half between one and two blocks,it drops by a factor of four between two and three blocks.Its rate of decrease diminishes from then on.This indicates that an individual's use of a park changes character radically if he lives more than three blocks away.

For more precision let us examine the relationship between distance and the logarithm of P.Under normal circumstances,the frequency of access to a given center will vary according to some distance decay function,such as $P=Ae^{-Br}$,where A and B are constants,and r is the radius.This means that if behavior and motivation are constant with respect to distance,and we plot the log of P against the radius,we should get a straight line. Any aberration from the straight line will show us the threshold where one kind of behavior and motivation changes to another. This plot is shown below:

We see that the resulting curve is S-shaped.It starts going down at a certain angle,then gets much steeper,and then flattens out again.Apparently there is a threshold somewhere between 2 and 3 blocks,where people's behavior and motivation change drastically.

Those people who live in close proximity to a green follow a high intensity use function—it has a steep gradient and it is very sensitive to increasing distance.But those people who live far from a green appear to adopt a low intensity use function(indicated by a shallower gradient),and their behavior is not as sensitive to distance.It is as if those people with ready access to a green display a full,free responsiveness to it;while people far away have lost their awareness of it and have suffered a reduced sensitivity to the pleasures of the green—for

而更远的居民已经不再考虑距离的远近，他们对绿地给人以愉快的感受已淡薄了——对这些人来说，绿地不再是他们的邻里生活中须臾不可缺少的重要因素了。

很明显，在 2～3 个街区的半径范围内（3min 的步行距离），人们能满足自己去绿地的需要，但是超过这一距离，他们就难以满足这种需要了。

这种距离的干扰是难以预料的。我们知道，住在绿地附近的人相当频繁地去绿地，据推测，他们需要松弛筋骨，休息养神。居住在离绿地超过 3min 步行路程的人同样有此需要。但是，就他们的情况而言，距离妨碍他们满足自己的需要。这么看来，每一个人、每一幢住宅和每一工作场所都必须位于 3min 路程到达公园的范围之内。

还有一个悬而未决的问题。为了满足这种需要，一块绿地究竟应该有多大呢？根据功能，这一问题是极易回答的。它的规模必须大到至少在它的中心，足以使你感到离开了拥挤喧闹之地来到了大自然的怀抱。据我们现在的估计，为了满足这种需要，一块绿地的面积应为 60000ft²，窄边的宽度至少为 150ft。

因此：

建造一个开阔的公共绿地，使其位于约 750ft 即 3min 步行路程内的每家每户和每一工作地点的市民都能到达。这意味着绿地应按 1500ft 的间距均匀地分散于全城。绿地的直径至少为 150ft，面积至少为 60000ft²。

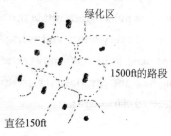

绿化区

1500ft的路段

直径150ft

thees people,the green has ceased to be a vital element in their neighborhood life.

Apparently,within a two to three block radius(a three-minute walking distance)people are able to satisfy their need for access to a green,but a greater distance seriously interferes with their ability to meet this need.

This inference is rather unexpected.We know that people who are close to a green go to it fairly often, presumably because they need the relaxation.The people who live more than three minutes walk from the green also need the relaxation,presumably.But in their case the distance prevents them from meeting their need.It seems then,that to meet this need,everyone—and that means every house and every workplace—must be within three minutes of such a park.

One question remains.How large must a green be in order to satisfy this need?In functional terms this is easy to answer.It must be large enough so that,at least in the middle of it,you feel that you are in touch with nature,and away from the hustle and bustle.Our current estimates suggest that a green should be as much as 60,000 square feet in area,and at least 150 feet wide in the narrowest direction in order to meet this requirement.

Therefore:

Build one open public green within three minutes'walk— about 750 feet—of every house and workplace.This means that the greens need to be uniformly scattered at 1500-foot intervals,throughout the city.Make the greens at least 150 feet across,and at least 60,000 square feet in area.

∞∞∞

对古树要倍加爱护，精心照料——**树荫空间**（171）；
要使绿地形成一个或一个以上的围合的户外正空间，四
周有树木、围墙或建筑物环绕，但无公路或汽车——**户
外正空间**（106）、**花园围墙**（173）；也许要划出绿地的
一部分供社区特殊需要之用——**圣地**（66）、**墓地**（70）、
地方性运动场地（72）、**动物**（74）和**在公共场所打盹**
（94）……

Pay special attention to old trees,look after them—TREE PLACES(171);shape the green so that it forms one or more positive room—like spaces and surround it with trees,or walls,or buildings,but not roads or cars—POSITIVE OUTDOOR SPACE(106),GARDEN WALL (173);and perhaps set aside some part of the green for special community functions—HOLY GROUND(66),GRAVE SITES(70),LOCAL SPORTS(72),ANIMALS(74),SLEEPING IN PUBLIC(94)...

模式61 小广场**

……本模式形成**活动中心**（30）的内核：本模式也有助于形成小路交叉口的活动中心。如果该活动中心选址合理，仅仅因其存在，居民就会十分频繁地使用它。本模式还有助于形成**散步场所**（31）、**工作社区**（41）和**易识别的邻里**（14），通过各种活动人们聚集在小广场上。但重要的是：在任何情况下，小广场不能太大。

❀❀❀

城镇需要广场。广场是城镇所拥有的最大的公共空间。但当广场太大时，人们会感到空空荡荡。

很自然，在每一公共街道活动最频繁的那些重要的活动中心都有一块扩大的地方。正是这样一些扩大的地方逐

61 SMALL PUBLIC SQUARES**

...this pattern forms the core which makes an ACTIVITY NODE(30):it can also help to generate a node,by its mere existence,provided that it is correctly placed along the intersection of the paths which people use most often.And it can also help to generate a PROMENADE(31),a WORK COMMUNITY(41),an IDENTIFIABLE NEIGHBORHOOD (14),through the action of the people who gather there.But it is essential,in every case,that it is not too large.

৪৩৫৪

A town needs public squares;they are the largest,most public rooms,that the town has.But when they are too large,they look and feel deserted.

It is natural that every public street will swell out at those important nodes where there is the most activity.And it is only these widened,swollen,public squares which can accommodate the public gatherings,small crowds,festivities,bonfires,carnivals,speeches,dancing,shouting,mourning,which must have their place in the life of the town.

But for some reason there is a temptation to make these public squares too large.Time and again in modern cities,architects and planners build plazas that are too large. They look good on drawings;but in real life they end up desolate and dead.

Our observations suggest strongly that open places

步变成广场，从而使许多活动才得以在广场上进行，如举行公共集会、狂欢节、篝火晚会、化妆游行、三五成群的人扎堆聊天、发表演说、跳舞、大声喧闹、举行悼念活动等，这些活动在城镇的生活中具有十分重要的意义。

但是，由于某种原因，目前存在着一种一味扩大广场的不良倾向。在现代社会中，建筑师和规划师一再建造规模过大的广场。这些广场在图纸上看来很出色，但在现实生活中效果不佳：显得荒凉而又死气沉沉。

我们的观察有力地表明，有意用作广场的开阔空间应当是很小的。一般说来，我们已经发现，广场的直径为60ft左右时，就算物尽其用了。因为这样规模的广场大家都愿常去，结果成为令人向往的活动场所。广场的直径超过70ft，就开始呈现出荒凉和令人不悦的气氛。就我们所知，圣马可广场和特拉法尔格广场算是例外，它们位于大城镇中心，游人拥挤，热闹非凡。

这些观察可能具有何种为实用而设计的基础呢？首先，我们根据模式**行人密度**（123）认识到，当一片空地的面积超过每人 $300ft^2$ 时，人们就会有荒凉之感。

利马的两个广场：一个小而生气勃勃，另一个大而荒凉寂寞
The squares in lima:one small and dive,the other huge and deserted

据此，一个直径为100ft的广场，如果游人少于33名，就会显得空空荡荡。市内几乎没有什么地方你可确保总有

intended as public squares should be very small.As a general rule,we have found that they work best when they have a diameter of about 60 feet—at this diameter people often go to them,they become favorite places,and people feel comfortable there.When the diameter gets above 70 feet,the squares begin to seem deserted and unpleasant.The only exceptions we know are places like the Piazza San Marco and Trafalgar Square,which are great town centers,teeming with people.

What possible functional basis is there for these observations? First,we know from the pattern,PEDESTRIAN DENSITY(123),that a place begins to seem deserted when it has more than about 300 square feet per person.

On this basis a square with a diameter of 100 feet will begin to seem deserted if there are less than 33 people in it.There are few places in a city where you can be sure there will always be 33 people.On the other hand,it only takes 4 people to give life to a square with a diameter of 35 feet,and only 12 to give life to a square with a diameter of 60 feet.Since there are far far better chances of 4 or 12 people being in a certain place than 33,the smaller squares will feel comfortable for a far greater percentage of the time.

The second possible basis for our observations depends on the diameter.A person's face is just recognizable at about 70 feet;and under typical urban noise conditions,a loud voice can just barely be heard across 70 feet.This may mean that people feel half-consciously tied together in plazas that have diameters of 70 feet or less—where they can make out the faces and half-hear the talk of the people around them;and this feeling of being at one with a loosely knit

33 名游人。另一方面，只要有 4 人就可使一个直径为 35ft 的广场生气勃勃，只要有 12 人就可使一个直径为 60ft 的广场获得相同效果。任何地方 4 名或 12 名游人去广场的机会总是会大大超过 33 名游人的机会，所以广场越小，感到心情舒畅的游人的比例就越大。

我们观察所依据的可能的第二个基础是直径的大小。一个人的脸在 70ft 左右的距离内才能辨认清楚；在一个典型的闹市区，一个洪亮的声音在直径为 70ft 的范围内才能听到。这兴许意味着，在直径为 70ft 或小于 70ft 的广场内人们会部分意识地感觉到大家联系在一起，他们能相互认出面孔，并能部分听清别人的谈话；而这种感觉在联系松散的更大的广场内是绝不会有的。菲利普·蒂埃尔（"An Architectural and Urban Space Sequence Notation". unpublished ms., University of California, Department of Architecture, August 1960, p.5）和汉斯·布卢门菲尔德（"Scale in Civic Design," Town Planning Review, April 1953, pp.35 ~ 46）对此都作过类似的描述。例如，布卢门菲尔德提供了以下数据：相距 70ft 或 80ft，可以看清对方的脸；相距约 48ft，可以就像看"肖像画"一样看清对方表情丰富的脸部细节了。

我们自己的非正式实验得出下面的结果：两个视力正常的人相距 75ft 可以舒适地进行相互联系。他们能提高嗓门谈话，他们能看清对方脸部表情的一般轮廓。75ft 这一最大值是极端可靠的。反复实验得出相同的距离值 ±10%。相距 100ft，交谈就不舒适了，对方的脸部表情再也看不清了。任何东西在 100ft 以外，就模糊得无法辨认了。

因此：

建造一个比你原先想象中小得多的广场。通常它的直径不大于 45 ~ 60ft，绝不超过 70ft。这一点仅适用于它的横向宽度。它的纵向长度当然可以再长些。

square is lost in the larger spaces.Roughly similar things have been said by Philip Thiel ("An Architectural and Urban Space Sequence Notation." unpublished ms.,University of California,Department of Architecture,August 1960,p.5) and by Hans Blumenfeld ("Scale in Civic Design," Town Planning Review,April 1953,pp.35-46).For example,Blumenfeld gives the following figures:a person's face can be recognized at up to 70 or 80 feet;a person's face can be recognized as "a portrait," in rich detail,at up to about 48 feet.

Our own informal experiments show the following results. Two people with normal vision can communicate comfortably up to 75 feet.They can talk with raised voice,and they can see the general outlines of the expression on one another's faces.This 75 foot maximum is extremely reliable.Repeated experiments gave the same distance again and again,±10 per cent.At 100 feet it is uncomfortable to talk,and facial expression is no longer clear.Anything above 100 feet is hopeless.

Therefore:

Make a public square much smaller than you would at first imagine;usually no more than 45 to 60 feet across,never more than 70 feet across.This applies only to its width in the short direction.In the long direction it can certainly be longer.

❧❦

An even better estimate for the size of the square:make a guess about the number of people who will typically be there(say,P),and make the area of the square no greater than 150 to 300P square feet—PEDESTRIAN DENSITY(123);ring

对广场的规模有一个更佳的估计值：估测将去广场的具有代表性的人数（P），并使广场面积不大于150～300Pft^2——**行人密度**（123）；在广场四周有活动角落，人们常聚集在那里——**袋形活动场地**（124）；在广场周围兴建建筑物，以便使广场具有一定的形状，并可从内向外看到其他更大的空间——**户外正空间**（106）、**外部空间的层次**（114）、**建筑物正面**（122）、**能坐的台阶**（125）；为了使广场的中心像它的边缘一样有用，使**空间中心有景物**（126）……

the square around with pockets of activity where people congregate—ACTIVITY POCKETS (124);build buildings round the square in such a way that they give it a definite shape,with views out into other larger places—POSITIVE OUTDOOR SPACE(106),HIERARCHY OF OPEN SPACE(114),BUILDINGFRONTS(122),STAIR SEATS(125);and to make the center of the square as useful as the edges,build SOMETHING ROUGHLY IN THE MIDDLE(126)...

模式62　眺远高地*

62 HIGH PLACES*

...according to FOUR-STORY LIMIT(21),most roofs in the community are no higher than four stories,about 40 or 50 feet. However,it is very important that this height limit be punctuated,just occasionally,by higher buildings which have special functions.They can help the character of the SMALL PUBLIC SQUARES(61) and HOLY GROUND(66);they can give particular identity to their communities,provided that they do not occur more frequently than one in each COMMUNITY OF 7000(12).

⠀⠀⠀⠀⠀⠀⠀⠀⠀⠀⠀⠀❧⟨⟩❧

The instinct to climb up to some high place,from which you can look down and survey your world,seems to be a fundamental haman instinct.

The tiniest hamlets have a dominating landmark—usually the church tower.Great cities have hundreds of them. The instinct to build these towers is certainly not merely Christian;the same thing happens in different cultures and religions,all over the world.Persian villages have pigeon towers; Turkey,its minarets;San Gimignano,its houses in the form of towers;castles,their lookouts;Athens,its Acropolis;Rio,its rock.

These high places have two separate and complementary functions.They give people a place to climb up to,from which they can look down upon their world.And they give people a place which they can see from far away and orient themselves toward,when they are on the ground.

Listen to Proust:

Combray at a distance,from a twenty-mile radius,as we

……根据**不高于四层楼**（21）在社区内大多数的屋顶不得超过四层楼高，约 40ft 或 50ft。但非常重要的是：这一高度极限偶尔会被更高的具有特殊功能的建筑物所打破。它们就是眺远高地，它们有助于**小广场**（61）和**圣地**（66）形成各自的特征。如果在每个 **7000 人的社区**（12）仅有一个眺远高地，就可赋予社区以特殊的个性。

80CR

攀登高地以便俯视和纵览风光，这似乎是人的基本天性。

就连最小的村庄也都有明显而突出的标记——通常为教堂尖塔。大城市往往有数以百计的教堂尖塔。当然，建造这些尖塔的天性不是基督教徒所独有的，而在全世界各种不同的文化和宗教中都表现出这种人所共有的天性。波斯人的村庄有鸽子塔；土耳其有伊斯兰教寺院的尖塔；圣吉米那诺有呈塔形的住宅；城堡有瞭望台；雅典有古希腊的卫城；里约热内卢有岩石。

这些眺远高地具有两种功能：区分功能和仰慕功能。人们登上这些高地，可以居高临下，极目四望，饱览风景。它们也是人们从远处地平线上仰慕的对象和确定行走方位的标志。

下面请听普劳斯特的描写吧：

我们总是一年一度在复活节前的一周内去遥远的科姆布雷教堂游玩，在离它 20ft 的半径范围内，我们从铁路线上就看得见它了。虽然它仅是一所教堂，但却是该城镇的缩影和代表。科姆布雷就是该城镇历史的见证，它为该城镇而矗立在地平线上。当游人来到科姆布雷附近时，他们都挤在它那又长又黑的、旷野上避风用的棚子周围，就像在那毛茸茸的、灰暗色的"包"式帐篷外，牧羊人集结着他的羊群似的……

当科姆布雷尚在视平线以下，人们从很远的地方就已能清晰地辨认出桑特—希莱尔教堂的尖顶，它以令人难以

used to see it from the railway when we arrived there every year in Holy Week,was no more than a church epitomising the town,representing it,speaking of it and for it to the horizon,and as one drew near,gathering close about its long,dark cloak,sheltering from the wind,on the open plain,as a shepherd gathers his sheep,the woolly grey backs of its blocking houses...

From a long way off one could distinguish and identify the steeple of Sainte-Hilaire inscribing its unforgetable form upon a horizon beneath which Combray had not yet appeared;when from the train which brought us down from Paris at master time my father caught sight of it,as it slipped into every fold of the sky in turn,its little iron cock veering continually in all directions,he could say: "Come,get your wraps together,we are there." (Marcel Proust,Swann's Way.)

High places are equally important,too,as places from which to look down:places that give a spectacular,comprehensive view of the town.Visitors can go to them to get a sense of the entire area they have come to;and the people who live there can do so too—to reassess the shape and scope of their surroundings.But these visits to the high places will have no freshness or exhilaration if there is a ride to the top in a car or elevator.To get a full sense of the magnificence of the view,it seems necessary to work for it,to leave the car or elevator,and to climb.The act of climbing,even if only for a few steps,clears the mind and prepares the body.

As for distribution,we suggest about one of these high places for each community of 7000,high enough to be seen throughout the community.If high places are less frequent,they tend to be too special,and they have less power as landmarks.

Therefore:

Build occasional high places as landmarks throughout

忘怀的形象矗立在那里。有一次我们在复活节假期从巴黎搭乘火车下行，我父亲从车厢内一眼就看见了桑特—希莱尔教堂，因为它高耸入云，在它的尖端有一只铁制的风信鸡不停地随风转动，变换方向。我父亲就会对它说："来吧，来取你的外套吧，我们终于来到了！"（Marcel Proust, Swann's Way.）

牛津：梦幻般美丽的塔尖之城
Oxford:the city of dreaming spires

凡是眺远高地，虽情况不同，都是同样重要的：当你登高远眺时，无限美景尽收眼底。游人登高，可以饱览风景；居民登高，可以重新评价周围环境的外貌和范围。如果游人乘坐汽车、电梯或架空缆车直达高地，就会感到索然无味和心中不快。为了观赏这一览无余的壮丽景色，看来不乘汽车、不乘电梯或空中缆车是必要的。游人拾级而上，其乐无穷，即使攀登寥寥数步，也会顿觉头脑清醒，身体舒服。

至于说到眺远高地的分布问题，我们认为，在每个7000人的社区中大概以一个为宜，其高度要求为全社区的男女老少都能看得见。如果眺远高地过少，就会显得过分特殊而不能成为人们的标记了。

因此：

在全市建造频繁出现的眺远高地作为标志。它们可能

the city.They can be a natural part of the topography,or towers,or part of the roofs of the highest local building—but,in any case,they should include a physical climb.

৪০৫জ

Elaborate the area around the base of the high place—it is a natural position for a SMALL PUBLIC SQUARE(61);give the stair which leads up to the top,openings with views out,so that people can stop on the stair,sit down,look out,and be seen while they are climbing—STAIR SEATS(125),ZEN VIEW(134),OPEN STAIRS(158)...

是自然地貌的一部分，也可能是塔或是该地区最高楼房的屋顶的一部分——但是，在任何场合下，它们都应当让人们自己攀登。

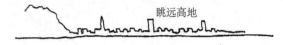

眺远高地

☙❦❧

精心设计眺远高地周围的地区——这是**小广场**（61）的天然位置；建造直达高地顶部的台阶，并设置观景的敞口，以便游人能在台阶上停下来坐着歇息，举目远眺，欣赏美景；当他们拾级而上时，也能相互看到身影——**能坐的台阶**（125）、**禅宗观景**（134）和**室外楼梯**（158）……

模式63　街头舞会*

63　DANCING IN THE STREET*

...several patterns have laid the groundwork for evening activity in public—MAGIC OF THE CITY(10), PROMENADE(31),NIGHT LIFE(33),CARNIVAL(58),SMALL PUBLIC SQUARES(61). To make these places alive at night,there is nothing like music and dancing;this pattern simply states the physical conditions which will encourage dancing and music to fill the streets.

⟡⟡⟡

Why is it that people don't dance in the streets today?

All over the earth,people once danced in the streets;in theater,song,and natural speech, "dancing in the street" is an image of supreme joy.Many cultures still have some version of this activity. There are the Balinese dancers who fall into a trance whirling around in the street;the mariachi bands in Mexico—every town has several squares where the bands play and the neighborhood comes out to dance;there is the European and American tradition of band—stands and jubilees in the park;there is the bon odori festival in Japan,when everybody claps and dances in the streets.

But in those parts of the world that have become "modern" and technically sophisticated,this experience has died.Communities are fragmented;people are uncomfortable in the streets,afraid with one another;not many people play the right kind of music;people are embarrassed.

Certainly there is no way in which a change in the environment, as simple as the one which we propose,can remedy these circumstances.But we detect a change in mood.The embarrassment and the alienation are recent developments, blocking a more basic need.And as we get in touch with these needs,things start

……前面的若干模式已为晚间的公共活动打好了扎实的基础——**城市的魅力**（10）、**散步场所**（31）、**夜生活**（33）、**狂欢节**（58）和**小广场**（61）。为了使这些地方在夜里喜气洋洋，没有什么东西比得上音乐和舞蹈了。本模式扼要叙述为激励人们在街头巷尾载歌载舞所必备的物质条件。

❧❧

今天人们不在街头巷尾载歌载舞原因何在？

许多人在街头巷尾翩翩起舞，曾一度风靡世界；剧院内的美妙歌声、无拘束的自然谈吐和"街头舞会"使欢乐的气氛达到极点。现在许多民族文化中仍有这种活动。例如，巴厘岛的舞蹈家在街头飞旋，竟至忘乎所以；在墨西哥有墨西哥街头乐队——在那里，每个城镇都有几个广场，街头乐队高奏乐曲，邻里居民随之起舞；在欧美的公园中有传统的音乐台和欢乐的佳节；在日本有佛教徒一年举行一次的盂兰盆节，他们在街上击掌跳舞，欢呼雀跃。

但在世界的另外一些已变得"现代化的"、技术高度发达的地方，这种欢乐的场面早已无影无踪了。现在的社区支离破碎；居民在街道上不会感到心情愉快，相互提心吊胆；演奏合适的乐曲的人寥若晨星；居民普遍有一种窘迫感。

当然，改变环境就能克服这种情况，但绝不会像我们所建议的那样简单。我们觉察到情绪的变化。这种窘迫感和相互疏远是现代社会发展的产物，抑制人们的基本需要。当我们着手设法满足人们的基本需要时，问题接踵而至。人们回忆应如何翩翩起舞；每人手执乐器；几百人组成许多小乐队。我们写到这里时，在旧金山、伯克利和奥克兰也正在展开一场关于"街头音乐家"的激烈争论。所谓"街头音乐家"就是当天气晴朗的时候，在街头巷尾和广场上自发演奏的乐队——在什么地方他们可获准演出？他们是

to happen.People remember how to dance;everyone takes up an instrument;many hundreds form little bands.At this writing,in San Francisco,Berkeley,and Oakland there is a controversy over "street musicians"—bands that have spontaneously begun playing in streets and plazas whenever the weather is good—where should they be allowed to play,do they obstruct traffic,shall people dance?

It is in this atmosphere that we propose the pattern.Where there is feeling for the importance of the activity reemerging,then the right setting can actualize it and give it roots.The essentials are straightforward:a platform for the musicians,perhaps with a cover;hard surface for dancing,all around the bandstand;places to sit and lean for people who want to watch and rest;provision for some drink and refreshment(some Mexican bandstands have a beautiful way of building tiny stalls into the base of the bandstand,so that people are drawn though the dancers and up to the music,for a fruit drink or a beer);the whole thing set somewhere where people congregate.

Therefore:

Along promenades,in squares and evening centers,make a slightly raised platform to form a bandstand,where street musicians and local bands can play.Cover it,and perhaps build in at ground level tiny stalls for refreshment.Surround the bandstand with paved surface for dancing—no admission charge.

❧✿❧

Place the bandstand in a pocket of activity,toward the edge of a square or a promenade—ACTIVITY POCKETS (124);make it a room,defined by trellises and columns—PUBLIC OUTDOOR ROOM(69);build FOOD STANDS(93)around the bandstand;and for dancing,maybe colored canvas canopies,which reach out over portions of the street,and make the street,or parts of it,into a great,half-open tent—CANVAS ROOFS(244)...

否妨碍交通？人们会伴着音乐的节奏而翩翩起舞吗？——这些就是正在争论不休的问题。

正是在这样一种气氛下，我们才提出本模式。凡是人们感到什么地方需要重现街头舞会，就可以适当安排，使它再现，并让它生根开花。显然，如下的物质条件是必不可少的：一个供音乐家们活动的平台，可能有顶篷；场地的中心地面坚硬，用作舞池，四周是音乐台；留出坐和靠的地方，供观看跳舞和休息的人用；供应饮料和点心（在墨西哥，许多饮食商亭十分巧妙地开设在音乐台附近，这样，无论跳舞的人还是演奏音乐的人都会被吸引过去，喝上一杯果汁或啤酒）；这一切都发生在人们聚集的地方。

因此：

沿着散步场所，在广场和夜市中心，建造一个略高出地面的平台，以便形成音乐台，供街头音乐家和地方乐队演奏用。平台有顶篷，建造一些室内外地坪相同的小型的饮食商亭。音乐台的四周铺砌舞池地面，免费开放。

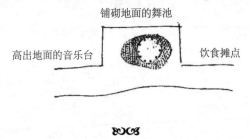

铺砌地面的舞池

高出地面的音乐台　　　　　　　　　　　饮食摊点

�

将音乐台设在袋形活动区内，向着小广场或散步场所的边缘——**袋形活动场地**（124）；用棚架或立柱围成一定的空间——**户外亭榭**（69）；在音乐台四周开设**饮食商亭**（93）；为了举办舞会，也许要支起彩色帆布顶篷，占用部分街道，并把整条街道或街道的若干地段用巨大的半敞式帐篷覆盖起来——**帆布顶篷**（244）……

模式64　水池和小溪*

64   POOLS AND STREAMS*

...the land,in its natural state,is hardly ever flat,and was,in its most primitive condition,overrun with rills and streams which carried off the rainwater.There is no reason to destroy this natural feature of the land in a town—SACRED SITES(24),ACCESS TO WATER (25)—in fact,it is essential that it be preserved,or recreated.And in doing so it will be possible to deepen several larger patterns—boundaries between neighborhoods can easily be formed by streams—NEIGHBORHOOD BOUNDARY(15),quiet backs can be made more tranquil—QUIET BACKS(59),pedestrian streets can be made more human and more natural—PEDESTRIAN STREET(100).

৪০০৪

We came from the water;our bodies are largely water;and water plays a fundamental role in our psychology.We need constant access to water,all around us;and we cannot have it without reverence for water in all its forms.But everywhere in cities water is out of reach.

Even in the temperate climates that are water rich,the natural sources of water are dried up,hidden,covered,lost. Rainwater runs underground in sewers;water reservoirs are covered and fenced off;swimming pools are saturated with chlorine and fenced off;ponds are so polluted that no one wants to go near them any more.

And especially in heavily populated areas water is scarce.We cannot possibly have the daily access to it which we and our children need,unless all water,in all its forms,is

……大地，就其自然状态而言，几乎从来不会是平坦的，在最原始的条件下，曾有过无数倾泻雨水的小河和溪流。所以，没有任何理由在城镇中毁坏大地的这种天然特征——**珍贵的地方**（24）和**通往水域**（25）——事实上，重要的是：这种自然特征应当保存下去或重新创造出来。这样，我们就会加深理解几个较大的模式的意义——小溪极容易成为邻里间的边界——**邻里边界**（15）；僻静区可能变得越发宁静——**僻静区**（59）；步行街可能变得更富有人情味和更天然成趣——**步行街**（100）。

෨෩

我们起源于水。我们的躯体之内大部分是水。水在我们的生理过程中起着重要作用。我们需要经常接触周围的水；我们如不珍惜所有各种形状的水域，我们就不会有水了。但在市内，各处居民难以接近水域。

甚至在雨水充沛的温带地区，天然水源也在干涸，或流入地下，或隐而不见，或完全消失。雨水经地下污水道流走；蓄水池被遮盖并用栅或篱围住；游泳池内氯的含量已达饱和状态，并围着栅栏；池塘已污染不堪，无人敢再涉足。

尤其在严重污染区水域寥寥无几。我们和孩子们都需要水域，但不可能天天去那里，除非各种不同形状的水域都暴露在外面，保持水质，并不断地从邻里无数的小水池、池塘、蓄水池和小溪的地区水系得到补充和更新水质。

可以从不同的角度来说明人和水之间的联系。生物学家亨德森观察到人的血液中金属盐的成分和海水的一模一样，因为我们都起源于大海。人类学家伊莱恩·摩根推测在上新世的干旱时期，我们都曾回到了大海，作为海洋哺乳动物在大海边缘的浅水域生活了长达 1000 万年之久。很明显，这一假设解释了许多有关人体之谜和人体适

exposed,preserved,and nourished in an endless local texture of small pools,ponds,reservoirs,and streams in every neighborhood.

There are various ways of expressing the connection between people and water.The biologist,L.J.Henderson,observed that the saline content of human blood is essentially the same as that of the sea,because we came from the sea.Elaine Morgan,an anthropologist,speculates that during the drought of the Pliocene era,we went back to the sea and lived 10 million years as sea mammals in the shallow waters along the edge of the ocean. Apparently,this hypothesis explains a great deal about the human body,the way in which it is adapted to water,which is otherwise obscure(The Descent of Woman,New York:Bantam Books,1973).

Furthermore,among psychoanalysts it is common to consider the bodies of water that appear in people's dreams as loaded with meaning.Jung and the Jungian analysts take great bodies of water as representing the dreamer's unconscious.We even speculate,in light of the psychoanalytic evidence,that going into the water may bring a person closer to the unconscious processes in his life.We guess that people who swim and dive often,in lakes and pools and in the ocean,may be closer to their dreams,more in contact with their unconscious,than people who swim rarely.Several studies have in fact demonstrated that water has a positive therapeutic effect;that it sets up growth experience.(For references,see Ruth Hartley et al.,Understanding Children's Play,Columbia University Press,New York:1964,Chapter V.)

All of this suggests that our lives are diminished if we cannot establish rich and abiding contact with water.But of course,in most cities we cannot.Swimming pools,lakes,and beaches are few in number and far away.And consider also the water supply.

应水的方式，但在其他方面至今仍然是解释不清的。(The Descent of Woman, New York : Bantam Books, 1973)

此外，精神分析学家普遍认为，人的睡梦中出现的水体含有某种意义。荣格和他的化验员们则认为巨大的水体表示做梦者的无意识状态。我们根据精神分析的证据，甚至可以推断"进入水中"会使做梦者更接近不省人事。我们猜测，常在湖泊、水池和海洋中游泳和潜水的人可能比很少游泳的人更容易做关于"水"的梦，更多地处于无意识状态。实际上，一些研究已经表明，水具有积极的疗效；水能恢复和增强人的体质。(For references, see Ruth Hartley et al., Understanding Children's Play, Columbia University Press, New York : 1964, Chapter V.)

所有这一切都表明，如果我们不能与水域建立起丰富多彩的、持久的联系接触，我们生活的范围就会日益缩小。当然，我们在大多数城市内却做不到这一点。游泳池、湖泊和海滩为数甚少且离得很远。而且还要考虑水的供应问题。现在我们只好旋开自来水龙头来接触水。我们认为处处有水是理所当然的事。可是，水的分配已经成为像处理水质的高超技术那样不可思议的了。只去接触当地蓄水池，只去理解水的循环——即水的局限性和神秘性，是无法满足人们精神上的需要的。

但是，构思出一个城镇内数以百计的水域分散在每家每户和每一工作地点附近是可能的。人可在水中游泳，可坐在水边濯足或晃动双脚尽情戏水。请想一下，例如流水：淙淙的小河，潺潺的溪流。如今却让铺砌好的路面覆盖在它们的上面，流水被迫"转入地下"。规划人员不是让建筑紧贴或靠近这些小溪流，而是让它们远远分开。他们仿佛在说："把变化莫测的大自然纳入一张合理的街道地图方格中是不可能的。"但是，我们有办法使建筑和水域，如池塘、水池、蓄水池、小河和溪流等保持接触。我

Our only contact with this water is to turn on the tap.We take the water for granted.But as marvelous as the high technology of water treatment and distribution has become,it does not satisfy the emotional need to make contact with the local reservoirs,and to understand the cycle of water:its limits and its mystery.

But it is possible to imagine a town where there are many hundreds of places near every home and workplace where there is water.Water to swim in,water to sit beside,water where you can dangle your feet.Consider,for example,the running water:the brooks and streams.Today they are paved over and forced underground.Instead of building with them,and alongside them,planners simply get them out of the way,as if to say:"the vagaries of nature have no place in a rational street grid." But we can build in ways which maintain contact with water,in ponds and pools,in reservoirs,and in brooks and streams. We can even build details that connect people with the collection and runoff of rain water.

Think of the shallow ponds and pools that children need.It is possible for these pools and ponds to be available throughout the city,close enough for children to walk to.Some can be part of the larger pools.Others can be bulges of streams that run through the city,where a balanced ecology is allowed to develop along their edges—ponds with ducks and carp,with edges safe enough for children to come close.

And consider the system of local and distribution reservoirs. We can locate local reservoirs and distribution reservoirs so that people can get at them;we might build them as kinds of shrines,where people can come to get in touch with the source of their water supply;the place immediately around the water an atmosphere inviting contemplation.These shrines could be set into the public space:perhaps as one end of a promenade,or as a

们甚至可以建造使人们和水接触的集蓄和排放雨水的构筑物。

要考虑到孩子们所需要的池塘和水池。水池和池塘要遍布全市，以便孩子们都能步行去玩，这一点是能办到的。其中一些可能是较大水池的一部分，另一些可能是流经市区河套的，沿岸可以发展平衡的生态环境：池塘里养鸭、养鲤鱼，池塘岸边要有安全设施，以防儿童失足落水。

还要考虑到当地蓄水池系统的分布情况。我们能把蓄水池配水池分布在居民能去的地方；我们也许会把它们建筑成各种圣地，居民们去那里可和供水水源保持接触；源头水清池秀，风景优美，素淡如画，雅趣横生，可令游人尽兴观赏。这些圣地可穿插在公共场所内：或位于散步场所的末端，或作为两个社区间公共用地的边界。

印第安人的台阶式井
Indian stepped well

再要考虑到各种可能的形式的流水。凡是在周围环境中被剥夺了每天和流水接触机会的人，都要出城到乡村去，经过长途跋涉就能观看到河中的流水，或坐在岸边凝视水面。儿童们见到流水欣喜若狂，他们不停地戏水，在水中尽兴打闹；时而把棍棒扔入水中，看着它们渐渐消失；时而把小纸船放在水面上，让它们漂浮远去；时而搅拌污泥，把水弄浑浊，看着它慢慢澄清。

boundary of common land between two communities.

And think of running water,in all its possible forms.People who have been deprived of it in their daily surroundings go to great lengths to get out of their towns into the countryside,where they can watch a river flow,or sit by a stream and gaze at the water. Children are fascinated by running water.They use it endlessly,to play in,to throw sticks and see them disappear,to run little paper boats along,to stir up mud and watch it clear gradually.

Natural streams in their original streambeds,together with their surrounding vegetation,can be preserved and maintained. Rainwater can be allowed to assemble from rooftops into small pools and to run through channels along garden paths and public pedestrian paths,where it can be seen and enjoyed.Fountains can be built in public places.And in those cities where streams have been buried,it may even be possible to unravel them again.

In summary,we propose that every building project,at every scale,take stock of the distribution of water and the access to water in its neighborhood.Where there is a gap,where nourishing contact with water is missing,then each project should make some attempt,on its own and in combination with other projects to bring some water into the environment.There is no other way to build up an adequate texture of water in cities:we need pools for swimming,ornamental and natural pools,streams of rain water,fountains,falls,natural brooks and streams running through towns,tiny garden pools,and reservoirs we can get to and appreciate.

Therefore:

Preserve natural pools and streams and allow them to run through the city;make paths for people to walk along them and footbridges to cross them.Let the streams form natural barriers in the city,with traffic crossing them only infrequently on bridges.

在原先河床上的天然河流及其周围的草木，能被保存并保持下去。天然雨水可以从屋顶汇集到小水池中，然后沿花园小径和公共人行道流过沟渠，以供行人观赏。在公共场所建造喷泉。就连那些城市内隐蔽的河流也可以重新挖通。

隐蔽的河流
The buried streams

总而言之，我们建议每一项建筑工程，不管规模大小如何，都应估计到水的分配以及邻里水域的通路。凡有漏洞的地方，接触水的希望就会成为泡影，所以，每项工程都应单独地或和其他工程一起注意设法把水引入周围环境。这是建立市内合理水网的唯一可行的途径，因为我们需要游泳池、人工构筑的和天然形成的水池、集雨水而成的小河、喷泉、瀑布、流经城镇的天然小溪和小河、小型的花园水池以及可供我们观赏的水库。

因此：

保存天然的水池和河流，并让河流流经全市；两岸修筑小路，供游人散步，并修筑人行桥梁，横跨河流之上。让河流成为市内的天然屏障，车辆只能在少量的桥梁上经过。

Whenever possible,collect rainwater in open gutters and allow it to flow above ground,along pedestrian paths and in front of houses.In places without natural running water,create fountains in the streets.

✂✃✂

If at all possible,make all the pools and swimming holes part of the running water—not separate—since this is the only way that pools are able to keep alive and clean without the paraphernalia of pumps and chlorine—STILL WATER(71). Sometimes,here and there,give the place immediately around the water the atmosphere of contemplation;perhaps with arcades,perhaps some special common land,perhaps one end of a promenade—PROMENADE(31),HOLY GROUND(66),ARCADES(119)...

无论何时，只要有可能，就把雨水收集在明沟内，并让它流经地面，并沿着人行道和宅前流过。没有天然流水的地方，就在街道上建造喷泉。

雨水

溪流

&oc&

　　如果确有可能，应使所有的水池和可供游泳的水湾都成为流水的一部分——不是分离的——因为这是使水池可能保持活水畅流、水质新鲜洁净而无须使用水泵设备和氯的唯一途径——**池塘**（71）。有时，在水域四周要处处造成一种引人入胜的气氛；也许有拱廊，也许有某种专用公共场地，也许有一端是散步场所——**散步场所**（31）、**圣地**（66）和**拱廊**（119）……

模式65　分娩场所

……生和死是人类社会的普遍现象，人是地方社区和邻里的一个组成部分——**7000 人的社区**（12）、**易识别的邻里**（14）和**生命的周期**（26）。至于说到分娩，每一组邻里都应能给予产妇的分娩过程以地方性的富有同情心的照料和护理。（原注：本模式发展到今天的样子在很大程度上应归功于朱迪思·肖的著作。我们在撰写本文时，她是伯克利加利福尼亚大学建筑系的研究生，一位有 3 个孩子的母亲。）

∞∞∞

把分娩过程视为疾病来处理，可能是一个健全社会的一个有益于健康的部分。这种说法看来可能性不大。

"怀孕并不处于像产妇产后可满怀希望恢复到'正常状态'的那种危急情况之中。……怀孕本身是发展到分娩的生理极限的一种高度活跃的、强有力的发展过程。"（I.H.Pearse and L.H.Crocker，The Peckham Experiment，New Haven：Yale University Press，1946，p.153）

现有的助产护理在大多数医院里都要遵循完备的程序。产妇生下一个婴儿被认为是患一场大病，需经住院恢复健康。临产妇被当作"病人"而进行外科检查，她们被全身消毒。她们的外阴部被擦拭干净，刮光消毒。她们身穿白色长外衣，躺在体检台上，让医生进行各种检查。忍着分娩阵痛的产妇，被一一安置在以墙或幕隔成的小室内，度过一段与世隔绝的时间。这段时间可能持续许多个小时。此时此刻，她的丈夫和孩子本可来安慰，但这是不被允许的。分娩通常在"分娩室"进行，那里有结构合理的"分娩台"。

65　BIRTH PLACES

...both birth and death need recognition throughout society,where people are,as part of local communities and neighborhoods—COMMUNITY OF 7000(12),IDENTIFIABLE NEIGHBORHOOD(14),LIFE CYCLE(26).As far as birth is concerned,each group of neighborhoods must be able to take care of the birth process,in local,human terms.(Note:The development of this pattern is due largely to the work of Judith Shaw,at this writing a graduate student in architecture at the University of California,Berkeley,and a mother of three children.)

∞⋙⋘

It seems unlikely that any process which treats childbirth as a sickness could possibly be a healthy part of a healthy society.

"Pregnancy is no state of emergency from which the mother may hopefully be returned to 'normality'after the birth of the child....It is a highly active,potent,developmental process of the family going forward to its natural culmination in delivery." (I.H.Pearse and L.H.Crocker,The Peckham Experiment,New Haven:Yale University Press,1946,p.153)

The existing obstetrics service in most hospitals follows a well outlined procedure.Having a baby is thought of as an illness and the stay in the hospital as recuperation.Women who are about to deliver are treated as "patients" about to undergo surgery.They are sterilized.Their genitals are scrubbed and shaved.They are gowned in white,and put on a table to be moved back and forth between the various parts of the hospital. Women in labor are put in cubicles to pass the time with virtually no social contact.This time can last for many hours.It

除了分娩台的特殊操作，分娩室还配备了和手术室一样的其他设备。分娩是离别的时刻而不是欢聚的时刻，产妇要经过长达 12h 才能被允许去抚摸她的新生婴儿。如果她服用了分娩镇定剂，甚至要经过更长的时间，才能看到她的丈夫。

目前正继续开展着一场微妙的运动：试图把分娩的本质重新看作一种自然现象。这一运动发展至今快 15 年了。在此运动中没有发生过大声疾呼的反对助产术和医院规章制度的行为，但悄悄地进行了一项宣传工作：出版了一些好的书籍，进行了与专职和非专职人员有关的口头宣传，莱奇联合会以及其他一些团体在乡村进行了宣传活动，他们最关心的是分娩和再出现乡村助产士。他们开始的努力目标是"自然的"正常分娩。该名称正被应用于促使"分娩"这一概念恢复为只指正常的生理现象。最近，他们的奋斗目标已扩大，包括已经改进了的分娩环境和妊娠保健家庭。(For an architectural slant, see Lewis Mumford, The Urban Prospect, New York: Harcourt Brace and World, 1968, p.25)

现在我们引证朱迪思·肖对一个良好的分娩场所所作的描述。她所描写的分娩场所可与小型护理室相媲美，也许它和地方保健中心有联系，必要时可和地方医院进行紧急联系：

给婴儿准备一只小筐……助产士要一直呆在那里提供产后护理……居住在那里的助产士应当有一套房间，包括卧室、起居室兼厨房和浴室……

餐室应当是公共的。每个婴儿也有一席之地，以便母亲给婴儿喂奶或照看……模式**农家厨房**（139）在这种建筑中能起到重要作用……一家人都可以来，不单是为了产妇生产，也是为了对产妇进行产前护理，学会自然分娩法，并尽可能学会婴儿护理，也可以促膝谈心，一般说来是为

is a time when father and children could be present to provide encouragement.But this is not permitted.Delivery usually takes place in a "delivery room" which has the proper "table" for childbirth.Except for the particular workings of the delivery table the room has the same properties as an operating theater.The birth becomes a time for separation rather than togetherness.It may be as long as 12 hours before the mother is even allowed to touch her baby,and if she was sedated for the delivery,even longer before she may see her husband.

For about fifteen years there has been a subtle movement to try and recapture the essence of childbirth as a natural phenomenon.There has been no loud protest against obstetricians and hospital rules,but a rather quiet one:several good books,word of mouth,concerned professionals and nonprofessionals,the La Leche League,a few groups around the country whose prime concern is with birth,and the reemergence of the nurse-midwife.The original effort of these people was aimed at "natural" childbirth,the name being applied in an attempt to bring the concept of childbirth back to a normal physiological occurrence.Lately the focus of the effort has been expanded to include an altered environment for childbirth and to include the family in a positive way.(For an architectural slant,see Lewis Mumford,The Urban Prospect,New York: Harcourt Brace and World,1968,p.25)

We quote now from Judith Shaw's description of a good birth place.She is describing a place comparable to a small nursing home,perhaps associated with a local health center,and with emergency connections to the local hospital:

A small basket for the baby would be provided....The nurse-midwife would be there always to provide postpartum care....The nurse-midwife,who lives in,would have a small suite containing a bedroom,sitting room—kitchen and bath...

了熟悉产妇行将分娩的场所。

分娩场所应能容纳全家人。他们能合住在一套房间里，产妇也在里面分娩……因为分娩是在全家居住的一套房间内发生的，婴儿、母亲和家人就能立即团圆。每套房间都应有自来水设备和一张简易床，供安放婴儿、擦洗婴儿并给婴儿进行第一次体检之用。

因此：

为产妇建造良好的当地分娩场所：该场所应是特别适于分娩，分娩既是自然的又是重大的一件事。该场所应是全家人都可去接受训练教育和进行产前护理的地方；丈夫和助产士都在那里，当产妇感到临产的阵痛和即将分娩时，就会帮助她顺利分娩。

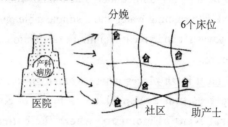

❧❧❧

使分娩场所包括各种房间，以便产妇产后可以和婴儿及家中其他人团圆——在一起同吃、同住和同煮饭烧菜——**中心公共区（129）、夫妻的领域（136）和农家厨房（139）**；辟出一部分私人花园供散步用——**半隐蔽花园（111）、花园围墙（173）**；就建筑形状来说，花园、停车场和周围的环境都要从**建筑群体（95）**开始考虑……

The eating place would be communal.Each baby would have a place too (his movable basket)so that the mother can bring her child with her to feed or to watch....The pattern FARMHOUSE KITCHEN(139)could play an important role in this building....families can come not only to have babies but have their prenatal care,learn methods of natural childbirth,possibly child care,maybe just to talk and in general to become familiar with the place they will come to for the delivery.

The birthing place should have accommodations for the entire family.They can occupy a suite in which they live and in which the mother gives birth to the baby....Since the delivery would take place in the family suite,the baby,mother,and the family can be together immediately.Each suite would have to be equipped with running water and a simple table on which to lay the baby,wash it and give it its initial examination.

Therefore:

Build local birth places where women go to have their children:places that are specially tailored to childbirth as a natural,eventful moment—where the entire family comes for prenatal care and education;where fathers and midwives help during the hours of labor and birth.

❧❧❧

Include rooms where after the birth the mother and her baby can stay together with the other members of the family— sleep together,eat together,cook together—COMMON AREAS AT THE HEART(129),COUPLE'S REALM (136),FARMHOUSE KITCHEN(139);provide apartly private garden to walk in—HALF-HIDDEN GARDEN(111),GARDEN WALL(173);for the shape of the building,gardens,parking,and surroundings,begin with BUILDING COMPLEX(95)...

模式66　圣地*

　　……我们业已说明一个完整的生命的周期以及生命的各个时期所经历的礼仪——**生命的周期**（26）；我们也介绍了一些土地应被闲置起来，因为它们具有明显的意义和重要性——**珍贵的地方**（24）。本模式将论述如何细致地组织这些地方周围的空间。组织这种空间十分重要，在某种程度上来说，它本身就能创造出神圣的气氛，也许，甚至会促使相关的礼仪过程慢慢出现。

∞∞∞

　　教堂和寺院是什么？当然，这是圣灵之所，也是圣徒顶礼膜拜和默默祈祷之地。但是，首先，从人的观点来看，它是一条门道。对于圣徒而言，一个人降生人世要经过教堂，离开人世也要经过教堂。而且，他在一生中的每个重要关头，都要一再经过教堂。

66   HOLY GROUND*

...we have defined the need for a full life cycle,with rites of passage between stages of the cycle—LIFE CYCLE(26);and we have recommended that certain pieces of land be set aside because of their importance and meaning—SACRED SITES(24).This pattern gives the detailed organization of the space around these places.The organization is so powerful,that to some extent it can itself create the sacredness of sites,perhaps even encourage the slow emergence of coherent rites of passage.

⊰✦⊱

What is a church or temple?It is a place of worship, spirit,contemplation,of course.But above all,from a human point of view,it is a gateway.A person comes into the world through the church.He leaves it through the church.And,at each of the important thresholds of his life,he once again steps through the church.

The rites that accompany birth,puberty,marriage,and death are fundamental to human growth.Unless these rites are given the emotional weight they need,it is impossible for a man or woman to pass thoroughly from one stage of life to another.

In all traditional societies,where these rites are treated with enormous power and respect,the rites,in one form or another,are supported by parts of the physical environment which have the character of gates.Of course,a gate,or gateway,by itself cannot create a rite.But it is also true that the rites cannot evolve in an

一个人从诞生起到青春期、结婚直至死亡，都要举行各种相应的仪式，这对人的发展是极为重要的。只有这些仪式赋予了人们所需要的浓厚的感情色彩，人们才可能从人生的一个阶段完全进入另一个阶段。

在所有的传统社会中，举行这些仪式都要花很大气力，怀着崇敬的心情去办。这些仪式要在物质环境中具有大门性质的地方以某种方式进行。当然，大门或门道，就其本身而言，是创造不出仪式的。但同样正确的是：仪式在特别被轻视或被视为微不足道的琐事的地方，也是无法发展的。医院不是举行洗礼的地方；殡仪馆不可能使人感到有葬礼的意义。

从功能的观点看，重要的是：每一个人当他自己或他的朋友通过这些重要关头时都有机会和他的伙伴进行社交。这种社交现在需要在公认举行仪式的某种圣灵之门的地方生根固定下来。

为了支持举行这些仪式，为了创造一种庄严肃穆和神圣的气氛，并使人感觉到和大地有着联系，而正是这种联系才使仪式具有重大意义，那么，这种"门道"必须具有什么样的外形或空间呢？

当然，它会在细节上因文化而异。无论什么东西确实被认为是神圣不可侵犯的——无论是大自然、上帝、特殊的地方、圣灵、圣物、大地本身，还是观念——神圣的东西在不同的文化中具有不同的形式，并需要不同的物质环境来支持。

可是，我们认为，从一种文化到另一种文化，有一个基本的特征是不变的。在一切文化中，似乎任何神圣的东西只不过是人们感到它是神圣的而已；它可望而不可及，或人们需经过重重回廊曲道、高高低低的台阶、许许多多的牌楼和大门，耐着性子一步步地接近它，才能渐渐揭示和看清它的真相。这样的例子层出不穷：北京的紫禁城；凡正式拜会罗马教皇的人都必须在 7 个等候室逐一等待；阿兹台克人（墨西哥印第安人——译者注）的献祭仪式以前是在有层层石级的金字塔上举行，他们上一个石级，就接近祭品一步；伊势神宫是日本最有名的圣地，它是一座大

environment which specifically ignores them or makes them trivial.A hospital is no place for a baptism;a funeral home makes it impossible to feel the meaning of a funeral.

In functional terms,it is essential that each person have the opportunity to enter into some kind of social communion with his fellows at the times when he himself or his friends pass through these critical points in their lives.And this social communion at this moment needs to be rooted in some place which is recognized as a kind of spiritual gateway for these events.

What physical shape or organization must this "gateway"have in order to support the rites of passage,and in order to create the sanctity and holiness and feeling of connection to the earth which makes the rites significant.

Of course,it will vary in detail,from culture to culture. Whatever it is exactly that is held to be sacred—whether it is nature,god,a special place,a spirit,holy relics,the earth itself,or an idea—it takes different forms,in different cultures,and requires different physical environments to support it.

However,we do believe that one fundamental characteristic is invariant from culture to culture.In all cultures it seems that whatever it is that is holy will only be felt as holy,if it is hard to reach,if it requires layers of access,waiting,levels of approach,a gradual unpeeling,gradual revelation,passage through a series of gates.There are many examples:the Inner City of Peking;the fact that anyone who has audience with the Pope must wait in each of seven waiting rooms;the Aztec sacrifices took place on stepped pyramids,each step closer to the sacrifice;the Ise shrine,the most famous shrine in Japan,is a nest of precincts,each one inside the other.

Even in an ordinary Christian church,you pass first

小建筑互套的、结构精巧别致的庙宇。

重重回廊曲道
Layers of access

即使在普通的基督教教堂内，你首先得穿过教堂的庭院，然后再穿过教堂的中堂；而在一些特殊场合下，你要在圣坛围栏以远的地方，才能进入环绕圣坛的高坛区，还有只有牧师本人才能进去的圣堂。祭祀用的面包外面裹着五层东西，就更难以接近它了。

这种层次重叠或大小围套的巧妙构思看来是符合人的基本心理状态的。我们认为，每个社区，不管居民的特殊信仰如何，甚至也不管他们是否已经有一种有组织的信仰，都需要有一片笼罩着神圣气氛的场所：你不能一步就登堂入室，必须移步重门，方得入内。当这样一个地方在社区中存在时，即使它不和任何特殊的信仰有着联系，我们认为，这种神圣的气氛也将以某种方式逐渐进入人们的共同生活之中。

因此：

在每一社区和邻里中，都要有一片容易识别的珍贵的地方作为圣地，并建成一系列叠套的教堂或寺庙的围墙；在每一叠套处均设门为标志；进门越多，越觉幽深，越有圣洁气氛，最幽深处是圣所，只有经过外面重重大门才能到达。

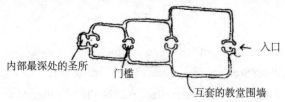

内部最深处的圣所　门槛　　　　　　　互套的教堂围墙　　　入口

through the churchyard,then through the nave;then,on special occasions,beyond the altar rail into the chancel and only the priest himself is able to go into the tabernacle.The holy bread is sheltered by five layers of ever more difficult approach.

This layering,or nesting of precincts,seems to correspond to a fundamental aspect of human psychology.We believe that every community,regardless of its particular faith,regardless of whether it even has a faith in any organized sense,needs some place where this feeling of slow,progressive access through gates to a holy center may be experienced.When such a place exists in a community,even if it is not associated with any particular religion,we believe that the feeling of holiness,in some form or other,will gradually come to life there among the people who share in the experience.

Therefore:

In each community and neighborhood,identify some sacred site as consecrated ground,and form a series of nested precincts,each marked by a gateway,each one progressively more private,and more sacred than the last,the innermost a final sanctum that can only be reached by passing through all of the outer ones.

⊱⊰

At each threshold between precincts build a gate—MAIN GATEWAYS(53);at each gate,a place to pause with a new view toward the next most inner place—ZEN VIEW(134);and at the innermost sanctum,something very quiet and able to inspire— perhaps a view,or no more than a simple tree,or pool—POOLS AND STREAMS(64),TREE PLACES(171)...

在重叠互套的教堂或寺庙的围墙之间，设一大门——**主门道**（53）；每一大门都是停顿处，从此又能窥见下一个更幽深的地方——**禅宗观景**（134）；最内部的圣所应造成一种寂静的气氛，使圣徒感到神的感召——它也许是一景物，或不过是一棵孤单单的树，或一池清水——**水池和小溪**（64）和**林荫空间**（171）……

在每一住宅团组和工作社区内，提供许多小片的公共土地，以各种不同的形式为同样的一些需要服务：

67. 公共用地

68. 相互沟通的游戏场所

69. 户外亭榭

70. 墓地

71. 池塘

72. 地方性运动场地

73. 冒险性的游戏场地

74. 动物

In each house cluster and work community,provide the smaller bits of common land,to provide for local versions of the same needs:

67.COMMON LAND

68.CONNECTED PLAY

69.PUBLIC OUTDOOR ROOM

70.GRAVE SITES

71.STILL WATER

72.LOCAL SPORTS

73.ADVENTURE PLAYGROUND

74.ANIMALS

模式67　公共用地**

　　……正如在邻里这一级内需要有公共用地一样——**近宅绿地**（60），在住宅团组和构成邻里的工作社区之内也是如此，需要有更小的、私密性更强的、为一些工作组和家庭共同使用的公共用地。这种公共用地实际上是任何住宅团组的心脏和灵魂。一旦对公共用地作出明确规定，住宅团组的私人住宅就会在其周围兴建——**住宅团组**（37）、**联排式住宅**（38）、**丘状住宅**（39）和**工作社区**（41）。

∞≪≫∞

　　没有公共用地，任何社会制度都无法继续存在。

　　工业化以前的社会中，在住宅之间和工作间之间，公共用地是自发存在的，所以过去在这方面就无须立论。通往住宅和工间的小路和街道是安全的社会空间，因此自动地起到了公共土地的作用。

67   COMMON LAND**

…just as there is a need for public land at the neighborhood level—ACCESSIBLE GREEN(60),so also, within the clusters and work communities from which the neighborhoods are made,there is a need for smaller and more private kinds of common land shared by a few work groups or a few families.This common land,in fact,forms the very heart and soul of any cluster.Once it is defined,the individual buildings of the cluster form around it—HOUSE CLUSTER (37),ROW HOUSES(38),HOUSING HILL(39),WORK COMMUNITY(41).

༄༅

Without common land no social system can survive.

In pre-industrial societies,common land between houses and between workshops existed automatically—so it was never necessary to make a point of it.The paths and streets which gave access to buildings were safe,social spaces,and therefore functioned automatically as common land.

But in a society with cars and trucks,the common land which can play an effective social role in knitting people together no longer happens automatically.Those streets which carry cars and trucks at more than crawling speeds,definitely do not function as common land;and many buildings find themselves entirely isolated from the social fabric because they are not joined to one another by land they hold in common.In such a situation common land must be provided,separately,and

在今天有大量轿车和卡车的社会里，在团结人民方面能起到有效社会作用的公共用地再也不会自发地出现。行驶速度比爬行略快的大大小小的汽车所经过的街道，肯定不再具有公共用地的功能了；并且，许多建筑都与社会的结构网络处于完全隔绝的状态，因为没有公共用地把它们连接在一起。在此情况下，就必须周密计划，分散地提供公共用地，它是一种社会需要，它的重要性可与街道相提并论。

公共用地具有两种特殊的社会功能。第一，它有可能使居民在他们的住宅和私人宅地以外感到心情舒畅，从而使他们感到自己和更大的社会体系保持联系——虽然不必感到和任何特定的邻居保持联系。第二，公共用地可用作居民集会的场所。

第一种功能是十分微妙的。当然，近邻的重要性在现代社会中比在传统的社会中小多了。这是因为人们在工作中、在学校里、在有关的小组会上常常见到朋友，因此，再也不用过分依赖邻居的友谊了。(See for instance, Melvin Webber, "Order in Diversity : Commtmity Without Propinquity," Cities and Space, ed.Lowdon Wingo, Baltimore : Resources for the Future, 1963 ; and Webber, "The Urban Place and the Nonplace Urban Realm," in Webber et al., Explorations into Urban Structure, Philadelphia, 1964, pp.79 ~ 153)

住宅间的公共用地不像过去用作友人见面的场所时那样重要了，这一点在某种程度上是正确的。但它仍有其重要性，这是因为它具有更深刻的心理功能，即使邻居间没有一丝一毫的关系。为了说明这种功能，请你设想一下，一个大深坑把你的住宅和城市分隔开了。每当你离开或进入家门时都必经此坑。你的住宅处于令人不安的孤立状态，而你，在这种住宅内，仅由于这个大深坑，就会和社会隔

with deliberation,as a social necessity,as vital as the streets.

The common land has two specific social functions. First,the land makes it possible for people to feel comfortable outside their buildings and their privae territory,and therefore allows them to feel connected to the larger social system— though not necessarily to any specific neighbor.And second, common land acts as a meeting place for people.

The first function is subtle.Certainly one's immediate neighbors are less important in modern society than in traditional society.This is because people meet friends at work,at school,at meetings of interest groups and therefore no longer rely exclusively on neighbors for friendship.(See for instance,Melvin Webber, "Order in Diversity:Community Without Propinquity," Cities and Space,ed.Lowdon Wingo, Baltimore:Resources for the Future,1963;and Webber, "The Urban Place and the Nonplace Urban Realm," in Webber et al.,Explorations into Urban Structure,Philadelphia,1964, pp.79-153)

To the extent that this is true,the common land between houses might be less important than it used to be as a meeting ground for friendship.But the common land between buildings may have a deeper psychological function,which remains important,even when people have no relation to their neighbors. In order to portray this function,imagine that your house is separated from the city by a gaping chasm,and that you have to pass across this chasm every time you leave your house,or enter it.The house would be disturbingly isolated;and you,in the house,would be isolated from society,merely by this physical fact.In psychological terms,we believe that a building without common land in front of it is as isolated from society as if it

离开来。从心理学的角度看，我们认为住宅前没有公共用地，将和前面有个大深坑一样与社会隔离。

现在有一种新的感情紊乱失调——集市恐怖症——在今天的城市中出现了。任何原因都会使患此症者害怕走出家门，甚至害怕出门寄信或到街角杂货店去，从字面上讲，他们害怕市场——**agora**。我们（毫无根据地）推测，这种感情紊乱会因缺乏公共用地而加剧，也会因人们感到无"权"待在他们的正门外这样一种环境而加剧。如果情况果真如此，那么集市恐怖症就是缺乏公共用地最具体的表现了。

公共用地的第二个社会功能是一目了然的。它为住宅团组的成员的流动和共同活动提供集会场所。为邻里服务的较大的公共用地——公园（社区设施）——并不符合上述条件。对整个邻里而言公园是很好的公共用地，但不能为住宅团组的共同功能提供基础。

刘易斯·芒福德写道：

即使每英亩有 12 户的住宅区——也许人们应当特别指出——在那里母亲们往往没有聚首见面的场所。倘有这种场所，她们就可以在风和日丽之日，来到大树下或藤架下一块儿做针线活或拉拉家常。那时，她们的小宝宝躺在婴儿车里熟睡，幼童在游戏坑四周挖土玩。也许查尔斯·赖利爵士的农村绿化区规划的最佳部分就是为这样的共同活动提供了场所：正如长岛森尼赛德的设计师斯坦和赖特先生早在 1924 年所做的那样。(The Urban Prospect，New York：Harcourt，Brace and World，1968，p.26)

公共用地应有多大呢？它必须大得足以发挥效用，必须能容纳孩子们做游戏和举行小型联欢会。必须有足够的土地是公共的，这样从心理学的角度看，私人土地不应压倒公共用地。我们猜测，邻里所需的公共用地的数量约占私人土地的 25％。绿化带的设计师正是把这一数据应用于典型的公共用地和绿化区。(See Clarence Stein，Toward

had just such a chasm there.

There is a new emotional disorder—a type of agoraphobia—making its appearance in today's cities. Victims of this disorder are afraid to go out of their houses for any reason—even to mail a letter or to go to the corner grocery store—literally, they are afraid of the marketplace—the agora. We speculate—entirely without evidence—that this disorder may be reinforced by the absence of common land, by an environment in which people feel they have no "right" to be outside their own front doors. If this is so, agoraphobia would be the most concrete manifestation of the breakdown of common land.

The second social function of common land is straightforward. Common land provides a meeting ground for the fluid, common activities that a house cluster shares. The larger pieces of public land which serve neighborhoods—the parks, the community facilities—do not fill the bill. They are fine for the neighborhood as a whole. But they do not provide a base for the functions that are common to a cluster of households.

Lewis Mumford:

Even in housing estates that are laid out at twelve families to the acre—perhaps one should say especially there—there is often a lack of common meeting places for the mothers, where, on a good day, they might come together under a big tree, or a pergola, to sew or gossip, while their infants slept in a pram or their runabout children grubbed around in a play pit. Perhaps the best part of Sir Charles Reilly's plans for village greens was that they provided for such common activities:as the planners of Sunnyside, Long Island, Messrs. Stein and Wright, had done as early as 1924.(The Urban Prospect, New York:Harcourt, Brace and World, 1968, p.26)

New Towns in America，Cambridge：M.I.T.Press，1966)

　　只有居民通力合作，才有可能在现存的邻里内通过封住街道的办法逐步实现本模式。

现已转变成邻里公共用地的伯克利大街
Berkeley street transformed to neighborhood commons

因此：

　　在住宅团组内划出 25% 以上的土地作为公共用地。这片土地距离共同使用它的各住户非常近。重要的是：谨防汽车，绝不能让汽车主宰这片土地。

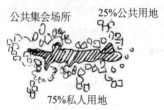

公共集会场所　　　25%公共用地

75%私人用地

᠗᠘᠓᠙

　　把公共用地围护起来，并使它具有充足的阳光——**朝南的户外空间**（105）、**户外正空间**（106）；以便有更小的、私密性更强的零星小块土地用作公共用地——**外部空间的层次**（114）；在公共用地之内提供公共设施——**户外亭榭**（69）、**地方性运动场地**（72）、**菜园**（177）；并把形状不同的和相邻的小块公共用地连成一个狭长的、相互沟通的游戏场地——**相互沟通的游戏场所**（68）。如果公路是**绿茵街道**（51），就可以是公共用地的一部分……

How much common land must there be?There must be enough to be useful,to contain children's games and small gatherings.And enough land must be common so that private land doesn't dominate it psychologically.We guess that the amount of common land needed in a neighborhood is on the order of 25 per cent of the land held privately.This is the figure that the greenbelt planners typically devoted to their cornons and greens.(See Clarence Stein,Toward New Towns in America,Cambridge:M.I.T.Press,1966)

With cooperation among the people,it is possible to build this pattern piecemeal,into our existing neighborhoods by closing streets.

Therefore:

Give over 25 per cent of the land in house clusters to common land which touches,or is very very near,the homes which share it.Basic:be wary of the automobile;on no account let it dominate this land.

❧❧

Shape the common land so it has some enclosure and good sunlight—SOUTH FACING OUTDOORS(105),POSITIVE OUTDOOR SPACE(106);and so that smaller and more private pieces of land and pockets always open onto it—HIERARCHY OF OPEN SPACE(114);provide communal functions within the land—PUBLIC OUTDOOR ROOM(69),LOCAL SPORTS (72),VEGETABLE GARDEN(177);and connect the different and adjacent pieces of common land to one another to form swaths of connected play space—CONNECTED PLAY(68). Roads can be part of common land if they are treated as GREEN STREETS(51)...

模式68 相互沟通的游戏场所*

……假设现在向住宅团组提供相互连接的公共土地——**公共用地**（67）。很有必要在此公共用地内为儿童们明确划出游戏场所，要使几块相邻空地的关系处理成能形成这样的游戏场所。

❧❧❧

如果1～5岁的儿童都不在一块儿痛痛快快地玩，则他们今后一生中将极有可能患上某种精神疾病。

儿童需要其他儿童。一些研究结果表明，儿童需要其他儿童的程度竟超过需要他们各自的母亲。从一些经验得出的证据表明，如果儿童几乎相互不接触，被迫在孤独中度过幼年时期，则他们在今后的岁月中很可能患精神病和神经病。

68   CONNECTED PLAY*

...suppose the common land that connects clusters to one another is being provided—COMMON LAND(67).Within this common land,it is necessary to identify play space for children and,above all,to make sure that the relationship between adjacent pieces of common land allows this play space to form.

᠄᠄᠄

If children don't play enough with other children during the first five years of life,there is a great chance that they will have some kind of mental illness later in their lives.

Children need other children.Some findings suggest that they need other children even more than they need their own mothers.And empirical evidence shows that if they are forced to spend their early years with too little contact with other children,they will be likely to suffer from psychosis and neurosis in their later years.

Since the layout of the land between the houses in a neighborhood virtually controls the formation of play groups,it therefore has a critical effect on people's mental health.A typical suburban subdivision with private lots opening off streets almost confines children to their houses.Parents,afraid of traffic or of their neighbors,keep their small children indoors or in their own gardens:so the children never have enough chance meetings with other children of their own age to form the groups which are essential to a healthy emotional development.

孤独……
Alone...

　　邻里内住宅之间的公共土地的布置实际上决定着儿童的游戏小组是否能形成，因此它对儿童的精神健康具有重大影响。在市郊典型的住宅团组内，住户都有一小块土地，且远离街道，所以几乎都把自己的孩子限制在住宅的小圈子里。凡是担心发生交通事故或担心邻居居心不良的父母，往往把幼童关在家里或花园里。结果，同龄的儿童就永远也不会有相互充分交往接触的机会，儿童游戏小组也就组织不起来了，而这样的游戏小组对儿童精神的健康发展是至关重要的。

　　我们将表明，如果每户都有通往某种安全的、相互沟通的公共用地，至少和 64 户人家发生接触交往，那么儿童们才可能相互串门。

　　首先，让我们考察一下有关这一问题的证据。最引人注目的证据摘自哈洛兄弟（the Harlows）有关猴子的著作。哈洛兄弟指出，在出生后前 6 个月离群独居的小猴子以后就没有能力和其他猴子一块儿过正常的群居生活、进行交配或玩耍：

　　猴子表现出来的反常行为在野生动物界里是罕见的。它们坐在笼子里，一动也不动地凝视着；或在笼内呆板地反复兜圈子；或用前爪和前臂抱着头，长时间地乱摇晃……这些猴子会咬伤或撕裂自己的皮肉直到鲜血淋淋……相似的精神失常症状在孤儿院里失去一切的儿童中以及在精神病院里孤独的青少年和成年人中都可以被观察到。（Henry F.Harlow and Margaret K.Harlow，"The

We shall show that children will only be able to have the access to other children which they need,if each household opens onto some kind of safe,connected common land,which touches at least 64 other households.

First,let us review the evidence for the problem.The most dramatic evidence comes from the Harlows'work on monkeys. The Harlows have shown that monkeys isolated from other infant monkeys during the first six months of life are incapable of normal social,sexual,or play relations with other monkeys in their later lives:

They exhibit abnormalities of behavior rarely seen in animals born in the wild.They sit in their cages and stare fixedly into space,circle their cages in a repetitively stereotyped manner,and clasp their heads in their hands or arms and rock for long periods of time...the animal may chew and tear at its body until it bleeds...similar symptoms of emotional pathology are observed in deprived children in orphanages and in withdrawn adolescents and adults in mental hospitals.(Henry F.Harlow and Margaret K.Harlow, "The Effect of Rearing Conditions on Behavior," Bull Menniger Clinic,26,1962,pp.213-214)

It is well known that infant monkeys—like infant human beings-have these defects if brought up without a mother or a mother surrogate.It is not well known that the effects of separation from other infant monkeys are even stronger than the effects of maternal deprivation.Indeed,the Harlows showed that although monkeys can be raised successfully without a mother,provided that they have other infant monkeys to play with,they cannot be raised successfully by a mother alone,without other infant monkeys,even if the mother is entirely normal.They conclude: "It seems possible that the

Effect of Rearing Conditions on Behavior," Bull Menniger Clinic,26,1962,PP.213～214)

众所周知,幼猴和幼儿一样,如果它们在成长中得不到母猴或代理母猴的照料,就会有上述缺陷。幼猴因相互隔离所造成的影响远远超过无母猴所造成的影响,这点尚未被人们所熟知。确实哈洛兄弟说过,尽管有些幼猴没有母猴,但只要有伴一块儿玩耍,仍能饲育成功,但如果没有伴在一起玩耍,即使有完全正常的母猴单独照料它们,也无法饲育成功。他们得出结论:"下面的情况似乎是可能的:幼猴——母猴感情体系是可以缺少的,而幼猴——幼猴感情体系在调节猴子今后生活的各方面是不可缺少的。"(Harry F.Harlow and Margaret K.Harlow,"Social Deprivation in Monkeys," Scientific American, 207, No.5, 1962, PP.136～146)

一只猕猴头6个月的生活与一个幼儿头3年的生活相符合。虽然现在没有正式的证据表明幼儿在头3年缺少接触会受到何种程度的损害——就我们所知,这一点还从未研究过——但是,在4～10岁的儿童之间的隔离状态所造成的后果已有确凿的证据。

赫尔曼·朗茨曾在美国陆军中进行了一次任意抽样调查,询问了1000名军人,他们由于精神错乱而早已被送进了精神病院。(Herman K.Lantz,"Number of Childhood Friends as Reported in the Life Histories of a Psychiatrically Diagnosed Group of 1,000," *Marriage and Family Life*, May 1956, PP.107～108) 美国陆军精神病医生们把其中的每个人进行分类,分成正常人、轻微精神性神经病患者和严重精神性神经病患者或精神病患者。接着,朗茨把他们列入3个范畴:他们报告他们在4～10岁期间,在任何一个有代表性的时刻,结交的朋友数。第一批人说他们有5个或5个以上的朋友,第二批人说他们平均约有2个朋友,

infant-mother affectional system is dispensable,whereas the infant-infant system is a sine-qua-non for later adjustment in all spheres of monkey life." (Harry F.Harlow and Margaret K.Harlow, "Social Deprivation in Monkeys," Scientific American,207,No.5,1962,pp.136-146)

The first six months of a rhesus monkey's life correspond to the first three years of a child's life.Although there is no formal evidence to show that lack of contact during these first three years damages human children—and as far as we know,it has never been studied—there is very strong evidence for the effect of isolation between the ages of four to ten.

Herman Lantz questioned a random sample of 1,000 men in the United States Army,who had been referred to a mental hygiene clinic because of emotional difficulties. (Herman K.Lantz, "Number of Childhood Friends as Reported in the Life Histories of a Psychiatrically Diagnosed Group of 1,000," *Marriage and Family Life*,May 1956, pp.107-108)Army psychiatrists classified each of the men as normal,suffering from mild psychoneurosis,severe psychoneurosis,or psychosis.Lantz then put each man into one of three categories:those who reported having five friends or more at any typical moment when they were between four and ten years old,those who reported an average of about two friends,and those who reported having no friends at that time. The following table shows the relative percentages in each of the three friendship categories separately.The results are astounding:

第三批人说他们没有任何朋友。下列的表格分别表示 3 种友谊范畴中的每一种的相对百分比。其结果是令人震惊的：

分类	5个或5个以上的朋友	大约2个朋友	没有任何朋友
正常人	39.5	7.2	0.0
轻微精神性神经病患者	22.0	16.4	5.0
严重精神性神经病患者	27.0	54.6	47.5
精神病患者	0.8	3.1	37.5
其他患者	10.7	18.7	10.0
合计	100.0	100.0	100.0

在他们童年时有 5 个或 5 个以上的朋友的人中，60.5％有轻微病情，而 27.0％有严重病情。在他们童年时没有任何朋友的人中，只有 5％有轻微病情，而 85％却有严重病情。

安娜·弗罗伊德的非正式报告令人信服地表明，幼儿之间的接触对儿童精神的发展会有多么巨大的影响。她描写了 5 个德国幼儿，当他们还是婴儿的时候，在集中营里，他们就失去了双亲。后来，他们在集中营内相互照顾，直到第二次世界大战结束为止，不久他们被遣返英国。（Anna Freud and Sophie Dann, "An Experiment in Group Upbringing," *Reading in Child Behavior and Development*, ed.Celia Stendler, New York, 1964, pp.122～140）她从社会和精神的角度描写了他们的成熟程度。读罢这篇报道，人们会感觉到，这些 3 岁的儿童是多么心心相印，他们相互需要的敏感性远远胜过现在的许多人。

由此可见，几乎可以断定：接触交往十分重要。缺少交往而达于极点，就会造成严重后果。除了我们在此引证的内容外，还有大量论述来自克里斯托弗·亚力山大的著作《城市的职能是维持人们的接触交往》。（Christopher Alexander："The City as a Mechanism for Sustaining Human

	5 or More Friends	About 2 Friends	No Friends
Normal	39.5	7.2	0.0
Mild psychoneurosis	22.0	16.4	5.0
Severe psychoneurosis	27.0	54.6	47.5
Psychosis	0.8	3.1	37.5
Other	10.7	18.7	10.0
	100.0	100.0	100.0

Among people who have five friends or more as children,60.5 percent have mild cases,while 27.0 per cent have severe cases.Among people who had no friends,only 5 per cent have mild cases,and 85 per cent have severe cases.

On the positive side,an informal account by Anna Freud shows how powerful the effect of contact among tiny children can be on the emotional development of the children.She describes five young German children who lost their parents during infancy in a concentration camp,and then looked after one another inside the camp until the war ended,at which point they were brought to England.(Anna Freud and Sophie Dann, "An Experiment in Group Upbringing," *Reading in Child Behavior and Development*,ed.Celia Stendler,New York,1964,pp.122-140) She describes the beautiful social and emotional maturity of these tiny children.Reading the account,one feels that these children,at the age of three,were more aware of each other and more sensitive to each other's needs than many people ever are.

It is almost certain,then,that contact is essential,and that lack of contact,when it is extreme,has extreme effects.A considerable body of literature beyond that which we have quoted,is given in Christopher Alexander: "The City as a Mechanism for Sustaining Human Contact," *Environment for Man*,ed.W.R.Ewald,Indiana

Contact," *Environment for Man*, ed.W.R.Ewald, Indiana University Press, Bloomington, 1967, pp.60 ~ 109)

如果我们认为，儿童和邻居的非正式接触是一条十分重要的经验，那么我们就会问：什么样的邻里会支持成立自发的游戏小组呢？我们认为答案是：邻里要有一块与儿童之家相连接的安全的公共用地，在这范围内儿童们可以相互接触交往。而重要的问题是：多少住户需要共同使用这种相互沟通的游戏场所呢？

所需住户的确切数字取决于这些住户的儿童数量。让我们假设：儿童约占已定人口数量的 1/4（略低于市郊住户的一般数字），并且这些儿童（包括青少年）的年龄从 0 ~ 18 岁分布均匀。粗略地说，某一学龄前儿童年龄为 x 岁，他将和年龄分别为 $x-1$ 或 $x+1$ 的儿童玩耍。为了达到一个合理的接触量并成立游戏小组，每一儿童必须至少和 5 个同龄儿童保持接触。统计分析表明，每个儿童有 95% 的机会同这样 5 个潜在游戏伙伴接触，所以他必须到达的住户范围为 64 户人家。

这一问题可以说明如下：在一个无限定的儿童人口中，对每一儿童来说，他有相当于总儿童人口的 1/6 的儿童年龄和他相当，有 5/6 的儿童年龄和他相差甚远。任意选出几个儿童为一组。该组包含 5 个或 5 个以上年龄相当的儿童。其概率为 $1-\sum_{k=0}^{4} P_{r,k}$，式中 $P_{r,k}$ 是超几何学分布。

如果我们问：在 $1-\sum_{k=0}^{4} P_{r,k} > 0.95$ 中 r 的最小值是多少？求出的结果为 54。

如果我们需要 54 个儿童，则我们需要的总人口为 $4 \times 54 = 216$。按每户 3.4 人计算，需有住户 64 户。

共同使用公共用地的 64 户住家是一个相当大的数量。事实上，面对这种要求，人们力图解决这一问题，办法是：

University Press,Bloomington,1967,pp.60-109.

If we assume that informal,neighborhood contact between children is a vital experience,we may then ask what kinds of neighborhoods support the formation of spontaneous play groups.The answer,we believe,is some form of safe common land,connected to a child's home,and from which he can make contact with several other children.The critical question is:How many households need to share this connected play space?

The exact number of households that are required depends on the child population within the households.Let us assume that children represent about onefourth of a given population(slightly less than the modal figure for suburban households),and that these children are evenly distributed in age from 0 to 18.Roughly speaking,a given preschool child who is x years old will play with children who are x-I or x or x+I years old.In order to have a reasonable amount of contact,and in order for playgroups to form,each child must be able to reach at least five children in his age range.Statistical analysis shows that for each child to have a 95 per cent chance of reaching five such potential playmates,each child must be in reach of 64 households.

The problem may be stated as follows:In an infinite population of children,one-sixth are the right age and five-sixths are the wrong age for any given child.A group of r children is chosen at random.The probability that this group of r children contains 5 or more right-age children in it is $1-\sum_{k=0}^{4} P_{r,\,k}$, where $P_{r,\,k}$ is the hypergeometric distribution.If we now ask what is the least r which makes $1-\sum_{k=0}^{4} P_{r,\,k}>0.95$ r turns out to be 54.

If we need 54 children,we need a total population of

将住户编成 10 ~ 12 户的住宅团组。但这起不了什么作用。住宅团组是一个有用的组合形式，它另有用途——**住宅团组**（37）和**公共用地**（67）。所以，住宅团组本身也解决不了儿童们使用相互沟通的游戏场所这一问题。还必须有一些安全的小路把许多块公共用地沟通起来才行。

沟通的小路
Connecting paths

因此：

将公共用地、小路、花园和桥梁布置在至少 64 户住家的范围内，并有一片狭长的、不横穿公路的土地把它们连接起来。把这片土地建成这些住户的孩子们使用的相互沟通的游戏场所。

ക౩⋈౩ക

务必把数个**住宅团组**（37）和**绿茵街道**（51）以及安全小道连接起来，才能建立相互沟通的游戏场所。在这游戏场内设置地方**儿童之家**（86），并确保儿童们能玩烂泥、观看动植物和到水塘做游戏——**池塘**（71）、**动物**（74）；保留一块废料堆放场，让他们利用废物制作各种东西——**冒险性的游戏场地**（73）……

4(54)=216,which at 3.4 persons per household,needs 64 households.

Sixty-four is a rather large number of households to share connected common land.In fact,in the face of this requirement,there is a strong temptation to try to solve the problem by grouping 10 or 12 homes in a cluster.But this will not work:while it is a useful configuration for other reasons—HOUSE-CLUSTER(37)and COMMON LAND(67)—by itself it will not solve the problem of connected play space for children.There must also be safe paths to connect the bits of common land.

Therefore:

Lay out common land,paths,gardens,and bridges so that groups of at least 64 households are connected by a swath of land that does not cross traffic.Establish this land as the connected play space for the children in these households.

৪০৫৪

Do this by connecting several HOUSE CLUSTERS(37) with GREEN STREETS(51)and safe paths.Place the local CHILDREN'S HOME(86)in this play space.Within the play space,make sure the children have access to mud,and plants,and animals,and water—STILL WATER (71),ANIMALS(74);set aside one area where there is all kinds of junk that they can use to make things—ADVENTURE PLAYGROUND(73)...

模式69 户外亭榭**

……在**主门道**（53）、**近宅绿地**（60）、**小广场**（61）、**公共用地**（67）、**步行街**（100）、**小路和标志物**（120）这些模式的公共用地内，至少要有一块地方供人们散步闲逛和"在外"抛头露面。为此，必须从公共用地划出一方空地，并稍加整饰以供使用。如果还没有一个较大的模式存在，本模式可起核心作用，并有助于较大的模式在其周围"结晶"。

❧❧❧

在现代城镇和邻里内，沿街几乎没有供人们悠闲自在地逗留几小时的地方。

人们寻觅街角啤酒店，把酒言欢，消磨时光：青少年，尤其是男孩，也选择街角闲逛，等待朋友。老人喜欢有一个特殊的天地，聚首聊天；儿童需要小块沙地，玩泥巴，摆弄植物，或在野外戏水；年轻的母亲们常常利用孩子们

69   PUBLIC OUTDOOR ROOM**

...the common land in MAIN GATEWAYS (53), ACCESSIBLE GREEN (60), SMALL PUBLIC SQUARES(61), COMMON LAND(67), PEDESTRIAN STREET(100),PATHS AND GOALS(120)needs at least some place where hanging out and being "out" in public become possible.For this purpose it is necessary to distinguish one part of the common land and to define it with a little more elaboration.Also,if none of the larger patterns exist yet,this pattern can act as a nucleus,and help them to crystallize around it.

৪০৫৪

There are very few spots along the streets of modern towns and neighborhoods where people can hang out, comfortably,for hours at a time.

Men seek corner beer shops,where they spend hours talking and drinking;teenagers,especially boys,choose special corners too,where they hang around,waiting for their friends. Old people like a special spot to go to,where they can expect to find others;small children need sand lots,mud,plants,and water to play with in the open;young mothers who go to watch their children often use the children's play as an opportunity to meet and talk with other mothers.

Because of the diverse and casual nature of these activities,they require a space which has a subtle balance of being defined and yet not too defined,so that any activity which is natural to the neighborhood at any given time can develop

游戏的机会相互攀谈。

人们要进行多种多样、不拘礼节的活动，所以需要这样一种空间，在其明确的和不太明确的限定之间达到十分微妙的平衡，以便邻里内任何一种天然合理的活动可在任何时刻自由开展，并有所创新。

例如，可以把未完工的室外空间留在那里，让附近居民自己动手完成，以满足他们最迫切的需要，如沙地、水龙头或儿童游戏设备——**冒险性的游戏场地**（73）；户外亭榭也许有台阶和坐位，男女青年可以在那里聚会——**青少年协会**（84）；或许有人会在一所面向这个地方的房屋内开设一个小酒吧间或咖啡店，它有拱廊，并使拱廊成为人们吃喝的场所——**饮食商亭**（93）；那里或许还有老年人酷爱的棋类游戏。

现代住宅建筑深受缺少这种空间之害。当社区提供室内活动空间时，利用者寥寥无几。人们不愿陷入他们未知的环境；同时在这种封闭式空间中所产生的缺点是它所固有的，以至于不会逐渐增强人们对它的兴趣。另一方面，空地总是不会那么封闭的，在它上面出现什么建筑不是一年半载之内的事，而要经过许多年。它几乎没有什么遮蔽，所以，没有理由"不在那里"休憩游玩。

户外亭榭所需要的就是一个足以规定范围的框架结构。人们见到它就会很自然地停步；不言而喻，他们的好奇心会怂恿他们去那里逗留。一旦社区的许多小组开始向往这种框架结构，如果他们获准，他们就将有良好的机会为他们自己创造出一个适合于他们活动的环境。

我们猜测，一个这样的小空间，它上面有顶盖，四周有立柱，但至少部分没有墙壁，这将正好提供一种"开敞性"和"封闭性"之间必要的平衡。

戴夫·蔡平和乔治·戈登以及来自俄亥俄州克利夫兰市的凯西·韦斯顿预备役军校学建筑的学生，共同建造的模式就是一个非常出色的例证。他们在一个地方精神病院

freely and yet has something to start from.

For example,it would be possible to leave an outdoor room unfinished,with the understanding it can be finished by people who live nearby,to fill whatever needs seem most pressing.It may need sand,or water faucets,or play equipment for small children—ADVENTURE PLAYGROUND(73);it may have steps and seats,where teenagers can meet— TEENAGE SOCIETY(84);someone may build a small bar or coffee shop in a house that opens into the area,with an arcade,making the arcade a place to eat and drink—FOOD STANDS (93);there may be games like chess and checkers for old people.

Modern housing projects especially suffer from the lack of this kind of space.When indoor community rooms are provided,they are rarely used.People don't want to plunge into a situation which they don't know;and the degree of involvement created in such an enclosed space is too intimate to allow a casual passing interest to build up gradually.On the other hand,vacant land is not enclosed enough.It takes years for anything to happen on vacant land;it provides too little shelter,and too little "reason to be there."

What is needed is a framework which is just enough defined so that people naturally tend to stop there;and so that curiosity naturally takes people there,and invites them to stay. Then,once community groups begin to gravitate toward this framework,there is a good chance that they will themselves,if they are permitted,create an environment which is appropriate to their activities.

We conjecture that a small open space,roofed,with columns,but without walls at least in part,will just about provide

的空地上及其周围的公共用地上建造了一系列的户外亭榭。根据采访人员的报导，这些地方引人注目地改变了精神病院的生活面貌，被吸引到户外去活动的人比平时多了许多。他们谈笑风生，欣喜之情溢于言表；那些一度为轿车所占据的户外空间，顷刻之间突然变得富有人情味了，它们不得不减速，缓慢地向前移动。

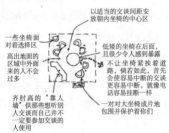

以适当的交谈间距安放朝内坐椅的中心区

一些坐椅面对着选择区

低矮的坐椅在后面，且很少令人感到暴露

高出地面的区域中外面来的人不会过多

不让坐椅紧挨着道路，倘若如此，首先会使容易中断的交谈更容易中断，就像电话容易挂断一样

齐肘高的"靠人墙"供那些想听别人交谈而自己并不一定要参加交谈的人使用

一对大坐椅成片地包围并保护着你们

戴夫·蔡平和乔治·戈登在俄亥俄州克利夫兰市设计的户外亭榭
Public outdoor room built by Chapin and Gordon in Cleveland, Ohio

蔡平和戈登以及他们的全体成员在该邻里中总共建造了 7 个户外亭榭。它们在造形上大同小异，因景观、朝向和规模的不同而处理略有变化。

我们从中世纪社会中也已发现本模式的一种早期形式。很明显，在 12 世纪和 13 世纪时，许多这样的户外亭榭分散布置在城镇中，它们曾是拍卖货物的平台、露天集会场所和商品展销地。它们与我们为邻里和工作社区所建议的户外亭榭十分相似。

在英国和秘鲁的户外亭榭
Outdoor rooms in England and Peru

the necessary balance of "openness" and "closedness."

A beautiful example of the pattern was built by Dave Chapin and George Gordon with architecture students from Case Western Reserve in Cleveland, Ohio. They built a sequence of public outdoor rooms on the grounds and on the public land surrounding a local mental health clinic. According to staff reports, these places changed the life of the clinic dramatically: many more people than had been usual were drawn into the outdoors, public talk was more animated, outdoor space that had always been dominated by automobiles suddenly became human and the cars had to inch along.

In all, Chapin and Gordon and their crew built seven public outdoor rooms in the neighborhood. Each one was slightly different, varying according to views, orientation, size.

We have also discovered a version of this pattern from medieval society. Apparently, in the twelfth and thirteenth centuries there were many such public structures dotted through the towns. They were the scene of auctions, open-air meetings, and market fairs. They are very much in the spirit of the places we are proposing forneighborhoods and work communities.

Therefore:

In every neighborhood and work community, make a piece of the common land into an outdoor room—a partly enclosed place, with some roof, columns, without walls, perhaps with a trellis; place it beside an important path and within view of many homes and workshops.

因此：

在每一邻里和工作社区，将一块公共用地建设成为户外亭榭——它部分封闭，上有某种屋顶而四周有立柱，但无墙壁，也许有花格墙；把户外亭榭设在重要的便道旁边，同时在许多住宅和车间的视野内。

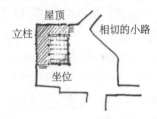

将户外亭榭设在有几条小路和它相切的地方，像任何其他的公共区一样——**中心公共区**（129）；在小路凸出的地方——**小路的形状**（121）；或在某一广场四周——**袋形活动场地**（124）；利用周围的**建筑物边缘**（160）去限定户外亭榭的一部分；户外亭榭四周有立柱，上有半棚式顶盖——**有围合的户外小空间**（163），和其他较小的户外空间一样，要造得小巧玲珑；也许，紧挨着它设一露天庭院——**有生气的庭院**（115）；**绕户外亭榭有拱廊**（119）或其他简易遮盖物——**帆布顶篷**（244）和供游人偶尔休憩的坐位——**能坐的台阶**（125）和**户外设座位置**（241）……

Place the outdoor room where several paths are tangent to it,like any other common area—COMMON AREAS AT THE HEART(129);in the bulge of a path—PATH SHAPE(121);or around a square—ACTIVITY POCKETS(124);use surrounding BUILDING EDGES(160)to define part of it;build it like any smaller outdoor room,with columns,and half-trellised roofs—OUTDOOR ROOM(163);perhaps put an open courtyard next to it—COURTYARDS WHICH LIVE(115),an ARCADE(119)around the edge,or other simple cover—CANVAS ROOFS(244),and seats for casual sitting—STAIR SEATS(125),SEAT SPOTS(241)...

模式70　墓地*

　　……根据**生命的周期**（26），人生从一个时期向另一个时期的转变，在每个社区内，必须是自然的和显而易见的。死亡也不例外。本模式有助于把死亡同每一邻里内的公共用地结合起来，并以此促成**易识别的邻里**（14）、**圣地**（66）和**公共用地**（67）。

୧୦ଓଃ

　　没有人能死而复生。在任何一个激励人们去生活的社会中活着的人纷纷谢世是司空见惯的现象。

70   GRAVE SITES*

...according to LIFE CYCLE(26)the transitions of a person's life must be available and visible in every community.Death is no exception.This pattern helps to integrate the fact of death with the public spaces of each neighborhood,and,by its very existence,helps to form IDENTIFIABLE NEIGHBORHOODS(14),and HOLY GROUND(66)and COMMON LAND (67).

৪০৪৪

No people who turn their backs on death can be alive.The presence of the dead among the living will be a daily fact in any society which encourages its people to live.

Huge cemeteries on the outskirts of cities,or in places no one ever visits,impersonal funeral rites,taboos which hide the fact of death from children,all conspire to keep the fact of death away from us,the living.If you live in a modern suburb,ask yourself how comfortable you would be if your house were next to a graveyard.Very likely the thought frightens you.But this is only because we are no longer used to it.We shall be healthy,when graves of friends and family,and memorials to the people of the recent and the distant past,are intermingled with our houses, insmall grave yards,as naturally as winter always comes before the spring.

In every culture there is some form of intense ceremony surrounding death,grieving for the dead,and disposal of the body. There are thousands of variations,but the point is always to give the community of friends left alive the chance to reconcile themselves to

市郊许多巨大的公墓，或人迹罕至的地方非个人的葬礼，向儿童隐瞒死亡之谜的忌讳，这些都说明我们这些活着的人在竭力回避死亡这一严峻的事实。如果你住在现代化的市郊，如果你的住宅就在墓旁，你扪心自问，是否会感到非常舒适。很可能这种想法会使你毛骨悚然。但这仅仅是因为我们不再习惯于此罢了。如果亲朋好友和至亲骨肉的坟墓，以及为悼念近期和远期的死者而建的纪念碑与我们的住宅混杂在一起，就像冬去春来那样自然时，我们的精神才将是健康的。

在每一种文化中，就丧事而言，都要举行某种隆重的葬礼，以便悼念死者，处理尸体。葬礼千差万别，但其主要用意是给社区内还活着的朋友以机会，使他们自己安于终将降临的死亡这一事实：生命稍纵即逝，只余一片空虚和乌有。

这些葬礼使人们经常感到人总有一死，这样就会使我们更加正视生和死的现实。当我们把这些感受和周围的环境以及每一个人的生活联系在一起，我们就能生活得充实并继续生活下去。但当环境和习俗不让我们去体验人终有一死的必然性，我们在生活中没有这种感受，我们的精神就是抑郁的、消沉的、缺少生气的。现在有大量的临床诊断的证据支持这种看法。

在一个提供上述证据的病例中记载了一个小男孩失去了他的外婆，周围的人为了"保护他的感情"，都告诉他，她刚刚"远去他乡"。这个男孩心神不安地意识到发生了什么意外，但在这种神秘的气氛中，他无法得知事情的真相，因而就不能充分地感受它。结果，他并没有受到保护，而是闷闷不乐，积郁成疾，患了严重的精神性神经病。在许多年之后，当他最后认识到并感受了他外婆已经死去这一事实后，他的顽症才被治愈。

这一病例和其他许多病例都非常清楚地告诉我们，一

the facts of death:the emptiness,the loss;their own transience.

These ceremonies bring people into contact with the experience of mortality,and in this way,they bring us closer to the facts of life,as well as death.When these experiences are integrated with the environment and each person's life,we are able to live through them fully and go on.But when circumstances or custom prevent us from making contact with the experience of mortality,and living with it,we are left depressed,diminished,less alive.There is a great deal of clinical evidence to support this notion.

In one documented case,a young boy lost his grandmother; the people around him told him that she had merely "gone away" to "protect his feelings." The boy was uneasily aware that something had happened,but in this atmosphere of secrecy,could not know it for what it was and could not therefore experience it fully.Instead of being protected,he became a victim of a massive neurosis,which was only cured,many years later,when he finally recognized,and lived through the fact of his grandmother's death.

This case,and others which make it abundantly clear that a person must live through the death of those he loves as fully as possible,in order to remain emotionally healthy,have been described by Eric Lindemann.We have lost the crucial reference for this work,but two other papers by Lindemann converge on the same point: "Symptomatology and Management of Acute Grief," *American Journal of Psychiatry*,1944,101,pp.141-148;and "A Study of Grief:Emotional Responses to Suicide," *Pastoral Psychology*,1953,4(39),pp.9-13.We also recommend a recent paper by Robert Kastenbaum,on the ways in which children explore their mortality: "The Kingdom Where Nobody Dies," *Saturday Review*,January 1973,pp.33-38.

个人为了保持健康的精神状态，必须尽可能充分经历他所热爱的亲人溘然长逝的情景。这一点已由埃里克·林德曼作了描述。我们已经遗失了供本书用的重要参考资料，但林德曼的其他两篇论文都集中表达了同样的观点：《症状学和极度悲痛之治疗》（"Symptomatology and Management of Acute Grief," *American Journal of Psychiatry*，1944，101，pp. 141～148）和《悲痛之研究：自杀的精神反应》（"A Study of Grief：Emotional Responses to Suicide," *Pastoral Psychology*，1953.4（39），pp.9～13）。我们再推荐罗伯特·卡斯顿鲍姆最近撰写的一篇论文，论述儿童们如何探索他们必然死亡的途径：《无人死亡的王国》（"The Kingdom Where Nobody Dies," *Saturday Review*，*January* 1973，pp. 33～38）。

加利福尼亚考尔麦的一片混凝土蜂窝状墓地。公墓的主管人说："亲属们将永远不会再来缅怀埋在这里的先人……此情此景真使他们含怨九泉……"

A concrete honeycomb graveyard in Colma,California.The superintendant of the cemetery said,"Families will never see the sinking...which so distressed them in older parts of the cemetery...."

在大工业城市，近百年来，殡葬仪式及其潜移默化的作用对活人已不复存在。以前极为简朴的悼念仪式现在已被奇形怪状的公墓和塑料花所替代，除了死亡这一现实，其余一切都被取代了。尤为明显的是，过去一个个的小墓

In the big industrial cities,during the past 100 years,the ceremonies of death and their functional power for the living have been completely undermined.What were once beautifully simple forms of mourning have been replaced by grotesque cemeteries,plastic flowers,everything but the reality of death.And above all,the small graveyards which once put people into daily contact with the fact of death,have vanished—replaced by massive cemeteries,far away from people's daily business.

What must be done to set things right?We can solve the problem by fusing some of the old ritual forms with the kinds of situations we face today.

1.Most important,it is essential to break down the scale of modern cemeteries,and to reinstate the connection between burial grounds and local communities.Intense decentralization:a person can choose a spot for himself,in parks,common lands,on his land.

2.The right setting requires some enclosure;paths beside the gravesites;the graves visible,and protected by low walls,edges,trees.

3.Property rights.There must be some legal basis for hallowing small pieces of ground—to give a guarantee that the ground a person chooses will not be sold and developed.

4.With increasing population,it is obviously impossible to go on and on covering the land with graves or memorials. We suggest a process similar to the one followed in traditional Greek villages.The graveyards occupy a fixed area,enough to cope with the dead of 200years.After 200 years,remains are put out to sea—except for those whose memory is still alive.

5.The ritual itself has to evolve from a group with some shared values,at least a family,perhaps a group that shares

地能每天使人接触死亡这一事实，如今已一一消失，代之而起的是巨大的、远离人们日常生活的公墓。

为了把事情纠正过来，我们必须做些什么呢？我们把一些古老的葬礼和今天我们所面临的种种情况巧妙结合，问题就不难解决。

1. 最重要的是要缩小现代公墓的范围，恢复墓地和地方社区之间的联系；同时要大力分散墓地：人人都可以在公园里、公共用地或私人土地上选择一块墓地。

2. 合理安排围栏。墓旁有小径，坟墓清晰可见，四周有矮墙、边缘或树木等绿篱围护。

3. 产权归属：必须具备一定的法律根据，以便使小块的墓地成为圣地，并保证任何人所选择的墓地将不被允许出卖或开发。

4. 随着人口的日益增加，覆盖这些小块墓地的坟墓或纪念碑绝不能无止境地增多。我们建议采用类似希腊村民的传统做法。墓地有一定面积，足以妥善保存死者遗体达200年之久。200年后，遗骸葬入大海——只有至今仍被人们所怀念的人才能例外。

5. 这种葬礼必须由具有共同道德标准的团体、至少家庭或同一宗教信仰的团体发展起来。有3条基本要求：死者生前的亲朋好友护送灵柩，列队徐徐经过街道；灵柩可为简易的松木棺材或骨灰盒；送葬者在墓旁集会，悼念亡魂。

因此：

绝不建造巨大的公墓；而要在全社区内划出一块块的小土地——公园的角隅、小路的部分地段、花园、主门道两侧——用作墓地。这里可举行仪式为死者建立墓碑。碑上铭刻碑文，或饰有纪念标志。为每一墓地装饰边缘，墓旁辟出一条小径和一方幽静之地，以便人们坐下歇息。根据习惯，这就是一块圣地。

a religious view.Three of the basics:friends carrying the casket through the streets in procession;simple pine coffins or urns;gathering round the grave.

Therefore:

Never build massive cemeteries.Instead,allocate pieces of land throughout the community as grave sites—corners of parks,sections of paths,gardens,beside gateways—where memorials to people who have died can be ritually placed with inscriptions and mementos which celebrate their Life. Give each grave site an edge,a path,and a quiet corner where people can sit.By custom,this is hallowed ground.

❧❧❧

If possible,keep them in places which are quiet—QUIET BACKS (59);and provide a simple seat or a bench under a tree,where people can be alone with their memories—TREE PLACES(171),SEAT SPOTS (241)...

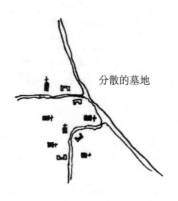

分散的墓地

⊷⊱⊰⊶

    如果可能的话，把墓地分散在幽静的地方——**僻静区**（59）；在树下设一简易坐位或放一条板凳，人们可以坐在那里独自回忆逝去的亲人——**绿荫空间**（171）和**户外设座位置**（241）……

模式71　池塘*

　　……模式**通往水域（25）、水池和小溪（64）**为全社区提供各种不同的水域。本模式有助于装饰点缀池塘——水池、河塘和水湾游泳区——并为儿童们提供安全的水滩。本模式也有益于**住宅团组（37）、工作社区（41）、保健中心（47）、公共用地（67）**和**地方性运动场地（72）**出现变化多端、风貌各异的公共空间。

⊱⊰

　　为了和水域保持接触，我们首先必须会游泳；为了让我们每天都能游泳，水池、河塘和水湾游泳区必须遍布全市，以便人人都能在几分钟之内到达一个游泳区。
　　我们在模式**水池和小溪（64）**中，已经解释了与水域保

71 STILL WATER*

...the patterns ACCESS TO WATER(25)and POOLS AND STREAMS(64)provide a variety of kinds of water throughout the community.This pattern helps to embellish the still waters—the pools and ponds and swimming holes—and provide them with a safe edge for children.It also helps to differentiate the public space in HOUSE CLUSTER(37),WORK COMMUNITY(41),HEALTH CENTER(47),COMMON LAND(67),LOCAL SPORTS(72).

∞∞∞

To be in touch with water,we must above all be able to swim;and to swim daily,the pools and ponds and holes for swimming must be so widely scattered through the city,that each person can reach one within minutes.

We have already explained,in POOLS AND STREAMS (64),how important it is to be in touch with water—and how the ordinary water of an area can,if left open,be a natural component of the everyday ecology of a community.

In this pattern we go a little further,and put the emphasis on swimming.On the one hand,adults cannot have any substantial contact with water unless they can swim in it,and for this purpose the body of water must be large enough and deep enough to swim in.On the other hand,the highly chlorinated,private,walled,and fenced off swimming pools,which have become common in rich people's suburbs,work directly against the very forces we have

持接触是何等的重要，以及一个地区的普通水域，如果是开放的，就可能是社区日常生态平衡的一个天然组成部分。

在本模式中我们要进一步强调游泳这一点。一方面，如果成年人不能游泳，就不会和水域保持真正的名副其实的接触。为此，水体必须有足够的宽度和深度以保证游泳者的安全。另一方面，那些私人的、用氯化物高度消毒的、环有围墙或护栏的游泳池，在富人们居住的郊区如今已成为十分普通的了。这样的游泳池直接违背我们在**水池和小溪**（64）中所描述的那些要求和做法，并使人感到来此游泳毫无意义，因为这是私人的，水中还含有大量的防腐灭菌剂。

我们认为，除非人人每天想去游泳就能游泳，否则游泳就不会占有恰当的位置。这就意味着，实际上在每一街区，几乎在每一住宅团组，至少在每一邻里，必须有一个游泳池，它离每幢住宅大约为100yd。

因此，我们试图在本模式中建立一个示范的"水湾游泳区"。它具有公共性质，所以它是公共场所，并非都是私人所有；它是十分安全的：深水处可游泳，在水边孩子们可以戏水而没有危险。

数百万年以来，儿童们在海滨、河边、湖畔平平安安地长大成人。而为什么今天的游泳池会如此危险呢？回答是简单的：危险出在水边上。

······水边······
...the edge...

通常，在水面和岸边之间的天然水边上，有一缓缓倾斜的凹凸不平的过渡区。当人们从岸上走入水中，就要经过这片水边，上面的各种东西、地形构造和生态环境呈现

described in POOLS AND STREAMS(64),and make the touch of water almost meaningless,because it is so private and so antiseptic.

We believe that the swimming cannot come into its proper place,unless everyone who wants to can swim every day:and this means,that,to all intents and purposes,there needs to be a swimming pool on every block,almost in every cluster,and at least in every neighborhood,within no more than about 100 yards of every house.

In this pattern,we shall therefore try to establish a model for a kind of "swimming hole" :public,so that it becomes a communal function,not a wholly private one;and safe,so that this public water can be deep enough for swimming without being dangerous to tiny children playing at the edge.

For millions of years,children have grown up in perfect safety along the edge of oceans,rivers,and lakes.Why is a swimming pool so dangerous?The answer depends simply on the edge.

As a rule,the natural edges between water and shore are marked by a slow,rough transition.There is a certain well—marked sequence of changes in materials,texture,and ecology as one passes from land to water.The human consequences of this transition are important:it means that people can walk lazily along the edge,without concern for their safety;they can sit at the edge and have their feet in the water,or walk along with the water around their ankles.

Children can play in the water safely when the edge is gradual.A baby crawling into a lake comes to no abrupt surprises;he stops when the water gets too deep,and goes back out again.It has even been shown that children teach themselves

着明显的变化序列。重要的是这种过渡区要富有人情味。这就是说人们可以在水边上懒洋洋地散步徜徉，而不必担心他们的人身安全；他们可以坐在水边，把双脚浸在水里，或在齐踝深的水中涉水而行。

当水边缓缓倾斜时，孩子们就能平安无事地戏水。一个胆怯的小孩爬到湖水里不会感到突如其来的惊讶：水太深了，令他止步不前，向后折回。事实早已表明，孩子们是在水塘四周的浅水区自由自在地戏水时才学会游泳的，而这一浅水区通往深水区的斜坡十分平缓。在这样一个水塘里，一些幼儿甚至在他们学会走路前就学会了游泳。即使在陡峭的多岩石的湖岸上，岩石也不会使人大吃一惊——因为离湖边稍远的沙地，就是通往岩石之路，当人们走向这陡岸时，岩石就会改变自己的倾斜角度和质感。

但是，凡是游泳池和任何一种有坚硬的人工堤岸的水域都不会有这种平缓的斜坡。一个小孩可能会以最快的跑步速度直奔而来，当他跳入水中浪花四溅时，他才突然发现他已在 6ft 深的水中了。

这种陡峭的、十分危险的水边，从心理学角度来看，不仅对儿童，对成年人也有影响。虽然成年人，就其本身来说，不会因滩地的陡峭而遇到什么危险——因为他们能应付自如——从生态学而言，这种不合时宜的水边的“陡峭性”往往会使人惊慌失措——从而破坏水面常有的安宁和平静。

因此，十分重要的是：在每一水边，无论是水塘的、湖泊的、游泳池的、河流的、还是运河的，都应有一个天然合理的梯度，它随人走向水边而不断变化，并从浅水区渐渐变化到深水区。

当然，一些深水区主要辟为游泳区；但深水区的水边必须不是直接可达的。而是在深水区一端，四周的水边必须用墙壁或栅栏围护，并要在那里建立安全岛，以便人们从容地下水游泳或潜水。

to swim when they are free to play around a pool with an extremely gradual slope toward the deep.In such a pool,some children have even learned to swim before they can walk.Even the rocks at the steep edge of a rock-bound lake are not that surprising—because the sandy earth further from the edge gives way to rocks,which change their angle and their texture,as one comes to the sharp edge.

But a swimming pool,and any kind of water with a hard and artificial edge,has none of this gradualness.A child may be running along,at top speed,when splash,suddenly he finds himself in six feet of water.

The abrupt edge,most serious for children,has its effects in psychological terms for adults too.Although they are not literally endangered by the edge—since they can learn its dangers—the presence of an ecologically wrong kind of abruptness is disconcerting—and destroys the peace and calm which water often has.

It is therefore essential that every water's edge,whether on a pond,or lake,or swimming pool,or river,or canal,be made so that it has a natural gradient,which changes as a person comes up to the edge,and goes on changing as the water is first very shallow,and then gradually gets deeper.

Of course,some deep water is essential for swimming;but the edge of the deep water must not be directly accessible. Instead,the edge around the deep end needs to be protected by a wall or a fence;and islands can be built there,for people to swim to,and to dive from.

因此：

在每一邻里内要有若干池塘——水池和河塘——供游泳用。使水池全时向大众开放，但是要使通往水池的入口只在浅水区一侧，并使池水逐步变深，起始水深为 **1 ~ 2in。**

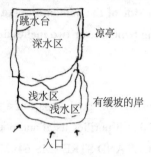

如有可能，可把水池安排为自然水系的一部分，以便天然流水净化水池的水质而不必用氯消毒——**水池和小溪**（64）。保证水池面南向阳——**朝南的户外空间**（105）。如有可能，则装饰水池边缘，建造一个小型户外凉亭或一棚架，人们可以在那里一边歇息，一边观看游泳——**户外亭榭**（69）、**棚下小径**（174）和**可坐矮墙**（243）……

Therefore:

In every neighborhood,provide some still water—a pond,a pool—for swimming.Keep the pool open to the public at all times,but make the entrance to the pool only from the shallow side of the pool,and make the pool deepen gradually,starting from one or two inches deep.

∞ℭ∞

If possible,arrange the pool as part of a system of natural running water,so that it purifies itself,and does not have to be chlorinated—POOLS AND STREAMS(64).Make sure the pool has southern exposure—SOUTH FACING OUTDOORS(105). If possible,embellish the edge of the pool with a small outdoor room or trellis,where people can sit and watch—PUBLIC OUTDOOR ROOM(69),TRELLISED WALK (174),SITTING WALL(243)...

模式72 地方性运动场地*

……在人们生活和工作的一切地区——尤其是**工作社区**（41），以及在**保健中心**（47）的预防条例所能照顾到的地区——都需要有运动场地和体育设施来加以完善。本模式说明这种健身运动的性质和分布方案。

᛭ᛞᚱᚷ

人体不会因使用而损坏，恰恰相反，不用才会损坏。

72   LOCAL SPORTS*

...all the areas where people live and work—especially the WORK COMMUNITIES(41)and the areas looked after by the preventive programs of the HEALTH CENTER(47)—need to be completed by provisions for sports and exercise.This pattern defines the nature and distribution of this exercise.

❧❧

The human body does not wear out with use.On the contrary,it wears down when it is not used.

In agricultural society,people use their bodies every day in many different ways.In urban society,most people use their minds,but not their bodies;or they use their bodies only in a routine way.This is devastating.There is ample empirical evidence that physical health depends on daily physical activity.

Perhaps the most striking evidence for the unbalance in our way of life comes from a comparison of the death rates between groups that have been able to live lives that include daily physical activity with those that have not.For example,in the age group 60 to 64,1 per cent of the men in the heavy exercise category died during the followup year,whereas 5 per cent of those in the no-exercise group died.(See P.B.Johnson et al.,*Physical Education,A Problem Solving Approach to Health and Fitness,*University of Toledo,Holt,Rinehart and Winston,1966)

There are very few modern societies where these facts are taken seriously.China and Cuba come to mind.In these societies,people work both with their hands and with their

在农业社会中，人们以各种不同的方式天天使用他们的躯体。在城市社会中，大多数人使用脑力而不是体力；或者他们只在日常工作中使用体力。这是十分有害的。大量的经验性证据表明，身体的健康取决于每天的体育活动。

也许，我们生活方式中不平衡的最触目惊心的证据就是两组人死亡率对比的结果。这两组人中一组人天天能进行身体锻炼，而另一组则不然。以 60 ～ 64 岁的人为例，在第二年中，每天坚持积极锻炼身体的死亡率为 1％，而不进行任何体育活动的死亡率则为 5％。(See P.B.Johnson et al., *Physical Education*, *A Problem Solving Approach to Health and Fitness*, University of Toledo, Holt, Rinehart and Winston, 1966)

认真考虑到这些事实的现代社会屈指可数。中国和古巴开始注意到这点。这两国的人民既从事体力劳动又从事脑力劳动，既有体力劳动的熟练技巧，又有脑力劳动的熟练技巧。医生往往深入施工现场治病；而建筑工人却常常参加行政会议。

在任何一个已经发展到这一阶段的社会中，人体的总体力萎缩症将不会发生。但是，在任何一个还未学会这种明智措施的社会中，有必要把分散运动场地作为临时解决办法，使体育锻炼可在近在咫尺的地方进行，即能够在紧邻每一住宅或工作地点开展起来。小块场地、游泳池、体育馆、运动场，必须像街角杂货店和餐厅那样多。地方性运动场地成为每一邻里和工作社区的一个天然组成部分是最理想不过的了。我们设想，这些设施是非营利性的中心，但获得使用者的支持，也许，与保健预防条例的精神一致，如英国佩卡姆先驱保健中心举办的游泳和跳舞活动——参阅**保健中心**（47）。

minds.Workdays embrace both kinds of skills.Doctors are as apt to be building houses as practicing medicine;and builders are often sitting in administrative sessions.

In any society which has reached this stage,the gross physical atrophy of human bodies will not occur.But in any society which has not learned this wisdom,it is necessary,as a kind of interim solution,to scatter opportunities for physical activity,so that they are close at hand,indeed next door,to every house and place of work.Small fields,swimming pools,gyms,game courts,must be as frequent as corner groceries and restaurants.Ideally,local sports would form a natural part of every neighborhood and work community.We imagine these facilities as nonprofit centers,supported by the people who use them,perhaps coordinated with a program of health prevention like the swimming and dancing at the Pioneer Health Center in Peckham—see HEALTH CENTER(47).

Sports also have a special life of their own,which cannot be duplicated.Throwing the ball around,shouting out,winning a crushing victory,losing a long drawn out match,getting a wild ball backon the net somehow,anyhow—these are moments that cannot be captured by a job of work.They are entirely different;perhaps they cater especially to what E.Hart calls the psychoemotional component of muscular activity.("The Need for Physical Activity," in S.Maltz,ed.,*Health Readings*,Wm. Brown Book Company,Iowa,1968,p.240)In any case,it is a kind of vitality that cannot bereplaced.

Therefore:

Scatter places for team and individual sports through every work community and neighborhood:tennis,squash,

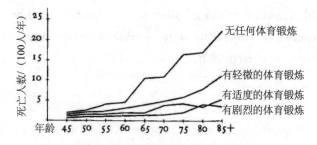

如果你有规律地进行体育锻炼，多半你将延年益寿
You will probably live longer if you exercise regularly

各种运动本身都具有一种不可替代的生命力。如到处
抛球；大声喊叫；赢得一次决定性的胜利；输了一场拖延
时间的比赛；拦网截住猛杀来的扣球等这些美好的时刻绝
非工作所能取代。体育活动与工作完全不同：体育活动也
许就是专门迎合伊·哈特称之为肌肉活动的心理感情成分
的那种东西（"The Need for Physical Activity," in S.Maltz,
ed., *Health Readings*, Wm.Brown Book Company, Iowa,
1968, p.240）在任何情况下，体育活动都具有不可取代的
生命力。

因此：

**把提供给运动队或个人进行体育活动的场地分散到每
个工作社区和每个邻里中去。这些场地可进行的运动项目
有：网球、壁球、乒乓球、游泳、台球、篮球、舞蹈、体
操……使这些运动成为过路人想要一睹为快的节目表演，
并以此吸引他们参加。**

table tennis,swimming,billiards,basketball,dancing,gym nasium...and make the action visible to passers-by,as an invitation to participate.

৪৩৫৪

Treat the sports places as a special class of recognizable simple buildings,which are open,easy to get into, with changing rooms and showers—BUILDING COMPLEX (95),BATHING ROOM(144);combine them with community swimming pools,where they exist—STILL WATER (71);keep them open to people passing—BUILDING THOROUGHFARE(101),OPENING TO THE STREET (165),and provide places where people can stop and watch-SEAT SPOTS(241),SITTING WALL(243)...

把地方性运动场地作为特种可识别的简易建筑来处理，它们应当是开放的、容易进入的，并设有各种不同的运动室和淋浴室——**建筑群体**（95）、**浴室**（144）；把这些运动场地和社区游泳池相结合——**池塘**（71）；并使它们向所有过路的人开放——**有顶街道**（101）、**向街道的开敞**（165），并向人们提供可止步观赏体育活动的地方——**户外设座点**（241）和**可坐矮墙**（243）……

模式73 冒险性的游戏场地

73 ADVENTURE PLAYGROUND

...inside the local neighborhood,even if there is common land where children can meet and play—COMMON LAND(67),CONNECTED PLAY(68);it is essential that there be at least one smaller part,which is differentiated,where the play is wilder,and where the children have access to all kinds of junk.

୫୦୧୫

A castle,made of cartons,rocks,and old branches,by a group of children for themselves,is worth a thousand perfectly detailed,exactly finished castles,made for them in a factory.

Play has many functions:it gives children a chance to be together,a chance to use their bodies,to build muscles,and to testnew skills.But above all,play is a function of the imagination.A child's play is his way of dealing with the issues of his growth,of relieving tensions and exploring the future. It reflects directly the problems and joys of his social reality. Children come to terms with the world,wrestle with their pictures of it,and reform these pictures constantly,through those adventures of imagination we call play.

Any kind of playground which disturbs,or reduces,the role of imagination and makes the child more passive,more the recipient of someone else's imagination,may look nice,may be clean,may be safe,may be healthy—but it just cannot satisfy the fundamental need which play is all about.And,to put it

……在地方邻里之内，即使儿童们有相互见面和玩耍的地方——**公共用地**（67）、**相互沟通的游戏场所**（68），但重要的是至少还应有一个更小的、与众不同的地方，以便儿童们的游戏更具有野趣：他们在那里可以钻进各种破破烂烂的东西里面去。

⚜

一群儿童为他们自己用硬纸板、岩石和枯枝等而筑成的一座古堡，抵得上 1000 座细部完善、装修精美、专为他们在工厂中制造的城堡。

游戏具有多种功能：游戏向儿童们提供聚集和锻炼身体的良好机会，他们的肌肉结实起来了，运动技巧改进了。但游戏的主要功能是培养儿童的想象力。儿童们的游戏就是处理他们自己的成长、松弛紧张的情绪以及探索未来的一种独特的方式。它直接反映出儿童社会现实中的苦恼和欢乐。儿童们接触这个世界，并和这个世界的形象作斗争，他们通过想象中的冒险性的游戏不断地改造着它的形象。

任何一种游戏场地，只要它起到干扰或减弱培养想象力的不良作用，并使儿童变得更加消极，更加容易接受别人的想象力，虽然可能看起来是挺好的，干净的和有益于健康的——但其实，这恰恰不能满足游戏的基本要求。坦率地说，这种游戏场地纯粹是浪费时间和金钱。以抽象的雕塑装饰的大型游戏场，正如沥青铺面的游戏场和供儿童攀爬的立体方格铁架一样糟糕。它们不仅仅是枯燥无味的，毫无生气的，而且也是无用的。它们所发挥的功能与儿童们的最基本的需要完全是风马牛不相及的。

在小城镇和乡村要注意灵活发展这种冒险性的和富于想象力的游戏场地，以便儿童们能接触到各种原料、空间和一种多少可以理解的环境。在城市里，解决这一问题已

bluntly,it is a waste of time and money.Huge abstract sculptured playlands are just as bad as asphalt playgrounds and jungle gyms.They are not just sterile;they are useless.The functions they perform have nothing to do with the child's most basic needs.

This need for adventurous and imaginative play is taken care of handily in small towns and in the countryside,where children have access to raw materials,space,and a somewhat comprehensible environment.In cities,however,it has become a pressing concern.The world of private toys and asphalt playgrounds does not provide the proper settings for this kind of play.

The basic work on this problem has come from Lady Allen of Hurtwood.In a series of projects and publications over the past twenty years,Lady Allen has developed the concept of the adventure playground for cities,and we refer the reader,above all,to her work.(See,for example,her book,*Planning for Play*,Cambridge:MIT Press,1968.)We believe that her work is so substantial,that,by itself,it establishes the essential pattern for neighborhood playgrounds.

Colin Ward has also written an excellent review, "Adventure Playgrounds:A Parable of Anarchy," *Anarchy 7*,September 1961. Here is a description of the Grimsby playground,from that review:

At the end of each summer the children saw up their shacks and shanties into firewood which they deliver in fantastic quantities to old age pensioners.When they begin building in the spring, "it's just a hole in the ground—and they crawl into it." Gradually the holes give way to two—storey huts.Similarly with the notices above their dens.It begins with

是刻不容缓的事情。私人的玩具和沥青铺面的游戏场不可能为这种冒险性的游戏提供合适的场地。

关于冒险性的游戏场地，霍特坞的艾伦夫人写了重要的著作。她在过去 20 年内设计的一系列工程和发表的许多论著中不断发展了这一概念，我们首先要向读者推荐她的著作。(See, for example, her book, *Planning for Play*, Cambridge : MIT Press, 1968.) 我们认为，她的著作内容丰富充实，本身就创立了邻里游戏场地这一重要模式。

不准入内游戏
No playing

（照片中的文字为：无旱冰场和自行车道；狗不得入内；严禁投入小圆石；这是你的游戏场地，请爱护它，保持清洁。纽约市园林局）

科林·沃德也写了一篇十分出色的评论《冒险性的游戏场地：杂乱无章之比喻》("Adventure Playgrounds : A Parable of Anarchy," *Anarchy 7*, September 1961。) 下面是这篇评论描述格利姆斯比游戏场地的精彩的一段：

每临夏末，儿童们把他们的各种各样的简陋的小木屋锯成一大堆一大堆的柴火，分头送给领取养老金的退休老人。春天来了，他们就开始造新的棚屋。"这简直是一个地下洞穴，他们往里爬呀爬。"不久，这样的洞逐渐多起来了，都通往一座又一座的二层棚屋。同样，在这些陋室的墙上

nailing up "Keep Out" signs.After this come more personal names like "Bughold Cave" and "Dead Man's Cave," but by the end of the summer they have communal names like "Hospital" or "Estate Agent." There is an everchanging range of activities due entirely to the imagination and enterprise of the children themselves...

Therefore:

Set up a playground for the children in each neighborhood. Not a highly finished playground,with asphalt and swings,but a place with raw materials of all kinds—nets,boxes,barrels,trees,ropes,simple tools,frames,grass,and water—where children can create and recreate playgrounds of their own.

✿

Make sure that the adventure playground is in the sun—SUNNYPLACE(161);make hard surfaces for bikes and carts and toy trucks and trolleys,and soft surfaces for mud and building things—BIKE PATHS AND RACKS(56),GARDEN GROWING WILD(172),CHILD CAVES (203);and make the boundary substantial with a GARDEN WALL(173)or SITTING WALL(243)...

糊满了各种布告之类，并钉上"切勿入内"的牌子。接着是一连串的个人命名，比如"爬虫穴"和"死人洞"等，但在夏末，这些洞穴就更换成普通的名称了，如"医院"或"房地产中间商"等。儿童们完全出于他们自己的想象力和冒险精神来搞游戏活动，所以他们的活动范围一直在不断变化……

因此：

为每一邻里的儿童们建立一个游戏场地。它不是装修一新的、铺有沥青地面和设有秋千架的游戏场地，而是一块堆放各式各样原料的地方——上面有为数众多的网、盒子、琵琶桶、树木、绳索、简易工具、框架、草和水——儿童们在这里可以创造出或再创造出他们自己的游戏场地。

各种破旧的物件

✽✽✽

确保冒险性的游戏场地阳光充足——**有阳光的地方**（161）；既有硬质地面供自行车、手推车、玩具卡车和玩具有轨电车用，又有软质地面供儿童们玩泥巴和造东西——**自行车道和车架（56）花园野趣**（172）、**儿童猫耳洞**（203）；并使其有实在的边界，可采用**花园围墙**（173）或**可坐矮墙**（243）……

模式74　动物

74 ANIMALS

...even when there is public land and private land for individual buildings—COMMON LAND(67),YOUR OWN HOME(79),there is no guarantee that animals can flourish there.This pattern helps to form GREEN STREETS(51)and COMMON LAND(67)by giving them the qualities they need to sustain animal life.

၈၁၉ၶ

Animals are as important a part of nature as the trees and grass and flowers.There is some evidence,in addition, which suggests that contact with animals may play a vital role in a child's emotional development.

Yet while it is widely accepted that we need "parks" — at least access to some kind of open space where trees and grass and flowers grow—we do not yet have the same kind of wisdom where sheep,horses,cows,goats,birds,snakes,rabbits,deer,chickens,wild—cats,gulls,otters,crabs,fish,frogs,beetles,butterflies,and ants are concerned.

Ann Dreyfus,a family therapist in California,has told us about the way that animals like goats and rabbits help children in their therapy.She finds that children who cannot make contact with people,are nevertheless able to establish contact with these animals.Once this has happened and feelings have started to flow again,the children's capacity for making contact starts to grow again,and eventually spreads out to family and friends.

But animals are almost missing from cities.In a city there are,broadly speaking,only three kinds of animal:pets,vermin,and animals in the zoo.None of these three provides the emotional

……即使在有公共用地和个人住宅的私人土地——**公共用地**（67）、**自己的家**（79）之时，也无法保证各种动物能在那里生息繁衍。本模式有助于形成**绿茵街道**（51）和**公共用地**（67），并使它们获得为维持动物生命所必需的特性。

⊰⊱

飞禽走兽犹如花草树木，是大自然的一个重要组成部分。除此之外，还有证据表明，经常和各种动物保持接触对儿童的精神发展起到十分重要的作用。

目前，当我们大家公认需要"公园"——到有花草树木的开阔空间去——之际，我们仍未明智地认识到，对绵羊、马、牛、山羊、鸟、蛇、野兔、鹿、鸡、野猫、鸥、水獭、螃蟹、鱼、蛙、甲虫、蝴蝶和蚂蚁等都要一视同仁地关心。

安·德赖弗斯，加利福尼亚的一位家庭治疗学家，告诉我们说，像山羊和兔子这样的动物在协助治疗儿童疾病方面起的作用非常明显。她发现，不合群的儿童，却能和这些动物打交道。一旦如此，儿童的感情重新开始活动，儿童的交往能力重新开始发展，最后逐步发展到和全家人以及朋友们密切来往。

但是，动物在城市中几乎销声匿迹了。一般说来，在城市中只有三种动物：玩赏动物、有害动物和动物园中的动物。三者中无一能满足精神需要也不能提供其所需的生态联系。宠物是令人愉悦的，但是太人性化了，不再过野生的无羁绊的生活了。并且它们几乎不会使人体验到动物的本性。有害动物——老鼠、蟑螂——城市特有的害人动物栖息在又脏又乱十分恶劣的生态环境中，它们自然而然地被视为人类的大敌。大多数人除了偶尔被好

sustenance nor the ecological connections that are needed.Pets are pleasant,but so humanized that they have no wild free life of their own.And they give human beings little opportunity to experience the animalness of animals.Vermin—rats,cockroaches—are animals which are peculiar to cities and which depend ecologically on miserable and disorganized conditions,so they are naturally considered as enemies. Animals in the zoo are more or less inaccessible to most of the human population—except as occasional curiosities.Besides,it has been said that animals living under the conditions which a zoo provides are essentially psychotic—that is,entirely disturbed from their usual mode of existence—so that it is probably wrong to keep them there—and certainly they can in no way recreate the missing web of animal life which cities need.

It is perfectly possible to reintroduce animals into the natural ecology of cities in a useful and functioning sense, provided that arrangements are made which allow this and do not create a nuisance.

Examples of ecologically useful animals in a city: horses,ponies,donkeys—for local transportation and sport. Pigs—to recycle garbage and for meat.Ducks and chickens—as a source of eggs and meat.Cows—for milk.Goats—milk. Bees—honey and pollination of fruit trees.Birds—to maintain insect balance.

There are essentially two difficulties to overcome.(1)Many of these animals have been driven out of cities by law because they interrupt traffic,leave dung on the street,and carry disease. (2)Many of the animals cannot survive without protection under modern urban conditions.It is necessary to make specific provisions to overcome these difficulties.

Therefore:

Make legal provisions which allow people to keep any

奇心驱使，是不大会接近动物园内的动物的。还有，据说，生活在动物园条件下的动物实际上患有精神病——即它们原来的生存方式完全遭到破坏——所以把它们豢养在那里也许是错误的——当然，它们绝不能再重建正在日益消失的动物生活之网，而这一点正是城市所需要的。

向里看和往外瞧——有何差别
Looking in or looking out—what's the difference

就用途和功能而言，把动物重新引回到城市的自然生态环境之中是完全可能的。但有一条件：我们要安排适当，要使这种做法不至于造成混乱。

在生态上有用的动物的例子如：马——赛马用的马、驴——可用于地方运输和体育运动；猪——把食物下脚料转化成肉；鸭和鸡——蛋、肉之源；奶牛——产奶；山羊——产奶；蜜蜂——酿蜜并为果树的花朵传授花粉；鸟——使昆虫保持平衡。

从本质上看，现有两个困难亟待克服：（1）许多动物根据法律已被驱逐出城市，因为它们妨碍交通，在街道上排泄粪便以及传播疾病。（2）许多动物在现代城市条件下，如不加以保护，就无法再生存下去。为解决这些困难而作出若干特别的规定是完全必要的。

· 因此：

制定法律条款，明确规定允许居民在自己私人的一

animals on their private lots or in private stables.Create a piece of fenced and protected common land,where animals are free to graze,with grass,trees,and water in it.Make at least one system of movement in the neighborhood which is entirely asphalt-free— where dang can fall freely without needing to be cleaned up.

ꕥꕥ

Make sure that the green areas—GREEN STREETS(51), ACCESSIBLE CREEN(60)—are all connected to one another to form a continuous swath throughout the city for domestic and wild animals.Place the animal commons near a children's home and near the local schools,so children can take care of the animals— CHILDREN'S HOME (86);if there is a lot of dung,make sure that it can be used as a fertilizer—COMPOST(178)...

within the framework of the common land,theclusters and the work communities,encourage transformation of the smallest independent social institutions:the families,workgroups,and gathering places.

First,the family,in all its forms;

75.THE FAMILY
76.HOUSE FOR A SMALL FAMILY
77.HOUSE FOR A COUPLE
78.HOUSE FOR ONE PERSON
79.YOUR OWN HOME

小片土地上或在私人的马厩和牛棚里饲养任何动物。开辟一块公共土地，四周由栅栏围护，里面水草丰茂，绿树成荫，允许居民自由放牧。至少在完全没有沥青地面的邻里内建立一种流动系统：准许动物在那里随地排泄粪便而无须清除。

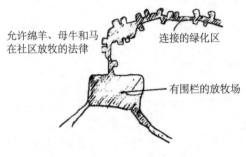

允许绵羊、母牛和马在社区放牧的法律

连接的绿化区

有围栏的放牧场

确保绿化区——**绿茵街道**（51）、**近宅绿地**（60）——相互连接成一片连续不断的贯穿全市的狭长地带，作为家养动物和野生动物的活动场所。把动物公共用地设在儿童之家和地方中小学校附近，以便儿童们能照料动物——**儿童之家**（86）；如果有许多动物粪便，保证把它们作为肥料来使用——**堆肥**（178）……

在公共用地、住宅团组和工作社区的范围内，鼓励改造最小的独立的社会机构：家庭、工作小组和集合场所。

首当其冲的是家庭及其一切形式；

75. 家庭

76. 小家庭住宅

77. 夫妻住宅

78. 单人住宅

79. 自己的家

模式75　家庭*

　　……现在假定，你已决定为你自己建造一幢住宅。如果你选址合理，这幢住宅就会有助于形成住宅团组，或联排式住宅，或丘状住宅——**住宅团组**（37）、**联排式住宅**（38）、**丘状住宅**（39）——或它能促使工作社区生气勃勃——住宅与**其他建筑间杂**（48）。本模式下面的叙述会给你提供某种关于住户本身的社会性质的重要信息。如果你遵照本模式行事而获得成功，则这将有益于修补你社区中的**生命的周期**（26）和**户型混合**（35）。

75　THE FAMILY*

...assume now,that you have decided to build a house for yourself.If you place it properly,this house can help to form a cluster,or a row of houses,or a hill of houses—HOUSE CLUSTER(37),ROW HOUSES(38),HOUSING HILL(39)— or it can help to keep a working community alive—HOUSING IN BETWEEN(48).This next pattern now gives you some vital information about the social character of the household itself. If you succeed in following this pattern,it will help repair LIFE CYCLE(26)and HOUSEHOLD MIX(35)in your community.

৪০০৪

The nuclear family is not by itself a viable social form.

Until a few years ago,human society was based on the extended family:a family of at least three generations,with parents,children,grandparents,uncles,aunts,and cousins,all living together in a single or loosely knit multiple household.But today people move hundreds of miles to marry,to find education,and to work.Under these circumstances the only family units which are left are those units called nuclear families:father,mother,and children.And many of these are broken down even further by divorce and separation.

Unfortunately,it seems very likely that the nuclear family is not a viable social form.It is too small.Each person in a nuclear family is too tightly linked to other members of the family;any one relationship which goes sour,even for a few hours,becomes critical;people cannot simply turn away toward uncles,aunts,grand—children,cousins,brothers.

核心家庭不是一种单独可行的社会形式。

直到若干年前，人类社会还是以大家庭为基础：一个家庭至少三世同堂，父母、子女、祖父母、叔伯姑婶、堂兄弟姐妹，共同居住在一个单一的或联结松散的大家庭内。但是，今非昔比，人们纷纷迁往数百英里之外的地方去结婚，或是求学，或是找工作。在这些情况下，留下来的只有称为核心家庭的家庭：父亲、母亲和儿女。其中许多核心家庭由于离婚和分居而进一步破裂。

不幸的是，核心家庭看来很可能不是一种可行的社会形式。它太小了，它的成员相互结合得太紧密了：任何一种关系搞僵了，哪怕是儿时发生的，也会变得一发不可收拾。他们不能简单地出走，到叔伯、姑婶、孙儿女、堂兄弟姐妹和兄弟那里去。纠缠这个核心家庭的重重困难似卷线越绕越紧；儿女们成为各种依附和恋母情结精神病的牺牲品；相互依赖的双亲，最终被迫分离。

菲利普·斯莱脱描述了美国家庭的这种情况，他发现在家庭的成年人中，尤其在妇女中，有一种可怕的郁闷压抑感。在他们的周围简直没有足够数量的人，没有足够的共同活动，能使他们平时感到家庭感情的深度和丰富多彩。(Philip E.Slater, The Pursuit of Loneliness, Boston: Beacon Press, 1970, p.67, and throughout.)

看来，重要的是一家人的周围至少有 12 人，在他们飞黄腾达或家道衰落之际，得到他们所需的慰藉和关系。因为以血缘关系为基础的旧式的大家庭似乎已经过时——至少在目前是如此——如果有小家庭、一对对夫妻和单身男女以自愿的方式组成 10 人左右的"家庭"时，大家庭才会再重现。

奥尔德斯·赫克斯利在他的最后一部著作《岛屿》中

Instead,each difficulty twists the family unit into ever tighter spirals of discomfort;the children become prey to all kinds of dependencies and oedipal neuroses;the parents are so dependent on each other that they are finally forced to separate.

Philip Slater describes this situation for American families and finds in the adults of the family,especially the women,a terrible,brooding sense of deprivation.There are simply not enough peoplearound,not enough communal action,to give the ordinary experience around the home any depth or richness. (Philip E.Slater,*The Pursuit of Loneliness*,Boston:Beacon Press,1970,p.67,and throughout.)

It seems essential that the people in a household have at least a dozen people round them,so that they can find the comfort and relationships they need to sustain them during their ups and downs.Since the old extended family,based on blood ties,seems to be gone—at least for the moment—this can only happen if small families,couples,and single people join together in voluntary "families" of ten or so.

In his final book,Island,Aldous Huxley portrayed a lovely vision of such a development:

"How many homes does a Palanese child have?"

"About twenty on the average."

"Twenty?My God!"

"We all belong," Susila explained, "to a MAC—a Mutual Adoption Club.Every MAC consists of anything from fifteen to twenty-five assorted couples.Newly elected brides and bridegrooms,oldtimers with growing children,grandparents and great-grandparents—everybody in the club adopts everyone else.Besides our own blood relations,we all have our quota of deputy mothers,deputy fathers,deputy aunts and uncles,deputy

栩栩如生地描写了这种自愿组合的家庭的发展的情景:

"一个班伦族的孩子有多少个家庭?"

"平均 20 个上下。"

"20 个? 我的天哪!"

苏希拉解释道:"我们都属于相互收养俱乐部,简称MAC。每一个相互收养俱乐部由 15 ~ 25 对各式各样的夫妻组成。新选配的新娘和新郎、拖男带女的老前辈、祖父母、曾祖父母——俱乐部的每个成员都收养一个人。我们除了有自己的血缘关系的亲戚外,还有一定限额的代理父母、代理叔伯、代理姑婶、代理兄弟姐妹、代理婴儿、代理幼儿和代理青少年。"威尔摇了摇头。"以前一个家庭发展的地方现在有 20 个家庭在发展。"

"可是,你们的那种家庭以前发展成什么模样了⋯⋯"她振振有词地说着,仿佛读着一份食谱的说明似的。"取一名性无能的工薪族,"她继续滔滔不绝地说下去:"一名不满足的女性,两个或(如果情愿的话)三个小电视迷;将他们腌泡在弗洛伊德主义与冲淡的基督教教义的混合汁中;然后再将他们紧紧封在四室一套的公寓里,把他们泡在自己的汁中炖 15 年。而我们的配方是截然不同的:取 20 对性欲上满足的夫妻和他们的子女;加添同等数量的科学知识、直觉和幽默感,使他们沉浸于印度佛教之中,将他们放入一只敞开的大锅里,在露天用旺盛的爱之火无限期地煨熬。"

"你们的露天大锅产生了什么结果?"

"一种完全不同的家庭。限制不严,像你们的家庭那样,而且不是预先指定的,不是强迫的。这是一种包容性的、非预先指定的、自愿组合的家庭。它有 20 对父亲和母亲,8 名或 9 名前父和前母,40 名或 50 名各种不同年龄的孩子。"(Aldous Huxley,Island,New York:Bantam,1962,pp.89 ~ 90.)

就实体环境处理来说,安置一个大的自愿组合家庭必

brothers and sisters,deputy babies and toddlers and teenagers."

Will shook his head. "Making twenty families grow where only one grew before."

"But what grew before was your kind of family..." As though reading instructions from a cookery book, "Take one sexually inept wage slave," she went on, "one dissatisfied female,two or(if preferred)three small television addicts;marinate in a mixture of Freudism and dilute Christianity;then bottle up tightly in a fourroom flat and stew for fifteen years in their own juice.Our recipe is rather different:Take twenty sexually satisfied couples and their offspring;add science,intuition and humor in equal quantities;steep in Tantrik Buddhism and simmer indefinitely in an open pan in the open air over a brisk flame of affection."

"And what comes out of your open pan?" he asked.

"An entirely different kind of family.Not exclusive,like your families,and not predestined,not compulsory.An inclusive,unpredestined and voluntary family.Twenty pairs of fathers and mothers,eight or nine ex-fathers and ex-mothers,and forty or fifty assorted children of all ages." (Aldous Huxley, Island,New York:Bantam,1962,pp.89-90.)

Physically,the setting for a large voluntary family must provide for a balance of privacy and communality.Each small family,each person,each couple,needs a private realm,almost a private household of their own,according to their territorial need. In the movement to build communes,it is our experience that groups have not taken this need for privacy seriously enough.It has been shrugged off,as something to overcome.But it is a deep and basic need;and if the setting does not let each person and each small household regulate itself on this dimension,it is sure to

须使私密性和公共性处于平衡之中。每个小家庭、每个人、每对夫妇，均需私人领域，几乎均需私人住宅，根据他们的领域需求而作出合理安排。在建立群居村的运动中，我们的体会是：各种群体还未充分意识到这种私密性的需要。这一需要被视为要加以克服的某种东西而一直不予理睬。但这确是一种深刻而基本的需要。如果安排不当，不能让每个人、每个小家庭在这方面进行调整，必定会引起麻烦。因此，我们建议，单身男女、成对夫妻、年轻人和老年人——每个小群体——各自都有合法的独立住宅：在某些情况下，他们各自都有空间上分开的住宅和住所，至少有空间上分开的房间、成套的居室和楼层。

由此可见，私人领域与公共空间和公共设施是相辅相成的。最重要的公共用地是厨房和就坐用餐的地方。还有花园。每周至少有几个夜晚共同进餐，看来，这在维系这种群体方面起到首要的作用。在厨房一起烹调美味佳肴，共进晚餐，就创造了相互会面的机会。大家欢聚在一起，无拘束地畅谈，如安排照料孩子、修缮养护、工程项目等——参阅**共同进餐**（147）。

这就表明要有一个大家庭的空间——农家厨房，它恰好位于居住地的中心，即主要的十字交叉路口。在一天行将结束之时，人人都想去那里见一见面，闲聊几句。按照家庭的方式，这种厨房也可能是一所独立的、带有一个工作间和若干花园的建筑，或者是一幢住宅的一翼，或者是一栋两三层高的楼房的整个底层。有某种证据表明，孕育着巨大的、自愿组合的群体家庭的过程已正在社会中进行着。(Cf.Pamela Hollie, "More families share houses with others to enhance 'life style,' *Wall Street Journal*, July 7, 1972.)

一种刺激自愿组合家庭发展的途径：当有人转让或出售自己的住宅或房间或公寓时，他们首先应当通知他们周围的邻居。这些邻居有权去找自己的朋友来购置这一房地

cause trouble.We propose,therefore,that individuals,couples,people young and old—each subgroup—have its own legally independent household—in some cases,physically separate households and cottages,at least separate rooms,suites,and floors.

The private realms are then set off against the common space and the common functions.The most vital commons are the kitchen,the place to sit down and eat,and a garden.Common meals,at least several nights a week,seem to play the biggest role in binding the group.The meals,and taking time at the cooking,provide the kind of casual meeting time when everything else can becomfortably discussed:the child care arrangements,maintenance,projects—see COMMUNAL EATING(147).

This would suggest,then,a large family room-farmhouse kitchen,right at the heart of the site—at the main crossroads,where everyone would tend to meet toward the end of the day. Again,according to the style of the family,this might be a separate building,with workshop and gardens,or one wing of a house,or the entire first floor of a two or three story building.

There is some evidence that processes which generate large voluntary group households are already working in the society.(Cf.Pamela Hollie,"More families share houses with others to en-hance'life style,'"*Wall Street Journal*,July 7,1972.)

One way to spur the growth of voluntary families:When someone turns over or sells their home or room or apartment,they first tell everyone living around them—their neighbors.These neighbors then have the right to find friends of theirs to take the place—and thus to extend their "family." If friends are able to move in,then they can arrange for themselves how to create a functioning family,with commons,and so on.They might build a connection between the homes,knock out a wall,add a room.If the

产，他们用这种办法来扩大自己的"家"。如果朋友们能迁入此地，那么他们就能为自己妥善安排好去创造一种功效卓著的家庭，有公共用地等，不一而足。他们可在各家之间建造衔接空间，如打穿墙壁或增添一室。如果邻居在数月之内不能立即做成这笔房地产交易，那么，它就可归属于正式的市场。

因此：

建立法律手续，鼓励 8 ~ 12 人的群体来到一起组成群居家庭。从形态上讲，下列几点是重要的：

1. 为构成扩大了的家庭的群体和男女个人准备好私人的领域：夫妻的领域、私密性房间、小家庭的小住宅。

2. 准备好具有公共设施的公共空间：供烹饪、劳作、园艺活动和照料孩子用。

3. 在居住地的重要交叉路口，辟出一方空地，供群居家庭全体成员见面和游憩之用。

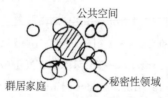

公共空间

群居家庭　　　秘密性领域

⌘

在较大的大家庭内，每一私人家庭都要不惜一切代价拥有它自己控制的地界分明的土地范围——**自己的家**（79）；根据私人家庭的性质处理私人的土地面积——**小家庭住宅**（76）、**夫妻住宅**（77）和**单人住宅**（78）；在它们之间建立公共空间，各种不同的较小的家庭成员能够在那里相互见面和共同进餐——**中心公共区**（129）、**共同进餐**（147）。为了建筑物造型的美观，花园、停车场和周围环境的建造都要着眼于**建筑群体**（95），并由此开始……

people immediately around the place cannot make the sale in a few months,then it reverts to the normal marketplace.

Therefore:

Set up processes which encourage groups of 8 to 12 people to come together and establish communal households. Morphologically,the important things are:

1.Private realms for the groups and individuals that make up the extended family:couple's realms,private rooms,subhouseholds for small families.

2.Common space for shared functions:cooking,working,gardening,child care.

3.At the important crossroads of the site,a place where theentire group can meet and sit together.

∞∞

Each individual household within the larger family must,at all costs,have a clearly defined territory of its own,which it controls—YOUR OWN HOME(79);treat the individual territories according to the nature of the individual households—HOUSE FOR A SMALL FAMILY(76),HOUSE FOR A COUPLE(77),HOUSE FOR ONE PERSON(78);and build common space between them,where the members of the different smaller households can meet and eat together—COMMON AREAS AT THE HEART(129),COMMUNAL EATING(147).For the shape of the building,gardens,parking,and surroundings,begin with BUILDING COMPLEX(95)...

模式76　小家庭住宅*

　　……根据模式**家庭**（75），每一核心家庭应当是较大的群体家庭中的一个成员的家庭。如果这是不可能的话，你就尽力而为。在为小家庭建造住宅的时候，尽可能形成某种较大的群体家庭，并和周围相邻的住家连接起来；在任何情况下，小家庭住宅至少成为**住宅团组**（37）的开端。

❧❧

　　在小家庭住宅内，儿童和成人之间的关系最为重要。

　　许多小家庭的住宅面积不大，没有玩具齐全的儿童室；他们也并不富裕阔绰，雇不起保姆，所以常常处于儿童的吵闹和纠缠不休的困境之中。孩子围着大人转是十分自然的，但他们的父母却没有心绪或能力使他们待在一个特定

76   HOUSE FOR A SMALL FAMILY*

...according to THE FAMILY(75),each nuclear family
ought to be a member household of a larger group household.
If this is not possible,do what you can,when building a house
for a small family,to generate some larger,possible group
household,by tying it together with the next door households;in
any case,at the very least,form the beginning of a HOUSE
CLUSTER(37).

ဆာဘ

**In a house for a small family,it is the relationship between
children and adults which is most critical.**

Many small households,not large enough to have a full
fledged nursery,not rich enough to have a nanny,find themselves
swamped by the children.The children naturally want to be
where the adults are;their parents don't have the heart,or the
energy,to keep them out of special areas;so finally the whole
house has the character of a children's room—children's
clothes,drawings,boots and shoes,tricycles,toy trucks,and
disarray.

Yet,obviously few parents feel happy to give up the calm
and cleanliness and quiet of the adult world in every square inch
of their homes.To help achieve a balance,a house for a small
family needs three distinct areas:a couple's realm,reserved
for the adults;a children's realm,where children's needs hold
sway;and a common area,between the two,connected to them
both.

的地方玩耍，结果，整个住宅终于具有儿童室的性质了：儿童的衣服、图画、靴子、鞋子、三轮自行车和玩具汽车等乱七八糟一堆，真是不可名状。

可是，很明显，几乎没有一对父母面对家里成人活动的领域没有清静、安宁和干净的方寸之地会感到高兴而喜形于色。为了有助于取得户内活动的平衡，小家庭住宅需要三个不同的区：专供成年人用的夫妻领域；由儿童支配的儿童领域；位于两领域之间而又与两领域相连接的公共区。

夫妻的领域应由若干房间组成，虽然房间只是它的一部分。正是这一领域供两个成年人即一对夫妻——不是父亲和母亲——使用。他们的其他生活是和儿童、朋友和工作有关的。夫妻领域必须表现出这就是成年人使用的地方。儿童们也可出入这一领域，但他们待在那里时，他们显然来到了成人区。参阅**夫妻的领域**（136）

儿童区必须视为儿童们的共享领域。参阅**儿童的领域**（137）。这里，重要的是在建造儿童的领域时，应考虑它是住宅内的一部分，并兼顾与其他两个区的平衡。还有，儿童领域的重要特征不是成人"被排挤在外"，当他们来到这个区时，他们就是在儿童领域里。

公共区起到成年人和儿童共享快乐的作用：共同进餐，促膝谈心，一块儿做游戏，也许一块儿洗澡，或在花园内栽花种草，还有其他。总之，凡使他们能陶醉于天伦之乐的有趣活动都可以在公共区内进行。公共区十分可能将大于小家庭住宅的另外两个区。

最后，要认识到本模式是不同于今天最小的家庭模式的。例如，目前流行着一种可与本模式比较的但又迥然不同的概念，即市郊的两区式住宅：卧室区和公共区。

The couple's realm should be more than a room,although rooms are a part of it.It is territory which sustains them as two adults,a couple—not father and mother.Other parts of their lives are involved with children,friends,work;there must be a place which becomes naturally an expression of them as adults,alone. The children come in and out of this territory,but when they are there,they are clearly in the adults'world.See COUPLE'S REALM (136).

The children's world must also be looked upon as territory that they share,as children,CHILDREN'S REALM(137);here,it is important to establish that this is a part of the house,in balance with the others.Again,the critical feature is not that adults are "excluded" but that,when they are in this world,they are in children's territory.

The common area contains those functions that the children and the adults share:eating together,sitting together,games,perhaps bathing,gardening—again,whatever captures their needs for shared territory.Quite likely,the common territory will be larger than the two other parts of the house.

Finally,realize that this pattern is different from the way most small family homes are made today.For example, a popular current conception,comparable to this,but quite different,is a suburban two part house:sleeping and commons.

Even though there is a "master bedroom" the sleeping part of the house is essentially one thing—the children are all around the master bedroom.This plan does not have the distinctions we are arguing for.

Here is a beautiful plan which does:

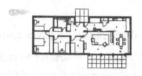

一幢典型的市郊两区式住宅
A typical suburban two part house

即使它有"主卧室",卧室区基本上也是一个区:儿童生活在主卧室周围。上列平面图不具有我们所讨论的那些特征。

下面是一幅非常出色的平面图:

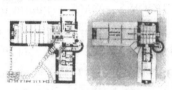

一幢三区式住宅——夫妻的领域在楼上
A three-part house—the couple's realm upstairs

因此:

使住宅具有三个不同的组成部分:夫妻的领域、儿童的领域和公共区。设想这三个区在规模上大致相似,而最大的是公共区。

双亲的领域　　公共区　　儿童的领域

இ୦୯ଓ

像任何一幢住宅一样,把小家庭住宅作为一个颇具特色的领域来处理——**自己的家**(79);根据下列的特殊模式——**中心公共区**(129)、**夫妻的领域**(136)和**多床龛卧室**(143)来建造三区式住宅,并按照**儿童的领域**(137)的要求,把公共区和多床龛卧室与儿童的领域连接起来……

Therefore:

Give the house three distinct parts:a realm for parents,a realm for the children,and a common area.Conceive these three realms as roughly similar in size,with the commons the largest.

✂✎✃

Treat the house,like every house,as a distinct piece of territory—YOUR OWN HOME(79);build the three main parts according to the specific patterns for those parts— COMMON AREAS AT THE HEART (129),COUPLE'S REALM(136),BED CLUSTER(143)and connect the common areas,and the bed cluster according to the CHILDREN'S REALM (137)...

模式77 夫妻住宅*

……再重复一遍，就理想的角度而言，每一对夫妻就是一个较大的群体家庭的一部分——**家庭**（75）。如果这一点无法实现，那你就竭力把这种夫妻住宅建造成这样的布局，即能和其他一些住宅联系起来，以便形成群体家庭的开端；如果这一点也实现不了，则夫妻住宅应至少形成**住宅团组**（37）的开端。

77   HOUSE FOR A COUPLE*

...again,ideally,every couple is a part of a larger group household—THE FAMILY(75).If this can not be so,try to build the house for the couple in such a way as to tie it together with some other households,to form the beginnings of a group household,or,if this fails,at least to form the beginnings of a HOUSE CLUSTER(37).

❧

In a small household shared by two,the most important problem which arises is the possibility that each may have too little opportunity for solitude or privacy.

Consider these forces:

1.Of course,the couple need a shared realm,where they can function together,invite friends,be alone together.This realm needs to be made up of functions which they share.

2.But it is also true that each partner is trying to maintain an individuality,and not be submerged in the identity of the other,or the identity of the "couple".Each partner needs space to nourish this need.

It is essential,therefore,that a small house be conceived as a place where the two people may be together but where,from time to time,either one of them may also be alone,in comfort,in dignity,and in such a way that the other does not feel left out or isolated.To this end,there must be two small places—perhaps rooms,perhaps large alcoves,perhaps a corner,screened off by a half-wall-places which are clearly understood as private

一对夫妻共同使用小住宅，可能产生的最大问题就是双方几乎都不存在隐居或拥有私密性的机会。

请考虑如下要点：

1. 当然，一对夫妻需要一个共同区，以便他们能共同活动，或邀请友人来访，共叙衷肠，或夫妻两人单独相聚在一起。这个共同区必须备有夫妻双方共同使用的设施。

2. 但是，同样正确的是：夫妻双方都力图保持而不是把自己的个性埋没在对方的个性中，或埋没在"夫妻"的共同个性中。因此，夫妻双方都需要空间来满足保持各自个性的这种需要。

由此可见，重要的是：夫妻住宅应当被设想成有如下特点：夫妻两人可以经常相处，也可以不时地独处，不但舒适，也保持了尊严，夫妻双方都不会感到被忽视或孤单寂寞。为此起见，必须有两个小地方——也许是房间，也许是大凹室，也许是由半矮墙遮挡的角落——这些地方清楚地被认为是私密性的，夫妻双方都可以在自己的私密地进行自己的一些活动。

再者，夫妻生活中私密性的平衡问题是十分微妙的。即使夫妻双方都有一个小天地和住宅若接若离，但双方都会感到在许多不同的场合被忽视了。而我们认为，本模式所建议的解决办法是有益的。只有在夫妻本身和其他的成年男女有某种亲密无间的、家庭似的友好关系时，这一问题才会彻底解决。那时，当夫妻一方需要私密性时，另一方就有可能就近和别的成年男女结交朋友。这一观念及其意义在模式**家庭**（75）中已论述过。

一旦隐退独处的需要得到满足，夫妻双方就会真正在一起相处；那时住宅才能成为夫妻山盟海誓、倾吐爱情的欢乐之所。

territories,where each person can keep to himself,pursue his or her own activities.

Still,the problem of the balance of privacy in a couple's lives is delicate.Even with a small place of one's own,tenuously connected to the house,one partner may feel left out at various moments.While we believe that the solution proposed in this pattern helps,the problem will not be entirely settled until the couple itself is in some close,neighborly,and family—like relationship to other adults.Then,when one needs privacy,the other has other possibilities for companionship at hand. This idea and its physical implications are discussed in the pattern,THE FAMILY(75).

Once the opportunity for withdrawal is satisfied,there is also a genuine opportunity for the couple to be together;and then the house can be a place where genuine intimacy,genuine connection can happen.

There is one other problem,unique to a house for a couple,that must be mentioned.In the first years of a couple's life,as they learn more about each other,and find out if indeed they have a future together,the evolution of the house plays a vital role.Improving the house,fixing it up,enlarging it,provides a frame for learning about one another:it brings out conflict,and offers the chance,like almost no other activity,for concrete resolution and growth.This suggests that a couple find a place that they can change gradually over the years,and not build or buy for themselves a "dream" home from scratch.The experience of making simple changes in the house,and tuning it to their lives,provides some grist for their own growth. Therefore,it is best to start small,with plenty of room for growth and change.

必须关注夫妻住宅所独有的另一个问题。夫妻双方共同生活的最初几年里，相互了解逐渐加深，他们会发觉，如果他们真能白头偕老，住宅的演变将起重要的作用。改进、修缮和扩大住宅，会加深夫妻双方的了解和感情。夫妻共同寻求具体的解决办法，使夫妻住宅不断完善。这表明，一对夫妻找到了安身立命之地，可以长年累月地逐步改变他们住宅的面貌，而不是为他们自己借款建造或购买"梦寐以求"的住宅。在住宅内部不断做些小变动，并使它和这对夫妻的生活和谐协调，这一经验是成功的，它给他们夫妻之间的感情发展带来了积极的有利因素。因此，夫妻住宅开始最好是小型的，但要留有足够的空间以备今后的变化和发展。

因此：

设想夫妻住宅由两种空间构成——夫妻的共同领域和个人的独居室。设想夫妻的共同领域为半公开的和半私密的；独居室为完全个人的和私密的。

夫妻的共同领域

夫妻独居室

⊰⊱

再说一遍，用户可以自己的某种方式把夫妻住宅作为一块具有特点的领域来处理——**自己的家**（79）。根据模式**夫妻的领域**（136），设计公共区，并给夫妻双方各安排一个独居室——**个人居室**（141）……

Therefore:

Conceive a house for a couple as being made up of two kinds of places—a shared couple's realm and individual private worlds.Imagine the shared realm as half-public and half-intimate;and the private worlds as entirely individual and private.

৪০০৪

Again,treat the house as a distinct piece of territory,in some fashion owned by its users—YOUR OWN HOME(79). Lay out the common part,according to the pattern COUPLE'S REALM(136),and give both persons an individual world of their own where they can be alone—A ROOM OF ONE'S OWN(141)...

模式78　单人住宅*

　　……单人住宅比其他任何住宅更有必要成为某种较大的家庭的一部分——**家庭**（75）。单人住宅或是和某种较大的群体家庭建造在一起，或是建成为附属于其他普通家庭住宅如**小家庭住宅**（76）或**夫妻住宅**（77）的辅助住所。

৪OCঙ

　　一旦单人住宅成为某一较大群体家庭的一部分时，首要的问题是简朴。

　　住宅市场有为数不多的住宅或公寓。它们是特意为单身独居者建造的。最为常见的是：决定独居的男女住在较大的、原先为夫妻两人或小家庭建造的住宅内。可是，对他们来说，这些较大的住房经常十分不紧凑、不实用，难以在其中生活，且也难以照料。最重要的原因是，这些住

78   HOUSE FOR ONE PERSON*

...the households with one person in them,more than any other,need to be a part of some kind of larger household—THE FAMILY (75).Either build them to fit into some larger group household,or even attach them,as ancillary cottages to other,ordinary family households like HOUSE FOR A SMALL FAMILY(76)or HOUSE FOR A COUPLE(77).

∞⋘

Once a household for one person is part of some larger group,the most critical problem which arises is the need for simplicity.

The housing market contains few houses or apartments specifically built for one person.Most often men and women who choose to live alone,live in larger houses and apartments,originally built for two people or families. And yet for one person these larger places are most often uncompact,unwieldy,hard to live in,hard to look after.Most important of all,they do not allow a person to develop a sense of self-sufficiency,simplicity,compactness,and economy in his or her own life.

The kind of place which is most closely suited to one person's needs,and most nearly overcomes this problem,is a place of the utmost simplicity,in which only the bare bones of necessity are there:a place,built like a ploughshare,where every comer,every table,every shelf,each flower pot,each chair,each log,is placed according to the simplest necessity,and supports

房不允许任何人在自己的生活中发展一种自给自足的、简朴的、紧凑的、经济实惠的意识。

非常朴实无华的、陈设简单的居室最能满足单身独居者的需要，从而单人住宅的问题也就随之解决。这种居室呈犁铧状，它的每个角落、每个书架、每个花盆、每把椅子、每一无知觉的摆设之物，完全按照最简单的需要来布置，直接满足居住者的生活所需。它们和谐协调，没有一样是不被需要的。

单人住宅的平面图将不同于其他住宅的平面图。主要因为它几乎不需要空间分隔，它只需是一个房间就行。单人住宅可能是一幢住所或一套一室户公寓，建造在平地上或设在较大的楼房内，成为群体家庭的一部分或一所独立的房屋。从本质上说，这简直就是一个四周有凹角区的中心空间。这些凹角区取代较大住宅中的房间；这些凹角区是用来安排床、浴室、厨房、工作间和入口的。

承认本书中许多模式都可被用于建造小型的单人住宅是颇为重要的。住宅规模小并不排斥丰富多彩的造型。诀窍就在于强化它和充实它；压缩这些模式，并使其表现力简单明了；对每一英寸土地要一当两用。如果这一点做得出色，单人住宅会使人感到是一妙不可言的连续性构件——就像调煮羹汤，使得宅内飘香；万籁俱寂，怡然独享。如果这一空间分隔成许多房间，就不会有这种情境了。

我们现在认为有必要唤起人们特别注意这一模式，因为在城市里建造这种小型的单人住宅几乎是不可能的——无法获得一块很小的宅基地。划区法规和银行事务严禁使用这样的小块土地；它们严禁把"标准的"土地分割成为单人住宅所需的小块宅基地。如果本模式能顺利发展，就必须修改这些法规。

the person's life directly,plainly,with the harmony of nothing that is not needed,and everything that is.

The plan of such a house will be characteristically different from other houses,primarily because it requires almost no differentiation of its spaces:it need only be one room.It can be a cottage or a studio,built on the ground or in a larger building,part of a group household or a detached structure. In essence,it is simply a central space,with nooks around it.The nooks replace the rooms in a larger house;they are for bed,bath,kitchen,workshop and entrance.

It is important to realize that very many of the patterns in this book can be built into a small house;small size does not preclude richness of form.The trick is to intensify and to overlay;to compress the patterns;to reduce them to simple expressions;to make every inch count double.When it is well done,a small house feels wonderfully continuous— cooking a bowl of soup fills the house;there is no rattling around.This cannot happen if the place is divided into rooms.

We have found it necessary to call special attention to this pattern because it is nearly impossible to build a house this small in cities—there is no way to get hold of a very small lot.Zoning codes and banking practices prohibit such tiny lots;they prohibit "normal" lots from splitting down to the kind of scale that a house for one person requires.The correct development of this pattern will require a change in these ordinances.

因此：

设想单人住宅为一种最简朴的居住空间：基本上是一居室的住所或一室户公寓，周缘有大小不等的凹室。如果单人住宅的内部空间得到充分利用，它的总面积可不超过 $300 \sim 400ft^2$。

凹室　　　　主室

⌘

还有一点：使单人住宅成为个人领域的一部分，并有自己的花园，不管它是多么小——**自己的家**（76）；使主室基本上成为一种农家厨房——**农家厨房**（139），并与四周的大小凹室相通。这些凹室专供休息、工作、洗澡、睡眠和更衣之用——**浴室**（144）、**窗前空间**（180）、**工作空间的围隔**（183）、**床龛**（188）和**更衣室**（189）；如果单人住宅是供老人或青少年使用的，则按**老人住所**（155）或**青少年住所**（154）建造……

Therefore:

Conceive a house for one person as a place of the utmost simplicity:essentially a one-room cottage or studio,with large and small alcoves around it.When it is most intense,the entire house may be no more than 300 to 400 square feet.

❧⊗❧

And again,make the house an individual piece of territory,with its own garden,no matter how small—YOUR OWN HOME(79);make the main room essentially a kind of farmhouse kitchen—FARMHOUSE KITCHEN(139),with alcoves opening off it for sitting,working,bathing,sleeping, dressing—BATHING ROOM(144),WINDOW PLACE(180),WORKSPACE ENCLOSURE(183),BED ALCOVE(188),DRESSING ROOM (189);if the house is meant for an old person,or for someone very young,shape it also according to the pattern for OLD AGE COTTAGE(155)or TEENAGER'S COTTAGE(154)...

模式79 自己的家＊＊

　　……根据**家庭**（75），每一独立住宅应当是较大的群体家庭的一部分。不管是否如此，每一独立住宅都必须有它自己完全控制的领域——**小家庭住宅**（76）、**夫妻住宅**（77）和**单人住宅**（78）；本模式只是描述每种独立住宅对这种领域的需要，它特别有助于形成较高密度的住宅团组，如**联排式住宅**（38）和**丘状住宅**（39）。分散的住宅在这些模式中不会有各住户自己的明确领域。

❧❧❧❧

　　金窝银窝，不如自己的草窝。只有住在自己的家里，人们才会感到舒适和愉快。各种租赁形式——无论是从私人房东还是从公共房管机构租赁的——都违背形成稳定的、自我康复的社区之自然过程。

79 YOUR OWN HOME**

...according to THE FAMILY(75),each individual household should be a part of a larger family group household.Whether this is so,or not,each individual household,must also have a territory of its own which it controls completely—HOUSE FOR A SMALL FAMILY(76),HOUSE FOR A COUPLE(77),HOUSE FOR ONE PERSON(78);this pattern,which simply sets down the need for such a territory,helps especially to form higher density house clusters like ROW HOUSES(38),HOUSING HILL(39),which often do not have well-defined individual territories for the separate households.

❧

People cannot be genuinely comfortable and healthy in a house which is not theirs.All forms of rental—whether from private landlords or public housing agencies—work against the natural processes which allow people to form stable,self-healing communities.

...in the imperishable primal language of the human heart house means my house,your house,a man's own house.The house is the winning throw of the dice which man has wrested from the uncanniness of universe;it is his defense against the chaos that threatens to invade him.Therefore his deeper wish is that it be his own house,that he not have to share with anyone other than his own family.(Martin Buber, *A Believing Humanism:Gleanings*,New York:Simon and Shuster,1969,p.93.)

This pattern is not intended as an argument in favor of "private property," or the process of buying and selling

出售房地产
Income property

……在人们心目中那些不可磨灭的基本词汇里,"住宅"一词意味着"我的住宅""你的住宅"和"某人的住宅"。住宅是人和神秘莫测的宇宙在孤注一掷的赌博中最后取胜的骰子点数。住宅是人反抗外来的侵袭和混乱的防御物。因此,人内心的最大愿望是住宅应当是他自己的,他应和自己的家庭而不是别人共同使用住宅。(Martin Buber,A Believing Humanism:Gleanings,New York:Simon and Shuster,1969,p.93.)

本模式不打算为"私人房地产"或买卖土地的法律手续辩护。的确,一切鼓励土地投机买卖的法律手续明显都是为了追逐利润,因而是有害的和破坏性的:它们诱使人们把住宅作为商品来处理,建造住宅就是为了"再出售",而不是去满足他们自身的需要。

正如投机和追求利润的动机不再能使造房的人把住宅造得适合他们自身的需要一样,房东和出租的房地产商也不是为了满足人们的需要。出租的住宅区总是最先变成贫民区。其中的道理是众所周知的、不言而喻的。(See,for example,George Sternlieb,The Tenement Landlord Rutgers University Press,1966)房东总是试图尽量少支出保养和维修费;房客也毫无热情来保养和维修——事实上,恰恰相反——住房条件改善会提高房租,从而增加房东的收入。像这种典型的出租房屋经过若干年后都会逐渐遭到彻底破坏。随后,房东又力图建造新的出租房屋——它们是用不着维修的——混凝土地面代替了花园;图案地板漆代替了地毯;叠层塑料贴面代替了木质表面;这是一种使新的单元住宅永久无须保养并使它们不再成为贫民区的尝试;但是这些住宅变得冷冰和缺乏独

land.Indeed,it is very clear that all those processes which encourage speculation in land,for the sake of profit,are unhealthy and destructive,because they invite people to treat houses as commodities,to build things for "resale," and not in such a way as to fit their own needs.

And just as speculation and the profit motive make it impossible for people to adapt their houses to their own needs,so tenancy,rental,and landlords do the same.Rental areas are always the first to turn to slums.The mechanism is clear and well known.See,for example,George Sternlieb,*The Tenement Landlord*(Rutgers University Press,1966).The landlord tries to keep his maintenance and repair costs as low as possible;the residents have no incentive to maintain and repair the homes—in fact,the opposite—since improvements add to the wealth of the landlord,and even justify higher rent.And so the typical piece of rental property degenerates over the years.Then landlords try to build new rental properties which are immune to neglect—gardens are replaced with concrete,carpets are replaced with lineoleum,and wooden surfaces by formica:it is an attempt to make the new units maintenance-free,and to stop the slums by force;but they turn out cold and sterile and again turn into slums,because nobody loves them.

People will only be able to feel comfortable in their houses,if they can change their houses to suit themselves,add on whatever they need,rearrange the garden as they like it;and,of course,they can only do this in circumstances where they are the legal owners of the house and land;and if,in high density multistory housing,each apartment,like a house,has a welldefined volume,in which the owner can make changes as he likes.

This requires then,that every house is owned—in some

创性了，结果又变成了贫民区，因为无人喜爱它们。

人们只有能把自己的住宅改变成适合于自身的需要，增添他们所需要的东西，并按自己的心愿重新布置花园，才会感到家里的舒适。当然，只有当他们是住宅和土地的合法所有者时才能做到这一点；如果在高密度的多层楼房内，每一套公寓就像一所住宅，都有明确规定的空间体量，所有者可按其心愿在公寓内作些变动。

由此可见，以下三点要求必须被满足：住宅必须以某种方式为居民所有；每所住宅不管是在平地上，还是在多层楼房内，都要有明确规定的空间体量，居民可以在宅内按自己的意愿自由改动；这就需要有一种阻止投机买卖房地产的所有制形式。

最近几年内，为解决每家都有一个"安乐窝"这一问题，已经提出几种解决办法。一种极端的办法如哈伯莱根的高密度"支撑体"系统说，即许多家庭可从公有的高层大厦内购买公寓住宅，并逐步改进完善自己的家。另一种极端的办法是田园群居村，即居民摒弃城市生活，到乡村去创建自己的家园。如果对房屋租赁进行些许修改，允许房客按他们的需要改进住宅，并在修缮保养过程中给予他们某种经济补贴，就能有助于这一问题的解决。这种修改是有益的，因为租赁住宅往往是通往取得住宅所有权的第一步；但是，除非房客在修缮保养和改进住宅方面投入的资金和劳务能以某种方式得到补偿，否则，出租房屋的衰颓败落和房客的经济能力下降的恶性循环将继续下去。(Cf.Rolf Goetze, "Urban Housing Rehabilitation," in Turner and Fichter,eds.,*The Freedom to Build*,New York:Macmillan，1972.)

在所有这些情况中的一个共同因素就是务必认识到，每家的"安乐窝"发展是否成功取决于下面两点：住户都必须拥有住宅和户外空间两者所需的明确规定的基地范围；住户完全有权控制基地如何发展。

fashion—by the people that live in it;it requires that every house,whether at ground level or in the air,has a well-defined volume within which the family is free to make whatever changes they want;and it requires a form of ownership which discourages speculation.

Several approaches have been put forward in recent years to solve the problem of providing each household with a "home." At one extreme there are ideas like Habraken's high density "support" system,where families buy pads on publicly owned superstructures and gradually develop their own homes. And at the other extreme there are the rural communes,where people have forsaken the city to create their own homes in the country.Even modified forms of rental can help the situation if they allow people to change their houses according to their needs and give people some financial stake in the process of maintenance.This helps,because renting is often a step along the way to home ownership;but unless tenants can somehow recover their investments in money and labor,the hopeless cycle of degeneration of rental property and the degeneration of the tenants'financial capability will continue.(Cf.Rolf Goetze, "Urban Housing Rehabilitation," in Turner and Fichter,eds.,The Freedom to Build,New York:Macmillan,1972.)

A common element in all these cases is the understanding that the successful development of a household's "home" depends upon these features:Each household must possess a clearly defined site for both a house and an outdoor space,and the household must own this site in the sense that they are in full control of its development.

Therefore:

Do everything possible to make the traditional forms of rental impossible,indeed,illegal.Give every household its own

因此：

尽量使传统的房屋租赁制实际上成为不合法的和行不通的。为每一家庭提供住宅和有足够空间的花园。要强调在控制方面的所有权而不是经济所有权。确实应优先考虑：建立居住者能控制他们的住宅和花园的所有制，防止经济上的投机买卖房地产。在任何情况下，要赋予居住者以合法的权利和真正的机会，以便他们能修缮并改进自己的住宅和住地。在高密度公寓的情况下，尤其要注意这条规则：建造私人公寓都必须有花园或种植蔬菜的露台；每一家庭都能按照自己的意愿建造和改造住宅或给住宅增建东西。

住宅　　　　　花园

控制权

❀❀❀

要从**建筑群体**（95）着手，通盘考虑住宅的形状。至于说到宅基地的形状，不要拘泥于泛泛的概念：正面宽度小而纵向深度相当可观。而要竭力使宅基地近似正方形，或甚至是一条沿街的长而浅的地带。所有这一切都是为了确立住宅和花园之间的正确关系——**半隐蔽花园**（111）……

工作组包括各种车间和办公室，甚至儿童们的学习小组：

80. 自治工作间和办公室

81. 小服务行业

82. 办公室之间的联系

83. 师徒情谊

84. 青少年协会

85. 店面小学

86. 儿童之家

home,with space enough for a garden.Keep the emphasis in the definition of ownership on control,not on financial ownership. Indeed,where it is possible to construct forms of ownership which give people control over their houses and gardens,but make financial speculation impossible,choose these forms above all others.In all cases give people the legal power,and the physical opportunity to modify and repair their own places.Pay attention to this rule especially,in the case of high density apartments:build the apartments in such a way that every individual apartment has a garden,or a terrace where vegetables will grow,and that even in this situation,each family can build,and change,and add on to their house as they wish.

&OC03

For the shape of the house,begin with BUILDING COMPLEX(95).For the shape of the lot,do not accept the common notion of a lot which has a narrow frontage and a great deal of depth.Instead,try to make every house lot roughly square,or even long along the street and shallow.All this is necessary to create the right relation between house and garden—HALF-HIDDEN GARDEN(111).

The workgroups,including all kinds of workshops and offices and even children's learning groups.

80.SELF-GOVERNING WORKSHOPS AND OFFICES

81.SMALL SERVICES WITHOUT RED TAPE

82.OFFICE CONNECTIONS

83.MASTER AND APPRENTICES

84.TEENAGE SOCIETY

85.SHOPFRONT SCHOOLS

86.CHILDREN'S HOME

模式80　自治工作间和办公室**

　　……办公室工作、工农业等各方面的工作都完全被**分散的工作点**（9）和**工业带**（42）所分散，并被组成为小型社区——**工作社区**（41）。本模式将明确类型不同的一切工作组织的基本性质，从而有助于产生这些较大的模式。

　　　　　　　　　　　　∞☙

　　如果人仅仅是机器中的一个嵌齿，那么谁也不会喜爱自己的工作了。

　　对自己的工作有全面的了解并对它的质量负责的人才会喜爱自己的工作。最重要的是：只有当富于人情味的自治小组来从事社会的各种工作时，人才能全面理解工作的

80   SELF-GOVERNING WORKSHOPS
AND OFFICES**

...all kinds of work,office work and industrial work and agricultural work,are radically decentralized by SCATTERED WORK(9),and INDUSTRIAL RIBBONS(42)and grouped in small communities—WORK COMMUNITY(41).This pattern helps to generate these larger patterns by giving the fundamental nature of all work organizations,no matter what their type.

❧✿❧

No one enjoys his work if he is a cog in a machine.

A man enjoys his work when he understands the whole and when he is responsible for the quality of the whole.He can only understand the whole and be responsible for the whole when the work which happens in society,all of it,is undertaken by small self-governing human groups;groups small enough to give people understanding through face-to-face contact,and autonomous enough to let the workers themselves govern their own affairs.

The evidence for this pattern is built upon a single, fundamental proposition:work is a form of living,with its own intrinsic rewards;any way of organizing work which is at odds with this idea,which treats work instrumentally,as a means only to other ends,is inhuman.Down through the ages people have described and proposed ways of working according to this proposition.Recently,E.F.Schumacher,the

意义并负起责任：自治小组以小见长：人们可以进行面对面的接触，相互了解，而自治程度却很高：工人们可以放手管理自己的工作。

本模式所依据的前提只有一个：工作是一种内心获得慰藉的生活方式；任何一种把工作组织起来的方式只要背离这一思想，就是以工具主义的态度来对待工作，就是仅仅把工作作为达到其他目的手段而已。这是不人道的。长久以来，人们根据这一前提早已叙述并建议了多种工作方式。近来，舒马赫，一位经济学家，对这种观点作了精彩的描述（E.F.Schumacher, "Buddhist Economics," Resurgence, 275 Kings Road, Kingston, Surrey, Volume 1, Number 11, January，1968）。请读下文：

佛教徒认为工作至少有三层含义：给人以机会去利用并发展其才华：在与别人合作共事中克服自我中心主义；提供社会所需的各种商品和服务。还有，由此观点而产生的良好成果不胜枚举。如果组织工人进行劳动的方式使工人感到毫无意义，厌烦、刺激神经和精神错乱，那近乎是犯罪。因为这表明对人的关心不如对商品的关心，这是一种邪恶的缺乏同情心的表现，也是一种醉心于这一世俗社会名利的昧心之举。同样，一味追求闲情逸致而无所事事，应被认为是对人类生存的基本真理之一，即劳逸结合的完全误解。劳和逸是同一生命过程中两个相辅相成的方面，是不能分离的，劳动的欢乐和闲暇的幸福是相互依存的。

根据佛教徒的观点，有两种机械化必须加以区分清楚：一种机械化是增强人的技艺和能力，而另一种机械化则是将人变成机器的奴隶，为机器服务。怎样把两者区别开来呢？安那达·库马拉斯瓦米，一位对现代西方和古代东方都有见地的老资格评论家说道："手艺工人本身，如果被允许这样做，总是能够辨出机器和手工工具之间的细

economist,has made a beautiful statement of this attitude(E. F.Schumacher, "Buddhist Economics," *Resurgence*,275 Kings Road,Kingston,Surrey,Volume I,Number Ⅱ ,January,1968).

The Buddhist point of view takes the function of work to be at least three-fold:to give a man a chance to utilize and develop his faculties;to enable·him to overcome his ego—centeredness by joining with other people in a common task;and to bring forth the goods and services needed for a becoming existence.Again,the consequences that flow from this view are endless.To organize work in such a manner that it becomes meaningless,boring,stultifying,or nerveracking for the worker would be little short of criminal;it would indicate a greater concern with goods than with people,an evil lack of compassion and a soul—destroying degree of attachment to the most primitive side of this worldly existence.Equally,to strive for leisure as an alternative to work would be considered a complete misunder standing of one of the basic truths of human existence,namely,that work and leisure are complementary parts of the same living process and cannot be separated without destroying the joy of work and the bliss of leisure.

From the Buddhist point of view,there are therefore two types of mechanization which must be clearly distinguished:one that enhances a man's skill and power and one that turns the work of man over to a mechanical slave,leaving man in a position of having to serve the slave.How to tell the one from the other? "The craftsman himself," says Ananda Coomaraswamy,a man equally competent to talk about the Modern West as the Ancient East, "the craftsman himself can always,if allowed to,draw the delicate distinction between the machine and the tool.The carpet loom is a tool,a contrivance for holding warp threads at a stretch for the pile

微差别。地毯织机是手工工具，它是系住拉伸的经线的机械装置，手艺工人用手指在经线间用绒毛编织地毯；但是，动力织机是机器，说它是文化的毁灭者，指的是它在编织地毯的过程中做了真正富有人情味的那部分工作。"所以，十分明显，佛教徒的经济学和现代唯物主义的经济学必定是截然不同的，因为佛教徒把文明的本质理解为：不在于增加必需品，而在于纯化人性。同时，人性主要是由人的劳动形成的。如果在有人的尊严和自由的条件下，人们适当进行劳动，就会纯化他们自己并同样纯化他们的产品。印度的哲学家兼经济学家库马拉巴对此作了如下概括：

"如果正确认识并恰如其分地评价劳动的本质，那么我们就会知道，劳动对于高级人员如同食物对于人体一样重要。劳动培育更加高级的人，并使他精力充沛，从而鞭策他生产出他所能生产的最佳产品。劳动指引他的自由意志沿正确的方向发展，并惩戒他的兽性，使之纳入进步的轨道。劳动显示出人的全部价值标准，并为人的个性发展提供一个良好的机会。"

与上述劳动方式呈鲜明对照的是过去 200 年间技术进步所形成的劳动方式。在这种劳动方式中，工人们被当作机器的零件而运转；他们生产无足轻重的零件，对工作的整体不负责任。我们可以公正地说，工人们对劳动的内在欢乐已无动于衷，显得十分淡漠。这是工业革命的主要后果。这种淡漠感在大的组织中尤其尖锐，在那里不露面的工人重复着无穷的奴役性劳动，他们生产着那些他们无法识别的产品并提供服务。

在这些组织中，工会一贯能从所有主手中夺取全部权利和利益，但证据表明，工人们基本上不满意他们的工作。例如，在汽车工业方面，每逢星期一和星期五缺勤率高达 15% ～ 20%；另有证据表明，"大量的酒精中毒者颇似

to be woven round them by the craftsmen's fingers;but the power loom is a machine,and its significance as a destroyer of culture lies in the fact that it does the essentially human part of the work." It is clear,therefore,that Buddhist economics must be very different from the economics of modern materialism,since the Buddhist sees the essence of civilization not in a multiplication of wants but in the purification of human character.Character,at the same time,is formed primarily by a man's work.And work,properly conducted in conditions of human dignity and freedom,blesses those who do it and equally their products.The Indian philosopher and economist J.C.Kumarappa sums the matter up as follows:

"If the nature of the work is properly appreciated and applied,it will stand in the same relation to the higher faculties as food is to the physical body.It nourishes and enlivens the higher man and urges him to produce the best he is capable of.It directs his freewill along the proper course and disciplines the animal in him into progressive channels.It furnishes an excellent background for man to display his scale of values and develop his personality."

In contrast to this form of work stands the style of work that has been created by the technological progress of the past two hundred years.In this style workers are made to operate like parts of a machine;they create parts of no consequence,and have no responsibility for the whole.We may fairly say that the alienation of workers from the intrinsic pleasures of their work has been a primary product of the industrial revolution. The alienation is particularly acute in large organizations,where faceless workers repeat endlessly menial tasks to create products and services with which they cannot identify.

In these organizations,with all the power and benefits that the unions have been able to wrest from the hands of the owners,there

俄罗斯人在他们的工厂中见到的酒鬼那般模样"（Nicholas von Hoffman,Washington Post）。如果工人们在劳动中没有发言权，而且劳动也不带有人情味，他们在劳动中是不会得到满足的。这就是事实。

近几年来，在现代工业中，工人对工作的不满已导致破坏行动和工人周转的加速。如俄亥俄州洛茨街的一座新的高度自动化的发动机总装配厂曾发生过破坏活动和关闭停产数星期。在过去的 7 年中，在 3 个最大的汽车制造公司，旷工人数已增加了一倍。工人周转速度也增加了一倍。一批工程师认为"在一些情况下，美国的工业把技术推进得太远了，从而把劳动中的最后一丁点儿的技巧都排斥掉了，它已经达到违背人性的地步"（Agis Salpukis, "Is the machine pushing man over the brink?" *San Francisco Sunday Examiner and Chronicle*, April 16，1972）。

也许，关于工作和生活之间的相互关系的最引人注目的经验性证据是 1972 年美国保健教育福利局秘书埃利奥特·理查森授意写的最新研究报告《美国的职业》。该报告指出，最佳的长寿预言家不是根据某人是否抽烟、是否常去请医生治病，而是根据他对自己的职业是否满意。该报告确认，工人们对职业不满意的两个主要因素：工人逐渐丧失独立性，任务日趋简单化、支离破碎和孤立状态。这两种不满情绪无论在现代工业中，还是在办公室业务中，都同样在不断蔓延。

但是，从人类的大部分历史来看，生产商品和提供服务是更具有个性的、自我调整的职业：那时各种职业都是创造性的、令人兴趣盎然的。而今天不再是那样，的确是毫无道理的。

例如，西摩·梅尔曼在《做决定和生产率》一文中把美国的底特律和英国的考文垂两个城市生产拖拉机的情况进行了比较。他对比了底特律的经营管理规则和考

is still evidence that workers are fundamentally unhappy with their work.In the auto industry,for example,the absentee rate on Mondays and Fridays is staggering—15 to 20 percent;and there is evidence of "massive alcoholism,similar to what the Russians are experiencing with their factory workers" (Nicholas von Hoffman,*Washington Post*).The fact is that people cannot find satisfaction in work unless it is performed at a human scale and in a setting where the worker has a say.

Job dissatisfaction in modern industry has also led to industrial sabotage and a faster turnover of workers in recent years.A new super-automated General Motors assembly plant in Lordstown,Ohio,was sabotaged and shut down for several weeks.Absenteeism in the three largest automobile manufacturing companies has doubled in the past seven years.The turnover of workers has also doubled.Some industrial engineers believe that "American industry in some cases may have pushed technology too far by taking the last few bits of skill out of jobs,and that a point of human resistance has been reached" (Agis Salpukis, "Is the machine pushing man over the brink?" *San Francisco Sunday Examiner and Chronicle*,April 16,1972).

Perhaps the most dramatic empirical evidence for the connection between work and life is that presented in the recent study, "Work in America," commissioned by Elliot Richardson,as Secretary of Health,Education and Welfare Department,1972.This study finds that *the single best predictor of long life is not whether a person smokes or how often he sees a doctor,but the extent to which he is satisfied with his job.*The report identifies the two main elements of job dissatisfaction as the diminishing independence of workers,and the increasing simplification,fragmentation,and isolation of tasks—both of

文垂的班组定额劳动制之后指出，班组定额劳动制生产了高质量的产品，并创造了英国工业中的最高工资额。"作决定过程的最大特点：一组工人运用手中掌握的最高权威作出决定的过程就是他们感情上相互共鸣的过程。"

其他的工程、实验和证据都表明，现代的工作都可以用这种方式组织起来，并可与现代最复杂的尖端技术和谐一致。这方面的情况已由亨尼厄斯、加森和蔡斯（See Workers'Control,New York:Vintage Books，1973）所收集。

还有另一例证取自特里斯特的《组织的选择》和赫勃斯特的《自治小组的功能》。这两位作者描写了德拉姆矿井的组织工作，几组矿工曾实施过这一方案。

混合采煤的组织工作可作为一个整体来描述。其中每一个工作组都对采煤工作面的操作总周期承担全部责任。工作组内每一成员的任务都不是固定不变的。相反，工作组根据不断提出的任务随时调配组员。他们在符合技术要求和安全要求的限度内，可以发挥才能，采取各种有效方法去组织生产和完成任务。

（实验表明）40～50人组成的工作组十分重要，它们作为自我调整和自我发展的社会有机体能使它们自己一直保持稳定的高生产率。（引自 Colin Ward，"The organization of anarchy," Patterns of Anarchy，Krimerman and Perry，eds.，New York:Anchor Books，1966，pp.349～351.）

我们认为，这些自治小组不仅是最有效率的，而且也是职业满意感的唯一可能的源泉。它们所提供的工作方式是唯一能滋养丰富人的内心世界并使他感到满足的工作方式。

which are rampant in modern industrial and office work alike.

But for most of human history,the production of goods and services was for a far more personal,self-regulating affair;when each job of work was a matter of creative interest.And there is no reason why work can't be like that again,today.

For instance,Seymour Melman,in *Decision Making and Productivity*,compares the manufacture of tractors in Detroit and in Coventry,England.He contrasts Detroit's managerial rule with Coventry's gang system and shows that the gang system produced high quality products and the highest wages in British industry. "The most characteristic feature of the decision—formulation process is that of mutuality in decision—making with fihal authority residing in the hands of the group workers themselves."

Other projects and experiments and evidence which indicate that modern work can be organized in this manner and still be compatible with sophisticated technology,have been collected by Hunnius,Garson,and Chase.See Workers' Control,New York:Vintage Books,1973.

And another example comes from the reports by E.L.Trist,Organizational Choice and P.Herbst,Autonomous Group Functioning.These authors describe the organization of work in mining pits in Durham which was put into practice by groups of miners.

The composite work organization may be described as one in which the group takes over complete responsibility for the total cycle of operations involved in mining the coal-face. No member of the group has a fixed work-role.Instead,the men deploy themselves,depending on the requirements of the ongoing group task.Within the limits of technological and safety requirements they are free to evolve their way of organizing and carrying out their task.

因此：

鼓励成立由 **5 ~ 20 名工人**参加的自治工作间和办公室。使每个工作组在有关组织、方式、与其他工作组的关系、雇用和解雇、工作进度安排等各个方面都完全自治。工作复杂的地方需要成立较大的组织。若干个这样的工作组可以联合起来，进行合作，以便生产出复杂的产品并提供服务。

自治工作间

❀❀❀

把工作组安排在它自己的一幢建筑物中——**办公室之间的联系**（82）、**建筑群体**（95）；如果工作组很大，如果它为公众服务，就把它分成若干个易于识别的自治小组，每一小组不超过 12 人——**小服务行业**（81）；在任何情况下，将一切工作都分配到小组，不是直接分配到合作工作组，就是分配到自治小组。每组成员都在公共的空间工作——**师徒情谊**（83）和**工作小组**（148）……

〔The experiment demonstrates〕the ability of quite large primary work groups of 40-50 members to act as self—regulating,selfdeveloping social organisms able to maintain.themselves in a steady state of high productivity.(Quoted in Colin Ward, "The organization of anarchy," *Patterns of Anarchy*,Krimerman and Perry,eds.,New York:Anchor Books,1966,pp.349-351.)

We believe that these small self-governing groups are not only most efficient,but also the only possible source of job satisfaction.They provide the only style of work that is nourishing and intrinsically satisfying.

Therefore:

Encourage the formation of self-governing workshops and offices of 5 to 20 workers.Make each group autonomous—with respect to organization,style,relation to other groups,hiring and firing,work schedule.Where the work is complicated and requires larger organizations,several of these work groups can federate and cooperate to produce complex artifacts and services.

೮೦೮೩

House the workgroup in a building of its own—OFFICE CONNECTIONS(82),BUILDING COMPLEX(95);if the workgroup is large enough,and if it serves the public,break it down into autonomous departments,easily identifiable,with no more than a dozen people each—SMALL SERVICES WITHOUT RED TAPE(81);in any case,divide all work into small team work,either direetly within the cooperative work-group or under the departments,with the people of each team in common space—MASTER AND APPRENTICES(83)and SMALL WORK GROUPS(148)...

模式81　小服务行业*

　　……为公众提供服务的办公室——**工作社区**（41）、**像市场一样开放的大学**（43）、**地方市政厅**（44）、**保健中心**（47）、**青少年协会**（84）都需有便民的附设部门。当然，这些小部门要逐步发展，一次发展一个，这样对逐渐形成这些较大的模式也是有益的。

৪৩৫৪

　　如果附设部门和公共服务机构太大就不起作用。它们一大就会失去人情味，就会变成官僚主义的东西；衙门作风就会盛行起来。

　　现有大量文献谈及衙门作风和官僚主义违情悖理的危害性 (See,for example,Gideon Sjoberg,Richard Brymer,and Buford Farris, "Bureaucracy and the Lower Class," Sociology and Social

81   SMALL SERVICES WITHOUT RED TAPE*

...all offices which provide service to the public—
WORK COMMUNITY(41),UNIVERSITY AS A
MARKETPLACE(43),LOCAL TOWN HALL(44),HEATH
CENTER(47),TEENAGE SOCIETY(84)need subsidiary
departments,where the members of the public go.And of course,
piecemeal development of these small departments,one department
at a time,can also help to generate these larger patterns gradually.

৪০৫৪

**Departments and public services don't work if they
are too large.When they are large,their human qualities
vanish;they become bureaucratic;red tape takes over.**

There is a great deal of literature on the way red
tape and bureaucracy work against human needs.See,for
example,Gideon Sjoberg,Richard Brymer,and Buford
Farris, "Bureaucracy and the Lower Class," Sociology
and Social Research,50,April,1966,pp.325-377;and Alvin
W.Gouldner, "Red Tape as a Social Problem," in Robert
Mertin,Reader in Bureaucracy,Free Press,1952,pp.410-418.

According to these authors,red tape can be overcome in
two ways.First,it can be overcome by making each service
program small and autonomous.A great deal of evidence
shows that red tape occurs largely as a result of impersonal
relationships in large institutions.When people can no longer
communicate on a face-to-face basis,they need formal
regulations,and in the lower echelons of the organization,these

Research,50,April,1966,pp.325 ～ 377;and Alvin W.Gouldner,"Red Tape as a Social Problem,"in Robert Mertin,Reader in Bureaucracy,Free Press， 1952， pp.410 ～ 418)。

根据这些作者的意见，衙门作风是可以克服的，其途径有二：第一，每个服务机构要小而自治。大量证据表明，衙门作风的滋生主要由于机构大以及人们之间的非个人的关系。当人们不能再面对面进行交际时，就需要正式的管理规章制度，但下级组织对规章制度往往是盲目地、狭隘地遵循着。

第二，改变业主对服务机构缺乏体谅的消极态度。大量证据表明，当业主对社会机构具有谅解精神的积极态度时，即使这些机构有官腔也就耍不起来。

因此，我们得出结论，任何一个服务机构人员总数不要超过 12 人（全体人员，包括秘书在内）。这一数字的根据是：12 人似乎是坐下来面对面讨论问题的最大数字。看来，人员再少些可能工作得更好。此外，每一服务机构都应是相对自治的，它只服从上级组织提出的一些简单的规章制度，而总体自治制应强调这一点。为了实现总体自治制，每一服务机构都应有一个完全由它自己管辖的地区；它有门通往有顶街道，并和其他服务机构在空间上完全分开。

本模式同样适用于市政厅和医疗中心的各个部门，以及福利中心的各地方机构。在大多数场合下，本模式要求对行政组织进行根本性的变动。不管实施起来困难有多大，我们认为，这些变动是必不可少的。

因此：

在任何一个有附设部门为公众提供服务的机构中：

1. 使每一服务机构或附设部门尽可能成为自治的。

2. 不准任何一个服务机构全体职工的总人数超过 12 名。

formal regulations are followed blindly and narrowly.

Second,red tape can be overcome by changing the passive nature of the clients'relation to service programs.There is considerable evidence to show that when clients have an active relationship with a social institution,the institution loses its power to intimidate them.

We have therefore concluded that no service should have more than 12 persons total(all staff,including clerks).We base this figure on the fact that 12 seems to be the largest number of people that can sit down in a face-to-face discussion.It seems likely that a smaller staff size will work better still. Furthermore,each service should be relatively autonomous— subject only to a few simple,coordinative regulations from parent organizations—and that this should be emphasized by physical autonomy.In order to be physically autonomous,each service must have an area which is entirely under its own jurisdiction;with its own door on a public thoroughfare,and complete physical separation from other services.

This pattern applies equally to the departments of a city hall,a medical center,or to the local branches of a welfare agency.In most of these cases the pattern would require basic changes in administrative organization.However difficult they may be to implement,we believe these changes are required.

Therefore:

In any institution whose departments provide public service:

1.Make each service or department autonomous as far as possible.

2.Allow no one service more than 12 staff members total.

3. 给每一服务机构在易识别的建筑内提供用房。

4. 使每一服务机构都能通往有顶街道。

可见的正面

公共有顶街道

12个人

❧❦

　　按**办公室之间的联系**（82）和**建筑群体**（95）的规定，为这些部门选址；如果公共通道在室内，则使它成为**有顶街道**（101），并使服务机构的正面成为可见的**各种入口**（102）；无论在什么地方这些服务机构都要以某种方式与社区的政治生活联系起来，并使它们和居民或用户所形成的特别小组混杂相间——**项链状的社区行业**（45）；根据**灵活办公室空间**（146）安排该部门的内部空间；并为三三两两地进行合作的人提供用房——**工作小组**（148）……

3.House each one in an identifiable piece of the building.

4.Give each one direct access to a public thoroughfare.

•••

Arrange these departments in space,according to the prescription of OFFICE CONNECTIONS(82)and BUILDING COMPLEX(95);if the public thoroughfare is indoors,make it a BUILDING THOROUGHFARE (101),and make the fronts of the services visible as a FAMILY OF ENTRANCES(102);wherever the services are in any way connected to the political life of the community,mix them with ad hoc groups created by the citizens or users—NECKLACE OF COMMUNITY PROJECTS(45);arrange the inside space of the department according to FLEXIBLE OFFICE SPACE(146);and provide rooms where people can team up in two's and three's—SMALL WORK GROUPS(148)...

模式82 办公室之间的联系*

……在任何一个工作社区和任何一个办公室，总会有各种不同的工作组。而决定这些工作组在空间上如何安排始终是十分重要的。哪些工作组应当相互靠拢？哪些应当相互分开？本模式回答这些问题。它的说明将极大地有助于**工作社区**（41）或**自治工作间和办公室**（80）或**小服务行业**（81）的内部布局的设计工作。

୨୦୯୬

如果一个办公室的两个部分离得太远，工作人员在其间来回行走的次数，将不会像他们所需要的那样频繁了；如果它们相距不止一层楼，上下楼层间就几乎不会有什么交往了。

现代建筑学方法常常包括近似矩阵法，它表示不同的工作人员在一个办公室或一个医院几部分之间来回行走的次数。近似矩阵法作出的假设总是不言而喻的：各部分紧挨着的地方一般都是工作人员在其间来回行走次数最多的地方。可是，如通常所说，这一概念是不切实际的，是完全行不通的。

上述概念是由泰勒追求高效率而提出的。他假定一个单位的各部门之间来回行走的工作人员越少，支出"浪费"在走动上的工资就越少。这种分析合乎逻辑的结论是：假如有可能，工作人员干脆就不要走动，应当整天呆板地坐在安乐椅里，打发单调无味的日子。

事实上，只有当工作人员的身心处于最佳的健康状态时，工作才最有效率。一个被迫整天坐在办公桌后面、连伸伸腿舒展一下身子都不许的人，将变成没有休息的、因而也

82　OFFICE CONNECTIONS*

...in any work community or any office,there are always various human groups—and it is always important to decide how these groups shall be placed,in space.Which should be near each other,which ones further apart?This pattern gives the answer to this question,and in doing so,helps greatly to construct the inner layout of a WORK COMMUNITY(41)or of SELF-GOVERNING WORKSHOPS AND OFFICES(80)or of SMALL SERVICES WITHOUT RED TAPE(81).

৪০৬৪

If two parts of an office are too far apart,people will not move between them as often as they need to;and if they are more than one floor apart,there will be almost no communication between the two.

Current architectural methods often include a proximity matrix,which shows the amount of movement between different people and functions in an office or a hospital.These methods always make the tacit assumption that the functions which have the most movement between them should be closest together. However,as usually stated,this concept is completely invalid.

The concept has been created by a kind of Taylorian quest for efficiency,in which it is assumed that the less people walk about,the less of their salary is spent on "wasteful" walking. The logical conclusion of this kind of analysis is that,if it were only possible,people should not have to walk at all,and should spend the day vegetating in their armchairs.

是不能工作的人，这样就毫无效率可言了。稍稍走动对你十分有益。这不仅对工作人员的身体健康有好处，而且也是让他们换换环境，思考一下别的问题，谈谈他们对上午工作中的某一细节或办公室内某一日常人事问题的看法的机会。

另外，如果某一工作人员行走同一路程来回许多趟，就会出现一个因行走路程长而感到费时烦人的临界点，从而也就无效率了。因为此时他容易发怒而无法克制，就会开始不来回行走了，因为路程太长，而且也太频繁。

只要在办公室里工作的人员并不感觉到他们必须行走的路程是烦人的，则办公室就会有效地发挥作用。行走的距离要短得使工作人员不会感到行走是一种烦恼——但行程不必再缩短。

行走的烦恼以行程和频率之间的关系为转移。你可以步行 10ft 到档案室去，一日数次而不觉厌烦；你偶尔步行 400ft 去档案室也不觉厌烦。在下列的图表中我们标绘出了行程和频率的各种不同组合的厌烦阈值。

这一图表是根据伯克利市政厅所进行的 127 次调查的结果而绘制的。工作人员被要求说明他们在工作周内必须按时行走的全部路程、频率，以及他们是否认为这种行走是令人厌烦的。

图表上的这条线表示各种距离的中位线，它是每一不同频率的厌烦阈值。我们定义中位线右侧的各种距离为厌烦距离。任何行走频率的厌烦距离即我们预言至少 50% 的人会开始感到厌烦的距离。

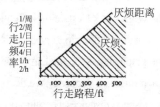

厌烦距离
Nuisance distances

The fact is that people work best only when they are healthy in mind and body.A person who is forced to sit all day long behind a desk,without ever stretching his legs,will become restless and unable to work,and inefficient in this way.Some walking is very good for you.It is not only good for the body,but also gives people an opportunity for a change of scene,a way of thinking about something else,a chance to reflect on some detail of the morning's work or one of the everyday human problems in the office.

On the other hand,if a person has to make the same trip,many times,there is a point at which the length of the trip becomes time—consuming and annoying,and then inefficient,because it makes the person irritable,and finally critical when a person starts avoiding trips because they are too long and too frequent.

An office will function efficiently so long as the people who work there do not feel that the trips they have to take are a nuisance.Trips need to be short enough so they are not felt a nuisance—but they do not need to be any shorter.

The nuisance of a trip depends on the relationship between length and frequency.You can walk 10 feet to a file many times a day without being annoyed by it;you can walk 400 feet occasionally without being annoyed.In the graph below we plot the nuisance threshold for various combinations of length and frequency.

The graph is based on 127 observations in the Berkeley City Hall.People were asked to define all the trips they had to make regularly during the work week,to state their frequency,and then to state whether they considered the trip to be a nuisance.

The line on the graph shows the median of the distances said to be a nuisance for each different frequency.We define distances to the right of this line as nuisance distances.The nuisance distance for any trip frequency is the distance at which we predict that at least 50per

直到现在我们关于近似矩阵法的讨论一直是以水平距离为基础的。怎样看待楼梯？垂直距离究竟起什么作用？或者，说得更为准确一些，什么是一段楼梯跑的水平距离的等值呢？试假设，根据近似图，两个部门之间相距不到 100ft。因某种原因，它们位于不同的楼层，相隔一层楼。两层楼之间的楼梯要占去 100ft 的多少呢？这一楼梯折合成水平距离后，两个部门之间的距离究竟有多远呢？

我们并不知道对这一问题的确切回答。但是，我们具有从玛丽娜·埃斯塔布鲁克和罗伯特·萨默未公开发表的研究报告中获得的某种间接证据。正如我们将会看到的那样，这份研究报告将表明，楼梯起着一种大得多的作用，它折合的水平距离比你想象中的长得多。

埃斯塔布鲁克和萨默调查研究了在一幢三层的大学楼内几个不同部门之间的工作人员是如何相互认识的。他们要求被调查者指名道姓地说出他们所熟悉的非本部门的人。他们调查的结果如下：

认识的人的百分比	非本部门的位置
12.2	在同一层楼
8.9	相隔一层楼
2.2	相隔二层楼

他们已经认识同一层楼非本部门的人达 12.2%，认识相隔一层楼非本部门的人达 8.9%，认识相隔二层楼非本部门的人则只有 2.2%。总而言之，各部门之间相隔二层楼或二层楼以上，实际上就连非正式的接触也不会有了。

很遗憾，我们在获知上述两位的研究成果之前就已完成了近似图的研究；所以我们一直未能说明水平距离和楼梯距离两者之间的关系。不过，十分清楚，一跑楼梯的长度一定等值于相当可观的水平距离：两跑楼梯的效果几乎

cent of all people will begin to consider this distance a nuisance.

So far, our discussion of proximity has been based on horizontal distances. How do stairs enter in? What part does vertical distance play in the experience of proximity? Or, to put it more precisely, what is the horizontal equivalent of one flight of stairs? Suppose two departments need to be within 100 feet of one another, according to the proximity graph—and suppose that they are for some reason on different stories, one floor apart. How much of the 100 feet does the stair eat up: with the stair between them, how far apart can they be horizontally?

We do not know the exact answer to this question. However, we do have some indirect evidence from an unpublished study by Marina Estabrook and Robert Sommer. As we shall see, this study shows that stairs play a much greater role, and eat up much more "distance" than you might imagine.

Estabrook and Sommer studied the formation of acquaintances in a threestory university building, where several different departments were housed. They asked people to name all the people they knew in departments other than their own. Their results were as follows:

Percent of people known:	When departments are:
12.2	on same floor
8.9	one floor apart
2.2	two floors apart

People knew 12.2 per cent of the people from other departments on the same floor as their own, 8.9 per cent of the people from other departments one floor apart from their own floor, and only 2.2 per cent of the people from other departments two floors apart from their own. In short, by the time departments are separated by two floors or more, there is

是一跑楼梯的 3 倍。据此我们推测，一跑楼梯的效果，就其相互作用的影响和人对距离的感觉来说，约等于 100ft 的水平距离；两跑楼梯的效果约等于 300ft 的水平距离。

因此：

为了确定各部门之间的距离，就要计算每天在两个部门之间的行走次数；从上列的图表中取"厌烦距离"；然后保证两个部门之间的总距离小于厌烦距离。把一跑楼梯的长度折算为 100ft 左右，把两跑楼梯折算为 300ft 左右。

两层楼的最大值

小于厌烦距离

⋇⋇⋇

使各部门的办公大楼符合**不高于四层楼**（21），并按**建筑群体**（95）设计其外形。使在上面各层的每个工作组都有楼梯直通外界——**步行街**（100）、**室外楼梯**（158）；如果工作组之间有内部走廊，把它扩大，使它起到街道的作用——**有顶街道**（101）；使每一工作组都容易被识别，并使它具有标志清晰的入口，以便工作人员易于发现从一个工作组到另一工作组的道路——**各种入口**（102）……

virtually no informal contact between the departments.

Unfortunately,our own study of proximity was done before we knew about these findings by Estabrook and Sommer;so we have not yet been able to define the relation between the two kinds of distance.It is clear,though,that one stair must be equivalent to a rather considerable horizontal distance;and that two flights of stairs have almost three times the effect of a single stair.On the basis of this evidence,we conjecture that one stair is equal to about 100 horizontal feet in its effect on interaction and feelings of distance;and that two flights of stairs are equal to about 300 horizontal feet.

Therefore:

To establish distances between departments,calculate the number of trips per day made between each two departments;get the "nuisance distance" from the graph above;then make sure that the physical distance between the two departments is less than the nuisance distance.Reckon one flight of stairs as about 100 feet,and two flights of stairs as about 300 feet.

❧❧❧

Keep the buildings which house the departments in line with the FOUR-STORY LIMIT(21),and get their shape from BUILDING COMPLEX(95).Give every working group on upper storys its own stair to connect it directly to the public world—PEDESTRIAN STREET(100),OPEN STAIRS(158);if there are internal corridors between groups,make them large enough to function as streets—BUILDING THOROUGHFARE(101);and identify each workgroup clearly,and give it a well-marked entrance,so that people easily find their way from one to another—FAMILY OF ENTRANCES(102)...

模式83 师徒情谊*

……社区的**学习网**（18）基于以下事实：即学习是分散的，每一活动都是分组进行的——而恰恰不是在教室内活动的。为了实现本模式，很有必要在工业企业、在办公室和工作社区普遍建立独立的工作组，使学习的过程成为可能。本模式说明所必需的处置办法，从而将对**自治工作间和办公室**（80）以及**学习网**（18）的形成大有裨益。

❦❧

学习的基本情况就是能者为师，即求学的人通过帮助懂行的干活而学得知识。

这是求知的最简单的方法，而且非常有效。据比较，只听课和从书本上学真是味同嚼蜡。但是，通过跟随能者实践来学习的这种方法几乎从现代社会中消失不见了。中小学和大学已经盛行并抽象出许多学习的途径。这些学习

83   MASTER AND APPRENTICES*

...the NETWORK OF LEARNING(18)in the community relies on the fact that learning is decentralized,and part and parcel of every activity not just a classroom thing.In order to realize this pattern,it is essential that the individual workgroups,throughout industry,offices,workshops,and work communities,are all set up to make the learning process possible.This pattern,which shows the arrangement needed,therefore helps greatly to form SELF-GOVERNING WORKSHOPS AND OFFICES(80)as well as the NETWORK OF LEARNING(18).

&OCS

The fundamental learning situation is one in which a person learns by helping someone who really knows what he is doing.

It is the simplest way of acquiring knowledge,and it is powerfully effective.By comparison,learning from lectures and books is dry as dust.But this situation has all but disappeared from modern society.The schools and universities have taken over and abstracted many ways of learning which in earlier times were always closely related to the real work of profess ionals,tradesmen,artisans,independent scholars.In the twelfth century,for instance,young people learned by working beside masters—helping them,making contact directly with every corner of society.When a young person found himself able to contribute to a field of knowledge,or a trade—he would prepare

途径在早期总是同专业人员、商人、手艺工人、独立的学者的实际工作密切联系的。例如，在 12 世纪，青年人就在师傅身旁边工作边学习——帮助他们和社会各个方面直接接触。当年轻人自以为他在某一领域或某一行中能有所建树时，他就要准备一件"样品"；在师傅同意后，他就成为这一行会的会员了。

亚历山大和戈德伯格的实验早已表明：一名"教师"让数人组成一个小组，也即"学生们"在实际中帮助自己做某件事或正要解决的某一问题，而不是他讲授某一抽象的或兴趣不大的课题——这种方法可能是最成功的。（Report to the Muscatine Committee,on experimental course ED.10X,Department of Architecture,University of California, 1966.）

如果这一点具有普遍意义——总而言之，如果学生当学徒能学得最好，并能助师傅一臂之力，就可得出结论：我们的中小学和大学，办公室和工业企业，都应提供有利的环境使这种师徒情谊融洽自然，以师傅为主带 6 名徒弟（不超过 6 名）共同工作，并由此产生密切的联系。

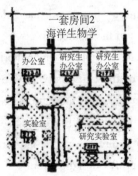

生物学实验室的师徒关系
Master-apprentice relationship for a biology laboratory

我们在俄勒冈大学分子生物学大楼获悉了本模式的一

a master "piece" ;and with the consent of the masters,become a fellow in the guild.

An experiment by Alexander and Goldberg has shown that a class in which one person teaches a small group of others is most likely to be successful in those cases where the "students" are actually helping the "teacher" to do something or solve some problem,which he is working on anyway—not when a subject of abstract or general interest is being taught.(Report to the Muscatine Committee,on experimental course ED.10X,Department of Architecture,University of California,1966.)

If this is generally true—in short,if students can learn best when they are acting as apprentices,and helping to do something interesting—it follows that our schools and universities and offices and industries must provide physical settings which make this master-apprentice relation possible and natural:physical settings where com-munal work is centered on the master's efforts and where half a dozen apprentices—not more—have workspace closely connected to the communal work of the studio.

We know of an example of this pattern,in the Molecular Biology building of the University of Oregon.The floors of the building are made up of laboratories,each one under the direction of a professor of biology,each with two or three small rooms opening directly off the lab for graduate students working under the professor's direction.

We believe that variations of this pattern are possible in many different work organizations,as well as the schools.The practice of law,architecture,medicine,the building trades,social services,engineering—each discipline has the potential to set

个实例。大楼的各层均设有许多实验室，每一实验室都有
一位生物学教授指导，每一实验室还附设 2 ～ 3 个小房间，
相互直接沟通，以供研究生在教授指导下进行工作之用。

我们认为，本模式的各种变形可能存在于许多不同的
工作组和学校中。在法律、建筑学、医学、房屋买卖、社
会服务和工程等各方面的实践中——每一门学科都有潜在
可能去创立自己的学习途径。因此，身体力行的实践者都
要在工作环境中按这些原则行事。

因此：

**在每一工作组、在工业企业和办公室，都要以做和学
相结合的方式来安排工作任务。把做每件工作都看成学习
的良机。为此，要按师徒的传统关系组织工作：支持这种
社会组织形式，把工作室划为供几组师徒共同工作、朝夕
相处的小空间。**

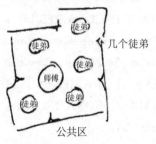

＊

把工作间布置成**半私密办公室**（152）或**工作空间的
围隔**（183）。工作组要小。给每一工作组以一公共区和一
公共集会场所，以及一处共同进餐的地方——**中心公共区**
（129）、**共同进餐**（147）、**工作小组**（148）和**小会议室**
（151）……

up its ways of learning,and therefore the environments in which its practitioners work,along these lines.

Therefore:

Arrange the work in every workgroup,industry,and office,in such a way that work and learning go forward hand in hand.Treat every piece of work as an opportunity for learning. To this end,organize work around a tradition of masters and apprentices:and support this form of social organization with a division of the workspace into spatial clusters-one for each master and his apprentices—where they can work and meet together.

෨൰ൟ

Arrange the workspaces as HALF-PRIVATE OFFICES (152) or WORKSPACE ENCLOSURES(183).Keep workgroups small,and give every group a common area,a common meeting space,and a place where they can eat together— COMMON AREAS AT THE HEART (129),COMMUNAL EATING(147),SMALL WORK GROUPS (148),SMALL MEETING ROOMS(151)...

模式84 青少年协会

......平衡的**生命的周期**（26）要求：从童年到成年的过渡应由比学校更灵活和更易于接受的青少年协会来处理。对这种单位进行定义的本模式能在**学习网**（18）中应运而生，并有助于**师徒情谊**（83）网的发展。现说明如下。

୫ୠୖଓ

青少年时期是童年走向成年的过渡时期。在传统社会中，各种满足过渡心理要求的礼仪会随之而来。但在现代社会中"高级中学"对此却完全不予理睬。

就我们所知，一个东非部落的例子是最激动人心的。为了成为一名硬汉子，该部落的每一个男孩都要出门远游两年，备尝千辛万苦，最后完成艰巨的任务——杀死一头狮子。在此期间，他漫游所到之处，各家族和村庄都要热情款待并照料他。他们认为这样做是义不容辞的，是礼仪的一部分。最后，他在历尽艰难险阻，并杀死了一头狮子，才会被大家公认是一个有大丈夫气概的男子汉。

在现代社会中，过渡并非如此直接和简单。其原因是错综复杂的，无法在此讨论。过渡的过程和时间已经大大扩展和复杂化了（See Edgar Friedenberg, *The Vanishing Adolescent*, Beacon Press, Boston, 1959 and *Coming of Age in America*, Random House Inc.N.Y., 1965）。青少年时期一般指 12 ～ 18 岁；前后相差 6 年而不是一两年。从童年到成年这一简单的性成熟的变化已让位于更广泛、更缓慢的变

84   TEENAGE SOCIETY

...the balanced LIFE CYCLE(26)requires that the transition from childhood to adulthood be treated by a far more subtle and embracing kind of teenage institution than a school;this pattern,which begins to define that institution,can take its place in the NETWORK OF LEARNING(18)and help contribute to the network of MASTERS AND APPRENTICES(83).

⊰⊱

Teenage is the time of passage between childhood and adulthood.In traditional societies,this passage is accompanied by rites which suit the psychological demands of the transition. But in modern society the "high school" fails entirely to provide this passage.

The most striking traditional example we know comes from an east African tribe.In order to become a man,a boy of this tribe embarks on a two year journey,which includes a series of more and more difficult tasks,and culminates in the hardest of all—to kill a lion.During his journey,families and villages all over the territory which he roams take him in,and care for him:they recognize their obligation to do so as part of his ritual. Finally,when the boy has passed through all these tasks,and killed his lion,he is accepted as a man.

In modern society,the transition cannot be so direct or simple.For reasons too complex to discuss here,the process of transition,and the time it takes have been extended and elaborated greatly.(See Edgar Friedenberg,*The Vanishing*

化。他或她在漫长的斗争中决定究竟"成为什么样的人"的时候，各人的自我本性就会在这种变化中显露出来。现在几乎没有一个人在重操他父亲的旧业；相反，在这个充满无限可能性的世界里，一切都得另起炉灶。自从工业革命以来，对这个世界来说，这一新的漫长的过程我们就称为青春期。

青春期这一过程唤起一种特别的希望。因为即将到来的成年期传统上是以生儿育女为标志的，一个扩大了的成年期是否会带来一种更加深刻的、不同的自我概念呢？

青春就是希望。但迄今为止，希望还是渺茫的。不同文化的青少年都有自己的青春期，同样都有自己的青春期问题。在技术发达的世界里，青春期会引起一系列的暴力行为，以极其相似的方式导致危机和绝境。少年犯罪，中小学生退学，青少年自杀、吸毒和私奔等触目惊心的事发生率很高。在这些情况下，甚至"正常的"青春期也是充满着忧虑，更不必说离通往更完整的和复杂的自我之大门还有多远。而对此我们在道德上和理智上往往麻木不仁。

就青春期问题而言，尤其是中学首当其冲。恰恰在青少年时期需要自由地、成群结队地将他们自己的活动联合起来，并反复探索成年人的世界——成年人的工作、爱情、科学、法律、习惯、旅行、游戏、交际和管理方法——之际，他们却被当作大孩子看待。他们在中学的责任性或权威不比幼儿园的小朋友大。他们只负责收拾整理自己的东西，负责学校乐队的演奏，也许还负责选举班长。可是这些事幼儿园里的孩子也都一样做。现在没有一种新的社会形式就是成人社会的缩影，而青少年在其中可以严肃地对

Adolescent,Beacon Press,Boston,1959 and *Coming of Age in America*,Random House Inc.N.Y.,1965).Teenage lasts,typically,from 12 to 18;six years instead of one or two. The simple sexual transformation,the change from childhood to maturity,has given way to a much vaster,slower change,in which the self of a person emerges during a long struggle in which the person decides "what he or she is going to "be". Almost no one does what his father did before him;instead,in a world of infinite possibilities,it has to be worked out from nothing.This long process,new to the world since the industrial revolution is the process we call adolescence.

And this process of adolescence calls up an extraordinary hope.Since coming of age traditionally marks the birth of self,might not an extended coming of age bring with it a more profound and varied self-conception?

That is the hope;but so far it hasn't worked that way.Every culture that has an adolescent period has also a complicated adolescent problem.Throughout the technically developed world,puberty sets off a chain of forces that lead,in remarkably similar ways,to crisis and impasse.High rates of delinquency,school dropout,teenage suicide,drug addiction,and runaway are the dramatic forms this problem takes.And under these circumstances even "normal" adolescence is full of anxiety and,far from opening the doors to a more whole and complicated self,it tends to benumb us morally and intellectually.

The institution of the high school has particularly borne the brunt of the adolescent problem.Just at the time when teenagers need to band together freely in groups of their own making and explore,step back from,and explore again,the adult world:its work,love,science,laws,habits,travel,play,com

照检查他们自己不断发育的成年期。而在这些条件下，青少年身上正在孕育着成年人的力量就会变成对他们的无责和无权威进行痛斥和报复。盲目的成年人极易把这种报复行为称为"少年犯罪"。

一个官方机构终于认识到了这一点。1973年12月美国全国中等教育改革委员会和基特林基金会合作时就得出以下结论：美国城市的中学简直不起作用；它们作为一种机构正处于崩溃之中。他们建议，中学对14岁以上的青少年应实行非义务教育，应给青少年提供在社会上充分就业的机会；中学的规模应急剧缩小，以避免让它们成为脱离社会的"世外桃源"。每个城市都应为青少年在地方商业和服务行业中提供当学徒的机会。而这种学徒工作应当被认为是青少年正式学习的一部分。

更准确地说，我们认为，我们应当鼓励青少年，即12～18岁的男女组成小型社会。在这种小型社会内，他们各有分工，但共同承担责任，如成年人在完全的成人社会中一样。他们必须共同负责，相互帮助，按年龄的大小和成熟的程度而具有不同的权利和权威。总而言之，他们的社会必须是成人社会的缩影，而不是假装成年的人为的成人社会。它有真正的实际内容：有真正的报酬、真正的不幸、真正的工作、真正的爱情、真正的友谊、真正的成就和真正的负责精神。为此，每个城镇必须建立一个或数个名副其实的青少年协会，部分是封闭式的，由成年人监督和协助的，但管理工作基本上是由成年人和青少年共同负责的。

munications,and governance,they get treated as if they were large children.They have no more responsibility or authority in a high school than the children in a kindergarten do.They are responsible for putting away their things,and for playing in the school band,perhaps even for electing class leaders.But these things all happen in a kindergarten too.There is no new form of society,which is a microcosm of adult society,where they can test their growing adulthood in any serious way.And under these circumstances,the adult forces which are forming in them,lash out,and wreak terrible vengeance.Blind adults can easily,then,call this vengeance "delinquency."

This has finally been recognized by an official agency. In December 1973 the National Commission on the Reform of Secondary Education,working with the Kittering Foundation,has come to the conclusion that the high schools in American cities are simply not working;that they are breaking down as institutions. They recommend that high school be non-compulsory after 14 years of age,and that teenagers be given many options for participation in society;that the size of high schools be reduced drastically,so that they are not so much a world apart from society;that each city provide opportunities for its young to work as apprentices in the local businesses and services,—and that such work be considered part of one's formal learning.

More specifically,we believe that the teenagers in a town,boys and girls from the ages of about 12 to 18,should be encouraged to form a miniature society,in which they are as differentiated,and as responsible mutually,as the adults in the full-scale adult society.It is necessary that they are responsible to one another,that they are able to play a useful role with respect to one another,that they have different degrees of power

因此：

实际上这是以一种成人社会模式的机构来取代"高级中学"。在这种社会中，学生担负起学习和社会生活的大部分责任，有明确规定的任务和训练方式。在学习和协会的社会结构方面，提供成人的指导；但要尽量使它们行之有效，而且由学生来掌握。

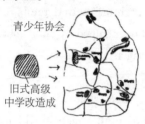

青少年协会

旧式高级中学改造成

∞∞

提供一个具有若干社会设施的中心地点和一本社区各班级学生的姓名地址录。在中心地点内为中小学生提供共同进餐和开展运动和游戏的各种机会，以及一个图书馆和一个学习网的辅导服务站。这个辅导服务站能告诉中小学生如何到分散在全城镇的各个班级、工作社区和家庭工作间——**学习网**（18）、**地方性运动场地**（72）、**共同进餐**（147）和**家庭工作间**（157）去；至于这些建筑物的形状，要从**建筑群体**（95）通盘考虑……

and authority according to their age and their maturity.It is necessary,in short,that their society is a microcosm of adult society,not an artificial society where people play at being adult,but the real thing,with real rewards,real tragedies,real work,real love,real friendship,real achievements,real responsibility.For this to happen it is necessary that each town have one or more actual teenage societies,partly enclosed,watched over,helped by adults,but run,essentially,by adults and the teenagers together.

Therefore:

Replace the "high school" with an institution which is actually a model of adult society,in which the students take on most of the responsibility for learning and social life,with clearly defined roles and forms of discipline.Provide adult guidance,both for the learning,and the social structure of the society;but keep them as far as feasible,in the hands of the students.

❧❧❧

Provide one central place which houses social functions, and a directory of classes in the community.Within the central place,provide communal eating for the students,opportunities for sports and games,a library and counseling for the network of learning which gives the students access to the classes,work communities,and home workshops that are scattered through the town—NETWORK OF LEARNING(18),LOCAL SPORTS(72),COMMUNAL EATING(147),HOME WORKSHOPS(157);for the shape of what buildings there are,begin with BUILDING COMPLEX(95)...

模式85　店面小学

……儿童之家（86）为社区的**学习网**（18）提供启蒙学习并形成学习的基础。随着儿童们的不断长大和具有更多的独立自主的见解，这些模式必须由大量的小型机构和学校来充实，而且还不是那种分布在社区现有设施之间的学校。

୫୦୧୫

六七岁上下的儿童通过动手干来培养自己渴望学习的热情，在家外的社区中崭露头角。如果安排得当，他们求知心切的需要就会直接引导他们去获得基本的技巧和养成学习的习惯。

85   SHOPFRONT SCHOOLS

...the CHILDREN'S HOME(86)provides the beginning of learning and forms the foundation of the NETWORK OF LEARNING(18)in a community.As children grow older and more independent,these patterns must be supplemented by a mass of tiny institutions,schools and yet not schools,dotted among the living functions of the community.

ॐ

Around the age of 6 or 7,children develop a great need to learn by doing,to make their mark on a community outside the home.If the setting is right,these needs lead children directly to basic skills and habits of learning.

The right setting for a child is the community itself,just as the right setting for an infant learning to speak is his own home.

For example:

On the first day of school we had lunch in one of the Los Angeles city parks.After lunch I gathered everyone,and I said, "Let's do some tree identification," and they all moaned. So I said, "Aw,come on,you live with these plants,you could at least know their names.What's the name of these trees we're sitting under?"

They all looked up,and in unison said "Sycamores." So I said, "What kind of sycamore?" and no one knew.I got out my Trees of North America book,and said, "Let's find out." There were only three kinds of sycamore in the book,only one on the West Coast,and it was called the California Sycamore.I

社区本身有责任适当安排儿童的学习，正如家庭有责任引导幼儿牙牙学语一样。

请看下面的例子：

儿童们开学的第一天，我们就在洛杉矶市的一个公园里吃了便餐。便餐之后我把大家召集在一起，我说："让我们来鉴定一下树木吧。"大家都唉声叹气。所以我又说："喔，得啦！你们和植物朝夕相处，至少能说出它们的名称。现在我们就坐在这些树下面哩，它们是什么树？"

他们都抬头看了看，然后异口同声地说："美国梧桐。"我又说："什么品种的美国梧桐？"无人知晓。于是我就取出一本书，名叫《北美树种》，我说："让我们在书中找一找。"这本书介绍了三种美国梧桐，只有一种是生长在西海岸的，称为加里福尼亚梧桐。我想，这个问题算是解决了，但我坚持还要继续提问。我说："为了确信无疑，我们根据书中的描写来检查一下这些美国梧桐。"接着我就开始朗读起来："树叶：6～8in。"我从一只匣子里拿出一条带尺，递给了杰夫，并说道："你去检查一下那些树叶。"结果他发现，梧桐树叶的长度真是6～8in。

我再翻开书接着读："成材的加利福尼亚梧桐高达30～50ft。我们将如何去检查呢？"这个问题引起了一阵热烈的讨论，最后决定：我——鲁希应靠着其中的一棵梧桐树直挺挺地站着，而他们尽量往后退，看着这棵梧桐估算一下它有几个"鲁希"那样高。一个简单的乘法运算折腾了一会儿，大家对它的高度心中大概有数了。现在每个人都参与了进去，都在动脑筋了。我就问他们："还有别的办法解决这道题吗？"埃里克是七年级的学生，懂得一点几何学，所以他告诉我们如何利用三角测量法求出高度。

我非常高兴，他们聚精会神地思索，我又打开书继续往下读。在一段的结尾，有一句无可争议的话："直径：1～3ft。"我又把带尺递给了他们，并说道："给我去量一

thought it was all over,but I persisted, "We better make sure by checking these trees against the description in the book." So I started reading the text, "Leaves:six to eight inches." I fished a cloth measuring tape out of a box,handed it to Jeff,and said, "Go check out those leaves." He found that the leaves were indeed six to eight inches.

I went back to the book and read, " 'Height of mature trees,30-50 feet.'How are we going to check that?" A big discussion followed,and we finally decided that I should stand up against one of the trees,they would back off as far as they could and estimate how many "Rusches" high the tree was.A little simple multiplication followed and we had an approximate tree height.Everyone was pretty involved by now,so I asked them "How else could you do it?" Eric was in the seventh grade and knew a little geometry,so he taught us how to measure the height by triangulation.

I was delighted just to have everyone's attention,so I went back to the book and kept reading.Near the bottom of the paragraph,came the clincher, "Diameter:one to three feet." So I handed over the measuring tape,and said, "Get me the diameter of that tree over there." They went over to the tree,and it wasn't until they were right on top of it that they realized that the only way to measure the diameter of a tree directly is to cut it down.But I insisted that we had to know the diameter of the tree,so two of them stretched out the tape next to the tree,and by eyeballing along one "edge" and then the other,they came up with eighteen inches.

I said, "Is that an accurate answer or just approximate?" They agreed it was only a guess,so I said, "How else could you do it?"

量那边的那棵梧桐的直径是多少。"他们向那棵梧桐走去，但还未走近它跟前，他们就认为，直接测量它的直径的唯一方法就是把它砍倒。但我却坚持，我们一定要测量出那棵梧桐的直径。两名学生贴近这棵树，拉直带尺，沿着它的这一"边"测量后，又在它的那一"边"测量，终于得出它的直径为18in。

我说："那是精确的答案，还是一个近似的估计？"他们同意那仅仅是一种猜测。我接着说："你们还能用别的什么办法计算出来呢？"

丹尼尔马上回答说："噢，您能绕着这棵梧桐树测量它的周边的全部长度，把它周边的长度在泥地上画成圆圈，然后再测那圆圈的直径。"他的话给我留下了深刻的印象。我说："你去测量吧。"我话音刚落，就转向其余的人问道："你们能用别的方法求解吗？"

埃里克早已证明是一个观察家，也许他正在观察这棵梧桐的两侧，开口道："嗯，您能测出它全部的周长，再除以2。"因为我认为，你们从错误中学到的东西和从成功中学到的一样多。"去试一试。"我说道。当时丹尼尔正在地上测量圆的直径。他在一个有点歪歪扭扭的圆上选择两个合适的点求算着它的直径呢，得出的答案分毫不差："18in。"我又把带尺给了埃里克，他测出这棵梧桐的周长为60in，除以2，得出它的直径为30in。他自然有点失望了。我告诉他："我喜欢你的想法，也许你的除数是错的。有没有比2更好的除数呢？"

迈克尔立即说："您就用3除。"他思索了一下，又抢先补充道："再减2。"

我说："好极了！现在你们有了一个公式，就用那边的那棵梧桐来检验它一下。"我指着那棵直径只有6in左右的梧桐树说。他们走到那棵梧桐树旁，测量了它的周长，除以3，再减去2，然后再按地上的圆圈来核对所求出的

Right off, Daniel said, "Well you could measure all the way around it, lay that circle out in the dirt, and then measure across it." I was really impressed, and said, "Go to it." Meanwhile, I turned to the rest of the group, and said, "How else could you do it?"

Eric, who turned out to be a visualizer and was perhaps visualizing the tree as having two sides, said, "Well, you could measure all the way around it, and divide by two." Since I believe you learn at least as much from mistakes as from successes, I said, "Okay, try it." Meanwhile, Daniel was measuring across the circle on the ground, and by picking the right points on a somewhat lopsided circle came up with the same answer, "Eighteen inches." So I gave the tape to Eric, he measured around the tree, got sixty inches, divided by two, and got thirty for the diameter. He was naturally a little disappointed, so I said, "Well, I like your idea, maybe you just have the wrong number. Is there a better number to divide with than two?"

Right off, Michael said, "Well you could divide by three," and then thinking ahead quickly added, "and subtract two."

I said, "Great! Now you have a formula, check it out on that tree over there," pointing to one only about six inches in diameter. They went over, measured the circumference, divided by three, subtracted two, and checked it against a circle on the ground. The result was disappointing, so I told them try some more trees. They checked about three more trees and came back. "How did it work?"

"Wall," Mark said, "Dividing by three works pretty well, but subtracting two isn't so good."

"How good is dividing by three?" I asked, and Michad replied, "It's not quite big enough."

数值。结果，他们大失所望。我告诉他们再去试试另外几棵梧桐。他们大概又测量了3棵，就走回来了。"这怎么搞的？"

"嗯，"马克说："除以3很对,但减2就不那么妥当了。"

"除以3，怎么个好法？"我问道。而迈克尔却答道:"这个数还不够大。"

"它应当是多大的数？"

"大约3.5。"丹尼尔答道。

"不！"迈克尔说："它更像是$3\frac{1}{8}$。"

在这一点上，有5个年龄为9～12岁的孩子一直在π值附近上下不超过1/100徘徊而未发现它。我激动得几乎不能自已。我想，我本来可以让他们把1/8换算成小数，但我太激动了。

"瞧！"我说："我要告诉你们一个秘密。有一个魔术般的数字。它非常奇特，有自己的名字。它叫做π。一旦你们知道这个魔数是多大，就能用它来求解任意圆了。不管圆是大是小，你们都可以从圆周求出直径，或从直径求出圆周。现在，下面我就讲解它是如何起作用的……"

我解释完之后，大家就一块儿逛公园了。我们一边游逛，一边猜测树木的直径，并求算它们的周长，或者一边测量树木的周长，再除上π，计算出它们的直径。后来，当我教会他们如何使用计算尺时，我向他们指出了π，并给他们出了一大串"树"的问题。以后，我在给他们进行复习全部内容时，我又利用电话电线杆和照明电线杆为例，使他们确信:π这一概念是不会消失于抽象数学的晦涩难懂之中的。我现在认识到，我过去在上大学之前，并未真正理解π，尽管我在高中读书时上过很好的数学课。但是现在，对这5个孩子来说，至少π是某种真实的东西；它"活在"许多树木和电线杆中（Charles W.Rusch, "Moboc:The Mobile Open Classroom," School of Architecture and Urban

"How big should it be?"

"About three and a half," said Daniel.

"No," said Michael, "It's more like three and an eighth."

At that point,these five kids,ranging in age from 9 to 12 were within two one hundredths of discovering$\pi$and I was having trouble containing myself.I suppose I could have extended the lesson by having them convert one-eighth to decimals,but I was too excited.

"Look," I said, "I want to tell you a secret.There's a magic number which is so special it has it own name.It's called π.And the magic is that once you know how big it is,you can take any circle,no matter how big or how small,and go from circumference to diameter,or diameter to circumference.Now here is how it works..."

After my explanation,we went around the park,estimating the circumferences of trees by guessing their diameter,or figuring the diameter by measuring the circumference and dividing by π.Later when I had taught them how to use a slide rule,I pointed out π to them and gave them a whole series of "tree" problems. Later still,I reviewed the whole thing with telephone poles and lighting standards,just to make sure that the concept of π didn't disappear into the obscurity of abstract mathematics.I know that I didn't really understand π until I got to college,despite an excellent math program in high school.But for those five kids at least,π is something real;it "lives" in trees and telephone poles.(Charles W.Rusch, "Moboc:The Mobile Open Classroom," School of Architecture and Urban Planning,University of California,Los Angeles,November 1973.)

A few children in a bus,visiting a city park with a teacher.That works because there are only a few children and

Planning, University of California, Los Angeles, November 1973.)

几个儿童和一位教师乘一辆小公共汽车一道去参观一个城市公园。那样做效果良好，因为只有几个儿童和一位教师呀！任何一个公立学校都能提供一位教师和一辆小公共汽车。但是，它们都不能提供学生和教师之间的低比例。因为学校的绝对规模吞没了全部的行政费用和通常开支，结果造成更高的学生率，而这一点在经济上是十分重要的。所以，即使每个人都知道，良好的教学秘密在于低的学生教师比，而学校要把这件事作为中心工作来抓是办不到的，因为学校会浪费金钱，且数额巨大。

但是，正如我们的例子所表明的那样，我们可以削减大而集中的学校通常的开支费用和降低学生教师比；很简单，把我们的学校变得更小就可行了。对教育采取如下措施——微型学校或店面小学——在美国的一些社区中已经在试验（See,for instance,Paul Goodman, "Mini-schools:a prescription for the reading problem," *New York Review of Books*, January 1968）。迄今为止，我们只知道这一实验的非系统的经验说明。但是，关于这些学校已有大量报道。也许，最令人感兴趣的报道首推乔治·丹尼森的《儿童的生活》一书了（New York: Vintage Book，1969）：

我愿意使大家清楚了解：我在对我们自己的教育过程和公立学校的教育过程进行对比时，并不是在竭力批评那些教师，他们在习以为常的环境里发现自己防不胜防，压到身上的沉重不堪的负担快使他们发疯了……准确地说，我的观点是：我们学校的亲切感和小的规模是值得广泛仿效的，因为这两点就能使接触交往具有人情味，从而就能治愈近十年来我们一直在频频不断地为之命名的社会弊病。

one teacher.Any public school can provide the teacher and the bus.But they cannot provide the low student—teacher ratio,because the sheer size of the school eats up all the money in administrative costs and over-heads—which end up making higher student ratios economically essential.So even though everyone knows that the secret of good teaching lies in low student—teacher ratios,the schools make this one central thing impossible to get,because they waste their money being large.

But as our example suggests,we can cut back on the overhead costs of large concentrated schools and lower the student-teacher ratio;simply by making our schools smaller.This approach to schooling—the mini-school or shopfront school—has been tried in a number of communities across the United States.See,for instance,Paul Goodman, "Mini-schools:a prescription for the reading problem," *New York Review of Books*,January 1968. To date,we know of no systematic empirical account of this experiment.But a good deal has been written about these schools. Perhaps the most interesting account is George Dennison's The Lives of Children(New York:Vintage Book,1969):

I would like to make clear that in constrasting our own procedures with those of the public schools,I am not trying to criticize the teachers who find themselves embattled in the institutional setting and overburdened to the point of madness....My point is precisely that the intimacy and small scale of our school should be imitated widely,since these things alone make possible the human contact capable of curing the diseases we have been naming with such frequency for the last ten years.

Now that "mini-schools" are being discussed(they have been proposed most cogently by Paul Goodman and Dr.Elliott Shapiro),it's worth saying that that's exactly what we were:the

既然"微型学校"（这种学校是由保罗·古德曼和埃利奥特·夏皮罗博士建议的，他们的建议很有说服力。）正处于讨论阶段，所以，在此值得一提的是这种学校恰恰就是我们所提倡的学校：第一类微型学校……

　　丹尼森通过节省集中化的学校的各项支出后发现，他能将学生教师比减小到 1/3！

　　23 个儿童有 3 个全时教师、一个半时教师（我自己），还有几名教师在规定的时间内来教唱歌、舞蹈和音乐。

　　公立学校的教师（30：1）将意识到我们已经进入纯粹奢侈的范围。可是，其中有一点必须重提：即这种奢侈是以每一儿童所耗资金略低于公立学校每一儿童所耗资金而换取的，因为管理费用大致相等反映不出公立学校巨大的基建投资或在服务质量上有巨大差别。不是我们的家庭付学费（几乎没有一个人付）；直截了当地说我的意思是，我们的金钱没有完全掷于庞大的行政费用、簿记、精致的建筑、维持费、提高人员素质和破坏财物等方面。

　　查尔斯·鲁希是灵活开放课堂的主任，也已同样发现：

　　……经削减建筑费用和那些不直接从事儿童工作的人的工资，学生教师比可以从大约 35/1 减小到 10/1。就此一举即可使公立学校许多亟待解决的问题无须额外费用就在学校或学校区内迎刃而解。（Rusch, "Moboc:The Mobile Open Classroom," p.7.）

　　因此：

　　建造小而独立的学校，一次一个，以便取代供 7 ~ 12 岁儿童上学的大规模的公立学校。学校要小，使它可以大大削减平时开支费用，并保持 1：10 的教师学生比。为这种学校在社区的公共地区内选择地址，使它有一个店铺门脸和 3 ~ 4 个房间。

first of the mini-schools...

By eliminating the expenses of the centralized school,Dennison found he was able to reduce the student-teacher ratio by a factor of three!

For the twenty-three children there were three full-time teachers,one part-time(myself),and several others who came at scheduled periods for singing,dancing,and music.

Public school teachers,with their 30 to 1 ratios,will be aware that we have entered the realm of sheer luxury.One of the things that will bear repeating,however,is that this luxury was purchased at a cost per child a good bit lower than that of the public system,for the similarity of operating costs does not refleet the huge capital investment of the public schools or the great difference in the quality of service.Not that our families paid tuition(hardly anyone did);I mean simply that our money was not drained away by vast administrative costs,bookkeeping,elaborate buildings,maintenance,enforcement personnel,and vandalism.

Charles Rusch,director of Moboc,Mobil Open Classroom, has made the same discovery:

...by eliminating the building and the salaries of all those persons who do not directly work with the children,the student/ teacher ratio can be reduced from something like 35/1 to 10/1.In this one stroke many of the most pressing public school problems can be eliminated at no extra cost to the school or school district. Rusch, "Moboc:The Mobile Open Classroom," p.7.

Therefore:

Instead of building large public schools for children 7 to 12,set up tiny independent schools,one school at a time. Keep the school small,so that its overheads are low and a

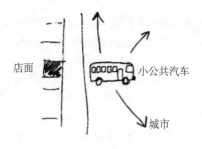

店面　　　小公共汽车

城市

⊗⟡⊗

　　将学校位于步行街——**步行街**（100）；靠近其他正在生产的车间——**自治工作间和办公室**（80）和一个可以步行到达的公园——**近宅绿地**（60）。使这种学校成为建筑物容易识别的一部分，而且也是**建筑群体**（95）的一部分；并在它的正面开一个美观而又坚固的敞口，以便它与街道连接——**向街道的开敞**（165）……

teacher—student ratio of 1:10 can be maintained.Locate it in the public part of the community,with a shopfront and three or four rooms.

◊◊◊

Place the school on a pedestrian street—PEDESTRIAN STREET (100);near other functioning workshops—SELF-GOVERNING WORKSHOPS AND OFFICES(80)and within walking distance of a park—ACCESSIBLE GREEN(60).Make it an identifiable part of the building it is part of—BUILDING COMPLEX(95);and give it a good strong opening at the front,so that it is connected with the street—OPENING TO THE STREET(165)...

模式86　儿童之家*

　　……每一邻里都有数百名少年儿童。他们，其中尤其是年幼无知的儿童，和外部世界的关系，由模式**市区内的儿童**（57）和**相互沟通的游戏场所**（68）来协助解决。可是，这些非常一般的、公共土地上的设施需要某种公共场所来支持。在公共场所，少年儿童可根据需要，逗留几小时或几天而不用父母陪同。本模式是为少年儿童提供的**学习网**（18）的一部分。

☜☞

　　照料少年儿童的任务比起那些娓娓动听的词句，如"临时看管孩子"和"关心孩子"，是一个更深刻的、更重要的社会问题。

86   CHILDREN'S HOME*

...within each neighborhood there are hundreds of children.The children,especially the young ones,are helped in their relation to the world by the patterns CHILDREN IN THE CITY(57)and CONNECTED PLAY(68).However,these very general provisions in the form of public land need to be supported by some kind of communal place,where they can stay without their parents for a few hours,or a few days,according to necessity.This pattern is a part of the NETWORK OF LEARNING(18)for the youngest children.

⊱⊰

The task of looking after little children is a much deeper and more fundamental social issue than the phrases "babysitting" and "child care" suggest.

It is true,of course,that in a society where most children are in the care of single adults or couples,the mothers and fathers must be able to have their children looked after while they work or when they want to meet their friends.This is what child care and babysitting are for.It is,if you like,the adult's view of the situation.

But the fact is that the children themselves have unsatisfied needs which are equally pressing.They need access to other adults beyond their parents,and access to other children;and the situations in which they meet these other adults and other children need to be highly complex,subtle,full of the same complexities and intensities as family life—not merely "schools" and "kindergarten" and "playgrounds."

千真万确，社会上大多数的儿童都得到单亲和双亲的关怀。做父母的在工作或会见朋友时都得请人照管自己的孩子。这就是"临时看护孩子"和"关心孩子"所指的具体内容。如果你愿意，就把它看成成人对这种情况所持的观点。

但事实是儿童们自己还有些需要未得到满足，这些需要也同样是迫切的。他们需要同自己父母以外的人接近，同别的儿童接近；他们遇见其他成人和其他儿童的环境必须是高度复杂的、难以捉摸的，充满着像家庭生活一样的错综复杂性和紧张性——而不只是像"学校""幼儿园"和"操场"那样的环境。

当我们考察儿童们的需要和成人的需要时，我们认识到儿童们所需要的是邻里中的一种新机构：儿童之家——这是一种儿童们共同使用的场所，它保证儿童们的身心安全，日夜为他们提供良好的照顾，并使他们能通过各种机会和社会活动而完全走向社会。

在某种程度上说，过去那些扩大的大家庭是专心于儿童们的这些需要的。在这样的大家庭中，各式各样的成人和年龄不同的少年伙伴对儿童们都具有积极的意义。这会使他们与更富有人情味的各种场所保持接触联系，并允许他们在与各种成人而恰恰不是双亲的交往中满足自己的需要。

然而，当这种大家庭已渐渐消失时，我们却继续坚持"培养子女只是家事，尤其是母亲的事"这样的观念。但是这种迂腐观念现在再也行不通了。下面是菲利普·斯莱特在论述一个小核心家庭悉心培育一两个孩子时所遇到的烦恼事：

新的双亲不会像他们自己的双亲那样沉湎于物质的占有和职业的自我扩张。他们会把自己做父母的虚荣心引向各个领域，使自己的子女跻身于杰出的艺术家、思想家和演奏家的行列。但是，以顽固不化的自我崇拜为核心的旧文化，在双亲和子女之间的关系"松绑"之前，是无法消灭的……

When we look at the children's needs,and at the needs of the adults,we realize that what is needed is a new institution in the neighborhood:*a children's home*—a place where children can be safe and well looked after,night and day,with the full range of opportunities and social activities that can introduce them,fully,to society.

To a certain extent,these needs were absorbed in the large,extended families of the past.In such a family,the variety of adults and children of other ages had a positive value for the children. It brought them into contact with more human situations,allowed them to work out their needs with a variety of people,not just two.

However,as this kind of family has gradually disappeared, we have continued to hold fast to the idea that child-raising is the job of the family alone,especially the mother.But it is no longer viable. Here is Philip Slater discussing the difficulties that beset a small nu-clear family focussing its attention on one or two children:

The new parents may not be as absorbed in material possessions and occupational self-aggrandizement as their own parents were.They may channel their parental vanity into different spheres,pushing their children to be brilliant artists,thinkers,and performers.But the hard narcissistic core on which the old culture was based will not be dissolved until the parent-child relationship itself is deintensified....

Breaking the pattern means establishing communities in which(a)children are not socialized exclusively by their parents,(b)parents have lives of their own and do not live vicariously through their children(*The Pursuit of Loneliness*,Boston:Beacon Press,1971,pp.141-142).

The children's home we propose is a place which "de-intensifies the parent-child relationship" by bringing the child into authentic social relationships with several other adults and

破坏这种旧模式意味着要建立社区。在这种社区中，(a) 儿童们的社会化不是专靠双亲实现的；(b) 双亲有他们自己的生活，并非要和子女形影不离地生活在一起 (The Pursuit of Loneliness, Boston: Beacon Press，1971，pp.141 ~ 142)。

我们所建议的儿童之家是一个"双亲和子女之间的关系松绑"的地方。儿童之家把去玩的儿童和其他的许多儿童和成年人建立相互可信赖的社会关系。

1. 从空间上讲，儿童之家是一个很大的可漫步的室内空间，并有一个规模适中的庭院。

2. 儿童之家要离儿童们自己的家不远，他们可以徒步走去。特伦斯·李已发现，步行或骑自行车去上学的少年儿童比乘公共汽车或坐轿车去的学到的东西要多。道理是简单而又惊人的。步行或骑自行车的儿童，依然和地面保持接触。因此，他们能在自己的脑海中形成一幅学校和家庭之间的认知地图。坐轿车的儿童，来去匆匆，从一地到另一地，仿佛坐在魔毯上似的，脑中模糊一片，无法有包括学校和家的认知地图。就全部的意图和目的来说，儿童们在学校里会感到若有所失；也许，他们甚至害怕已经失去自己的母亲。(T.R.Lee, "On the relation between the school journey and social and emotional adjustment in rural infant children," *British Journal of Educational Psychology*, 27 : 101, 1957.)

3. 有两三个成人充当核心职员，他们管理儿童之家；至少其中一人，最好多一人，住在那里，事实上，这样真正成了一批人的家；它夜里无须关门。

4. 双亲及其孩子们都加入了一个特殊之家。在此之后，儿童们在任何时候都可去儿童之家，并可在那里逗留一小时、一个下午，或在那里过夜。

5. 可以一小时起计时收费。我们假设每小时收费一美

many other children.

1.Physically,it is a very large,rambling home,with a good-sized yard.

2.The house is within walking distance of the children's own homes.Terence Lee was found that young children who walk or bike to school learn more than those who go by bus or car.The mechanism is simple and startling.The children who walk or bike,remain in contact with the ground,and are therefore able to create a cognitive map which includes both home and school.The children who are taken by car,are whisked,as if by magic carpet,from one place to the other,and cannot maintain any cognitive map which includes both home and school.To all intents and purposes they feel lost when they are at school;they are perhaps even afraid that they have lost their mothers.(T.R.Lee, "On the relation between the school journey and social and emotional adjustment in rural infant children," *British Journal of Educational Psychology*,27:101,1957.)

3.There is a core staff of two or three adults who manage the home;and at least one of them,preferably more,actually lives there.In effect,it is the real home of some people;it does not close down at night.

4.Parents and their children join a particular home.And then the children may come and stay there at any time,for an hour,an afternoon,sometimes for long overnight stays.

5.Payment might be made by the hour to begin with.If we assume$1 per hour as a base fee,and assume that a child might spend 20 hours a week there,the house needs about 30 member children to generate a monthly income of about $2500.

6.The home focuses on raising children in a big extended family setting.For example,the home might be the center of a local coffee

元作为收费标准，并再假设，每个儿童每周在那里待 20 小时，则儿童之家需有约 30 名儿童才能每月收入约 2500 美元。

6. 儿童之家将集中精力在一个扩展了的大家庭环境中来培养儿童。例如，儿童之家或许是地方咖啡座谈会的中心，每天有一些人来聚会，并和儿童们生活在一起。

7. 为了和这种气氛和谐一致，儿童之家是相对开放的，并有一条公共小路通过它的所在地。西尔弗斯坦已经指明，如果儿童之家的游戏区是向所有过路的成人和儿童开放的话，那么儿童们最初上学时所感到的与社会的"脱节"就能缩小。（Murray Silverstein, "The Child's Urban Environment," Proceedings of the Seventy-First National Convention of the Congress of Parents and Teachers, Chicago, Illinois，1967，pp.39～45.）

8. 为了保证少年儿童的安全，为了给他们充分的自由而又不完全失去对他们的控制，游戏区可略低于地面，并在其四周筑一堵围墙。如果围墙只有坐位的高度，这就会吸引行人去坐在矮围墙上，从而就给他们提供了观看儿童们做游戏的坐位，同时也给儿童们提供了和过路行人聊天攀谈的机会。

儿童之家这一模式已经在比我们此处想象中更为极端的形式中成功地进行了试验。在以色列的许多集体农场，儿童们都入集体托儿所进行培养，他们每周只有几小时能和父母亲热。这样极端的模式已经取得成功的事实，理应消除对我们所建议的、温和得多的模式是否切实可行的种种疑虑。

因此：

在每一邻里内建造一个儿童之家——少年儿童的第二家庭——一所可漫步的大屋或一个工作点——儿童们能在那里逗留一小时或两小时，乃至一个星期。至少有一个管理员居住在儿童之家的房屋内；儿童之家必须 24 小时开放；

klatch,where a few people meet every day and mix with the children.

7.In line with this atmosphere,the home itself should be relatively open,with a public path passing across the site. Silverstein has indicated that the child's sense of his first school being "separate" from society can be reduced if the play areas of the children's home are open to all passing adults and to all passing children.(Murray Silverstein, "The Child's Urban Environment," Proceedings of the Seventy—First National Convention of the Congress of Parents and Teachers,Chicago,Illin ois,1967,pp.39-45.)

8.To keep the young children safe,and to make it possible to give them this great freedom without losing track of them altogether,the play areas may be sunk slightly,and surrounded by a low wall.If the wall is at seat height,it will encourage people to sit on it—giving them a place from which to watch the children playing,and the children a chance to talk to passers-by.

The children's home pattern has been tried,successfully,in a far more extreme form than we imagine here,in many kibbutzim where children are raised in collective nurseries,and merely visit their parents for a few hours per week.The fact that this very extreme version has been successful should remove any doubts about the workability of the much milder version which we are proposing.

Therefore:

In every neighborhood,build a children's home—a second home for children—a large rambling house or workplace-a place where children can stay for an hour or two,or for a week.At least one of the people who run it must live on the premises;it must be open 24 hours a day;open to children of all ages;and it must be clear,from the way that it is run,that it is a second family for the children—not just a place where baby—sitting is available.

对各种不同年龄的少年儿童一律开放；同时也必须明确，从管理方式上来讲，儿童之家是儿童们的第二家庭——而并不是一个临时看管孩子的地方。

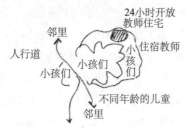

⊱⊰

把建筑作为许多相互联系的小建筑的集合体来处理——**建筑群体**（95）；铺一条重要的邻里小路，恰好穿过学校的建筑物，以便那些不在这所学校上学的儿童们通过集会能看见学校并了解学校——**有顶街道**（101）；使有顶街道附属于地方**冒险性的游戏场地**（73）；使教师的住宅成为学校建筑内部完整的一部分——**自己的家**（79）；并把公共区作为一个较大的家庭的家庭生活区来对待——**家庭**（75）和**中心公共区**（129）……

地方商店和聚集的场所：

87. 个体商店

88. 临街咖啡座

89. 街角杂货店

90. 啤酒馆

91. 旅游客栈

92. 公共汽车站

93. 饮食商亭

94. 在公共场所打盹

Treat the building as a collection of small connected buildings—BUILDING COMPLEX(95);lay an important neighborhood path right through the building,so that children who are not a part of the school can see and get to know it by meeting the children who are—BUILDING THOROUGHFARE(101);attach it to the local ADVENTURE PLAYGROUND(73);make the teachers'house an integral part of the interior—YOUR OWN HOME(79);and treat the common space itself as the hearth of a larger family—THE FAMILY(75),COMMON AREAS AT THE HEART(129)...

the local shops and gathering places:

87.INDIVIDUALLY OWNED SHOPS

88.STREET CAFE

89.CORNER GROCERY

90.BEER HALL

91.TRAVELER'S INN

92.BUS STOP

93.FOOD STANDS

94.SLEEPING IN PUBLIC

模式87　个体商店**

……**临街咖啡座**（88）、**街角杂货店**（89）和**商业街**
（32）上的所有个体商店和摊点以及**综合商店**（46）都应
受到法令的保护。法令应保证这些商店和摊点都应由个体
经营者自己掌管，而不应由不在此地的地主或连锁店、或
庞大的代销经营所控制。

⋘⋙

**当商店太大，或由不在此地的所有者所控制时，就会
变成易受影响的、平淡无奇的和无实际意义的了。**

利润动机造成一种倾向：商店的规模越来越大。但商
店变得越大，其服务就越缺少个性，而其他小商店就越难
生存下去。不用多久，各种商店在经济上几乎都受连锁店

87 INDIVIDUALLY OWNED SHOPS**

...the STREET CAFE(88)and CORNER GROCERY
(89)and all the individual shops and stalls in SHOPPING
STREETS(32)and MARKETS OF MANY SHOPS(46)
must be supported by an ordinance which guarantees that
they will stay in local private hands,and not be owned
by absentee landlords,or chain stores,or giant franchise
operations.

❧✣❧

**When shops are too large,or controlled by absentee
owners,they become plastic,bland,and abstract.**

The profit motive creates a tendency for shops to become
larger.But the larger they become,the less personal their service
is,and the harder it is for other small shops to survive.Soon,the
shops in the economy are almost entirely controlled by chain
stores and franchises.

The franchises are doubly vicious.They create the
image of individual ownership;they give a man who doesn't
have enough capital to start his own store the chance to run
a store that seems like his;and they spread like wildfire.But
they create even more plastic,bland,and abstract services.
The individual managers have almost no control over the
goods they sell,the food they serve;policies are tightly
controlled;the personal quality of individually owned shops
is altogether broken down.

或代销经营所控制。

代销经营是倍加恶劣的。它们创造出个人所有制的形象。它们给没有足够资本自己开店的人以经营商店的机会，商店看起来好像是经营者自己的。代销经营像野火一般到处蔓延。但是代销的服务变得更随心所欲、更平淡无奇而又抽象。个体经营者几乎无权控制他们为消费者服务而销售的物品和食品；各种政策受到严格控制：个体商店的个性品质已完全遭到破坏。

只顾赚钱的商店
Shop run for money alone

商店体现一种生活方式
Shop run as a way of life

只有当社区禁止各种形式的代销店和连锁店，限制社区商店的实际规模，并禁止不在此地的所有者拥有商店时，才可以重新找回这种个性品质。总而言之，社区必须做到力所能及的一切，使地方社区创造的财富掌握在地方社区的手中。

即使在这种情况下，如果租来的商店不是小规模的，本模式就不能保持原样了。全国性的大规模的代销经营兴起的最大原因之一就是开办商业的金融冒险，对于一般的个人来说风险是巨大的。一个个体商店的破产对店主个人可能是灾难性的。而且，这种情况之所以发生，很大程度是因为店主付不起租金。成百上千租金低的小商店将把开始开业时所冒的起始风险保持到最小。

在摩洛哥、印度、秘鲁和比较古老的城镇的最古老的地方，商店的面积往往不超过 $50ft^2$。这刚好是供一个人经营和存放若干商品的空间——但已足够大的了。

Communities can only get this personal quality back it they prohibit all forms of franchise and chain stores,place limits on the actual size of stores in a community,and prohibit absentee owners from owning shops.In short,they must do what they can to keep the wealth generated by the local community in the hands of that community.

Even then,it will not be possible to maintain this pattern unless the size of the shop spaces available for rent is small. One of the biggest reasons for the rise of large,nationally owned franchises is that the financial risks of starting a business are so enormous for the average individual.The failure of a single owner's business can be catastrophic for him personally;and it happens,in large part because he can't afford the rent.Many hundreds of tiny shops,with low rents,will keep the initial risk for a shop keeper who is starting,to a minimum.

Shops in Morocco,India,Peru,and the oldest parts of older towns,are often no more than 50 square feet in area. Just room for a person and some merchandise—but plenty big enough.

Therefore:

Do what you can to encourage the development of individually owned shops.Approve applications for business licenses only if the business is owned by those people who actually work and manage the store.Approve new commercial building permits only if the proposed structure includes many very very small rental spaces.

50平方英尺
Fifty square feet

因此：

尽量鼓励发展个体商店。对那些确实在经营和管理商店的店主，才批准他们申请营业执照。如果他们申请新的商业建筑许可证，只有当新建筑包含许多小的出租空间时，才予以批准。

所有主所拥有的

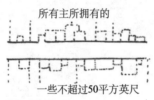

一些不超过50平方英尺

❀

把每个个体商店作为较大的**建筑群体**（95）的一个易识别的单位：至少使商店的某一部分成为人行道的一部分，以便行人沿街步行时走过商店——**向街道的开敞**（165）；商店内部布置要合理，各式商品要尽量开架摆放，且便于拿取——**室内空间形状**（191）、**厚墙**（197）和**敞开的搁架**（200）……

ॐ

Treat each individual shop as an identifiable unit of a larger BUILDING COMPLEX(95);make at least some part of the shop part of the sidewalk,so that people walk through the shop as they are going down the street—OPENING TO THE STREET(165);and build the inside of the shop with all the goods as open and available as possible—THE SHAPE OF INDOOR SPACE(191),THICK WALLS(197),OPEN SHELVES (200)...

模式88　临街咖啡座**

　　……邻里是由**易识别的邻里**（14）来加以说明；邻里的天然中心是**活动中心**（30）和**小广场**（61）。本模式和下面紧接着的几个模式都将阐明邻里和它的中心识别特征。

∞∞

　　临街咖啡座为城市提供一种独一无二的环境：人们可以懒洋洋地、无所顾忌地坐在那里观看街景，并注视熙来攘往的行人。

88   STREET CAFE**

...neighborhoods are defined by IDENTIFIABLE NEIGHBORHOOD(14);their natural points of focus are given by ACTIVITY NODES(30)and SMALL PUBLIC SQUARES(61).This pattern,and the ones which follow it,give the neighborhood and its points of focus,their identity.

୫୦ୠ୦ଓ

The street cafe provides a unique setting,special to cities:a place where people can sit lazily,legitimately,be on view,and watch the world go by.

The most humane cities are always full of street cafes.Let us try to understand the experience which makes these places so attractive.

We know that people enjoy mixing in public,in parks,squares,along promenades and avenues,in street cafes. The preconditions seem to be:the setting gives you the right to be there,by custom;there are a few things to do that are part of the scene,almost ritual:reading the newspaper,strolling,nursing a beer,playing catch;and people feel safe enough to relax,nod at each other,perhaps even meet.A good cafe terrace meets these conditions.But it has in addition,special qualities of its own:a person may sit there for hours—in public!Strolling,a person must keep up a pace;loitering is only for a few minutes.You can sit still in a park,but there is not the volume of people passing,it is more a private,peaceful experience.And sitting at home on

临街咖啡座在大多数富有人情味的城市到处可见。让我们尽量理解那使许多地方变得引人入胜的经验。

我们知道，人们在公开场合、在公园里、在广场上、沿着散步场所和林荫大道以及在临街咖啡座总是喜欢混杂在大庭广众之中。要做到这一点，似乎要具备一些先决条件：周围环境良好，根据习俗，你有权去消遣散心；要使读报、散步、品尝啤酒、轮流演唱成为街景的一部分；人们感到非常安全，轻松愉快，相互点头致意，甚至迎面相撞。良好的临街咖啡座的露台能满足这些条件。但是，除此之外，临街咖啡座还要有自己的特色：一个人可以在那里当众独坐几个小时。如要漫步街头，需保持一定步速；寸金光阴不可虚掷，闲荡只许片刻。你可以坐在没有大量来往游人的公园里，在更幽雅而又宁静的气氛中孤芳自赏。而你坐在家中的门廊上感受到的却又是另一番景象：围护得过多；并且也没有混杂的人群来往。而你在临街咖啡座的露台上静坐时，轻松自如，而且非常公开。作为一种体验和感受，它有多种特殊的可能性："也许下一个人……"这是一个危险的地方。

临街咖啡座支持的正是这种内心感受。它是城市引人入胜的地方之一，因为只有在城市里，我们才能形成它所需要的人口集中。但这一体验不必局限于城镇那些非常特殊的地方。在欧洲的许多大小城镇，在每一邻里内都有临街咖啡座——它们比比皆是，就像美国的加油站那样普遍。临街咖啡座的存在为社区提供了社交场所——它们变得像俱乐部了。这是人们最乐意去的地方。人们变得面熟了，在离你家不远的地方，有一个生意兴隆的临街咖啡座，这有多好啊！它大大有助于增加邻里的识别特征。临街咖啡座是刚来的邻里新人开始了解内情和会见在当地居住多年的友人的若干场所之一。

一个生意兴隆的临街咖啡座看来必须具备如下的要素：

1. 有一个已经形成的地方顾客群。即按名称、位置和

one's porch is again different:it is far more protected;and there is not the mix of people passing by.But on the cafe terrace,you can sit still,relax,and be very public.As an experience it has special possibilities; "perhaps the next person..." ;it is a risky place.

It is this experience that the street cafe supports.And it is one of the attractions of cities,for only in cities do we have the concentration of people required to bring it off.But this experience need not be confined to the special,extraordinary parts of town.In European cities and towns,there is a street cafe in every neighborhood—they are as ordinary as gas stations are in the United States.And the existence of such places provides social glue for the community.They become like clubs— people tend to return to their favorite,the faces become familiar. When there is a successful cafe within walking distance of your home,in the neighborhood,so much the better.It helps enormously to increase the identity of a neighborhood.It is one of the few settings where a newcomer to the neighborhood can start learning the ropes and meeting the people who have been there many years.

The ingredients of a successful street cafe seem to be:

1.There is an established local clientele.That is,by name,location,and staff,the cafe is very much anchored in the neighborhood in which it is situated.

2.In addition to the terrace which is open to the street,the cafe contains several other spaces:with games,fire,soft chairs,newspapers....This allows a variety of people to start using it,according to slightly different social styles.

3.The cafe serves simple food and drinks—some alcoholic drinks,but it is not a bar.It is a place where you are as likely

全体职员来说，它在邻里中处于不败之地。

2.除了露台向大街敞开之外，它还有其他的几处地方，供棋类比赛、放置火炉、软椅和阅读报纸……之用。因社会生活方式不尽相同，顾客可按自己的情况来进行活动。

3.它还要供应方便食品和饮料——一些酒精饮料，但它不是酒吧。你一清早就可去那里，开始你一天的生活，就像你晚上去喝一点临睡前的酒一样。

当临街咖啡座具备了这些条件时，大家就会承认它。它能为顾客的生活提供某种独特的东西：它提供了一个讨论伟大精神的场所——交谈、不重要的授课、半公开和半私密的学习和相互交流思想。

当我们过去为俄勒冈大学设计的时候，我们把在咖啡馆中和在类似咖啡馆的场所中的这种讨论的重要性与大学生在教室里接受的教育进行了比较。我们采访了30名大学生，以便评价商店和咖啡馆对他们在俄勒冈大学期间智力和感情的发展的影响。我们发现，"与咖啡馆内的一组大学生交谈"和"通过喝啤酒和大学生讨论"，他们的智力发展和感情发展的得分是高的，或比"各种考试"和"实验室研究"的得分高。很明显，大学生在商店和咖啡馆内的非正式活动和他们在学校进行的更为正式的教育活动一样，对他们自己的发展具有同等的重要性。

我们认为，这一现象是带普遍性的。我们在采访调查中设法抓住存在于邻里咖啡座中的这一普遍性——它对一切邻里都是极其重要的，它不只是对大学生的邻里来说是如此。临街咖啡座是大学生生活中的一个重要的部分。

因此：

在每一邻里内鼓励地方咖啡馆蓬勃发展。它们给人亲切感。它们有若干个房间，朝向一条繁忙的人行道敞开，人们在馆内可以坐着喝咖啡或饮酒，并可观看街上来去匆匆的行人。建造咖啡馆的正面，以便一系列的餐桌摆到馆外，直至街道。

to go in the morning,to start the day,as in the evening,for a nightcap.

When these conditions are present,and the cafe takes hold,it offers something unique to the lives of the people who use it:it offers a setting for discussions of great spirit— talks,two-bit lectures,half-public,half-private,learning,exchange of thought.

When we worked for the University of Oregon,we compared the importance of such discussion in cafes and cafe-like places,with the instruction students receive in the classroom.We interviewed 30 students to measure the extent that shops and cafes contributed to their intellectual and emotional growth at the University.We found that "talking with a small group of students in a coffee shop" and "discussion over a glass of beer" scored as high and higher than "examinations" and "laboratory study." Apparently the informal activities of shops and cafes contribute as much to the growth of students,as the more formal educational activities.

We believe this phenomenon is general.The quality that we tried to capture in these interviews,and which is present in a neighborhood cafe,is essential to all neighborhoods—not only student neighborhoods.It is part of their life-blood.

Therefore:

Encourage local cafes to spring up in each neighborhood. Make them intimate places,with several rooms,open to a busy path,where people can sit with coffee or a drink and watch the world go by.Build the front of the cafe so that a set of tables stretch out of the cafe,right into the street.

若干房间

餐桌

露台
读报栏

繁忙的人行道

　　在临街露台和座内之间的地段上建造一个宽敞的重要开口——**向街道的开敞**（165）；供附近公共汽车站或办公室用的露台面积为**等候场所**（150）的两倍；无论在咖啡座内还是在临街露台上都要摆放各种构思别致、结构精巧的桌椅——**各式坐椅**（251）；如果街道边缘处的露台处于受街道活动干扰的危险之中，则其高度就应降低——**能坐的台阶**（125）、**可坐矮墙**（243），也许还需要**帆布顶篷**（244）。至于建筑的外形、临街露台和周围环境，均要从**建筑群体**（95）出发考虑……

ℬⓍℭ

Build a wide,substantial opening between the terrace and the indoors—OPENING TO THE STREET(165);make the terrace double as A PLACE TO WAIT(150)for nearby bus stops and offices;both indoors and on the terrace use a great variety of different kinds of chairs and tables—DIFFERENT CHAIRS(251);and give the terrace some low definition at the street edge if it is in danger of being interrupted by street action—STAIR SEATS(125),SITTING WALL(243),perhaps a CANVAS ROOF(244).For the shape of the building,the terrace,and the surroundings,begin with BUILDING COMPLEX(95)...

模式89　街角杂货店*

　　……在任何一个社区内，**综合商场**（46）一般都能满足居民购买大部分生活必需品的需求。可是，如果没有许多较小的和分散得更广的商店来补充市场，并协助形成**易识别的邻里**（14）的天然识别特征，则**商业网**（19）就不会是完美无缺的。

⊰⊱

　　近来有一种假设：居民不再想步行去地方商店了。这种假设是完全错误的。

　　我们确实认为，居民不仅愿意步行去地方街角杂货店，而且也认为街角杂货店在任何一个健康的邻里中起着十分重要的作用。一部分原因是这杂货店对每一个人来说都较方便；另一部分原因是它有助于把邻里联结成为一个整体。

89   CORNER GROCERY*

...the major shopping needs,in any community,are taken
care of by the MARKET OF MANY SHOPS(46).However,the
WEB OF SHOPPING (19)is not complete,unless there are
also much smaller shops,more widely scattered,helping to
supplement the markets,and helping to create the natural
identity of IDENTIFIABLE NEIGHBORHOODS(14).

⊰⊱

**It has lately been assumed that people no longer want
to walk to local stores.This assumption is mistaken.**

Indeed,we believe that people are not only willing to walk
to their local corner groceries,but that the corner grocery plays
an essential role in any healthy neighborhood:partly because it
is just more convenient for individuals;partly because it helps
to integrate the neighborhood as a whole.

Strong support for this notion comes from a study by Arthur
D.Little,Inc.,which found that neighborhood stores are one of the
two most important elements in people's perception of an area as a
neighborhood (*Community Renewal Program*,New York:Praeger
Press,1966).Apparently this is because local stores are an important
destination for neighborhood walks.People go to them when they
feel like a walk as well as when they need a carton of milk.In this
way,as a generator of walks,they draw a residential area together and
help to give it the quality of a neighborhood.Similar evidence comes
from a report by the management of one of San Francisco's housing
projects for the elderly.One of the main reasons why people resisted

阿瑟·利特尔公司的研究报告对这种看法提供了强有力的支持。该研究报告发现，邻里商店是居民们把某一地区作为邻里的感知过程中两个最重要的因素之一。（*Community Renewal Program*,New York:Praeger Press，1966）。十分明显，这是因为地方商店是邻里内散步的重要目的地。当他们需要一纸盒奶时就到地方商店去购买，好像就是去散步似的。这样，地方商店作为"散步发生器"就吸引住一片居民区，并有助于该居民区获得邻里的性质。旧金山老人住宅建筑工程之一的管理报告也提供了类似的证据。据租赁经理谈，老人反对迁入市内新建的一些住宅的主要原因之一是这些工程的位置不在"商业区，而在商业区……每条街的拐角处都有商店。"（*San Francisco Chronicle*,August 1971.）

为了搞清楚居民愿意步行去商店的路程极限，我们在伯克利市的一家邻里商店内调查了 20 名顾客，我们发现，其中 80% 的顾客是步行来的，他们到该店的步行距离为 3 个街区或不足 3 个街区。其中一大半人前两天内曾来过这里。坐汽车来此店的顾客一般来自 4 个街区以外的地方。我们发现，本模式和我们考察过的邻里内的其他公共设施有相似之处。距商店的路程约 4 个街区或比 4 个街区更远，坐汽车去购物的顾客会比步行去的顾客多。如此看来，街角杂货店必须位于每一住户的散步距离之内，即相距 3 ～ 4 个街区，或 1200ft。

可是这些街角杂货店能够生存下去吗？它们受经济规模支配吗？多少顾客才能维持一个街角杂货店？我们查阅了电话簿的黄页分类，现在可以对街角杂货店的临界顾客人数作出估计。例如，人口为 75 万的旧金山市拥有 638 个邻里街角杂货店。这就是说，每 1160 人有一个。这一数据与贝利的估计相符合——参阅**商业网**（19）——同样也符合邻里的规模——参阅**易识别的邻里**（14）。

由此可见，一个街角杂货店能生存下去的条件是：在 3 ～ 4 个街区内有居民 1000 人——净密度至少每英亩 20 人，

TOWNS
城 镇
865

moving into some of the city's new housing projects,according to the rental manager,was that the projects were not located in "downtown locations,where...there is a store on every street corner." (*San Francisco Chronicle*,August 1971.)

To find out how far people will walk to a store we interviewed 20 people at a neighborhood store in Berkeley.We found that 80 percent of the people interviewed walked,and that those who walked all came three blocks or less.Over half of them had been to the store previously within two days.On the other hand,those who came by car usually came from more than four blocks away.We found the pattern to be similar at other public facilities in the neighborhoods that we surveyed. At distances around four blocks,or greater,people who rode outnumbered those who walked.It seems then,that corner groceries need to be within walking distance,three to four blocks or 1200 feet,of every home.

But can they survive?Are these stores doomed by the economics of scale?How many people does it take to support one corner grocery?We may estimate the critical population for grocery stores by consulting the yellow pages.For example,San Francisco,a city of 750,000,has 638 neighborhood grocery stores.This means that there is one grocery for every 1160 people,which corresponds to Berry's estimate—see WEB OF SHOPPING(19)—and corresponds also to the size of neighborhoods—see IDENTIFIABLE NEIGHBORHOOD(14).

It seems,then,that a corner grocery can survive under circumstances where there are 1000 people within three or four blocks—a net density of at least 20 persons per net acre,or six houses per net acre.Most neighborhoods do have this kind of density.One might even take this figure as a lower limit for a viable

或每英亩 6 幢住宅。大多数邻里均有此密度。人们甚至可以将这一密度值作为可实施的邻里的下限值，因为邻里为了其自身的社会团结，必须维持一个街角杂货店。

最后，邻里商店的成功与否取决于它的位置。我们早已说明，小的零售商店店主愿意支付的租金是直接随来往行人数量的多少而变化的。所以，位于街道拐角处的小店的租金一律高于位于街区中心的小店。

(Brian J.L.Berry,*Geography of Market Centers and Retail Distribution*,Prentice Hall，1967，p.49.)

因此：

在每一邻里的中心附近至少有一个街角杂货店。它们按人口密度分布，每隔 200～800yd 开设一个，因而每个街角杂货店可为 1000 人左右服务。将它们开设在有大量行人来往的街道拐角处。并将它们和住宅连接在一起，以便经营者能住在那里或邻近处。

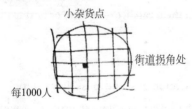

෧෧෨෩

防止代销经营，并通过法律防止出现大杂货店并吞街角杂货店——**个体商店**（87）。街角杂货店的内部要作为一个房间来处理，而且商品要摆放整齐——**室内空间形状**（191）、**厚墙**（197）和**敞开的搁架**（200）；要使街角杂货店具有醒目而又宽敞的入口，以便每个行人都能看见它——**主入口**（110）和**向街道的开敞**（165）。街角杂货店或是一小型建筑，或是某一较大建筑的一部分，它的形状要根据**建筑群体**（95）通盘考虑……

neighborhood,on the grounds that a neighborhood ought to be able to support a corner grocery,for the sake of its own social cohesion.

Finally,the success of a neighborhood store will depend on its location.It has been shown that the rents which owners of small retail businesses are willing to pay vary directly with the amount of pedestrian traffic passing by,and are therefore uniformly higher on street corners than in the middle of the block.(Brian J.L.Berry,*Geography of Market Centers and Retail Distribution*,Prentice Hall,1967,p.49.)

Therefore:

Give every neighborhood at least one corner grocery, somewhere near its heart.Place these corner groceries every 200 to 800 yards,according to the density,so that each one serves about 1000 people.Place them on corners,where large numbers of people are going past.And combine them with houses,so that the people who run them can live over them or next to them.

৪৩৫৪

Prevent franchises and pass laws which prevent the emergence of those much larger groceries which swallow up the corner groceries—INDIVIDUALLY OWNED SHOPS(87).Treat the inside of the shop as a room,lined with goods-THE SHAPE OF INDOOR SPACE(191),THICK WALLS(197),OPEN SHELVES(200);give it a clear and wide entrance so that everyone can see it—MAIN ENTRANCE(110),OPENING TO THE STREET(165).And for the shape of the grocery,as a small building or as part of a larger building,begin with BUILDING COMPLEX(95)...

模式90　啤酒馆

……有时有的邻里在一组邻里中起中心作用；在这样的邻里中，或在邻里之间的边界内——**邻里边界**（15）——或在一个成为大社区中心的散步场上——**散步场所**（31）和**夜生活**（33）——特别需要某种比临街咖啡座更大和更喧闹的场所。

☙❧

人们能在什么地方可以饮酒、唱歌、喧哗或宣泄他们内心世界的悲哀呢？

酒吧间是任何大社区的一个天然组成部分。在那里陌生人和朋友都是酒友。但屡见不鲜的是酒吧间都在蜕化变质，逐步成为仅仅是寂寞者聊以排忧解闷的场所而已。罗伯特·萨默对此作了描述（"Design for Drinking," Chapter

90   BEER HALL

...in an occasional neighborhood,which functions as the focus of a group of neighborhoods,or in a boundary between neighborhoods—NEIGHBORHOOD BOUNDARY(15)—or on the promenade which forms the focus of a large community—PROMENADE(31),NIGHT LIFE(33)—there is a special need for something larger and more raucous than a street cafe.

❧❧

Where can people sing,and drink,and shout and drink, and let go of their sorrows?

A public drinking house,where strangers and friends are drinking companions,is a natural part of any large community. But all too often,bars degenerate and become nothing more than anchors for the lonely.Robert Sommer has described this in "Design for Drinking," Chapter 8 of his book *Personal Space*,Englewood Cliffs,N.J.:Prentice-Hall,1969.

...it is not difficult in any American city to find examples of the bar where meaningful contact is at a minimum.V.S.Pritchett deschbes the lonely men in New York City sitting speechlessly on a row of barstools,with their arms triangled on the bar before a bottle of beer,their drinking money before them.If anyone speaks to his neighbor under these circumstances,he is likely to receive a suspicious stare for his efforts.The barman is interested in the patrons as customers—he is there to sell,they are there to buy....

8 of his book *Personal Space*,Englewood Cliffs,N.J.:Prentice-
Hall，1969.）他描写道：

　　……在美国任何城市，很容易找到这样的酒吧间，
在那里有意义的人际交往几乎一概不见了。普里切特
（V.S.Pritehett）描写了纽约市孤独寂寞者在酒吧间的冷漠
表情：他们坐在一排酒吧凳上，双臂呈三角形，面前放着
一瓶瓶的啤酒和他们的酒钱。在这种情况下，如果有人想
同他的邻座说上一句话，很可能，他得到的是一种怀疑的
目光。酒吧间的男招待对这些老主顾颇感兴趣——他在那
里卖啤酒，他们在那里买啤酒……

　　另一个观光的英国人也持同样的观点。他把美国酒
吧问说成是"勉强凑合的场所"；那里的气氛冷若冰镇啤
酒……当我请一位陌生人一块儿喝几杯时，他睁大眼睛
看着我，仿佛我疯了。在英国，如果一个陌生人……每
个人都会给另一人买点饮料。你会享受到相互交往的乐
趣，而且人人都会兴高采烈的……（Tony Kirby，"Who's
Crazy?" *The Village Voice*,January 26，1967，p.39.）

　　现在让我们根据这些英国小酒店的风格更仔细地研究
一下饮酒问题。饮酒有助于人们舒松筋骨，坦诚相见和轻
歌曼舞。但是，环境要布置得合理才会有这种效果。我们
认为这种环境要有两大特点：

　　1. 这种环境是人来人往的热闹所在。人们会在酒吧柜
台、舞池、炉火旁、投镖游戏处、卫生间、门厅和坐位之
间连续不断地来来往往；这些活动都集中在边上，结果形
成了连续不断的客流交叉现象。

　　2. 坐位布置的主要形式是四人一桌或八人一桌，都
布置在开敞的凹室内——就是说，规定酒桌供小批顾客使
用。凹室内有隔墙、立柱和帷幕——但凹室的两端是畅通无
阻的。

Another visiting Englishman makes the same point when he describes the American bar as a "hoked up saloon;the atmosphere is as chilly as the beer...When I asked a stranger to have a drink,he looked at me as if I were mad.In England if a guy's a stranger,...each guy buys the other a drink.You enjoy each other's company,and everyone is happy...." (Tony Kirby, "Who's Crazy?" *The Village Voice*,January 26,1967, p.39.)

Let us consider drinking more in the style of these English pubs.Drink helps people to relax and become open with one another,to sing and dance.But it only brings out these qualities when the setting is right.We think that there are two critical qualities for the setting:

1.The place holds a crowd that is continuously mixing between functions—the bar,the dance floor,a fire,darts,the bathrooms,the entrance,the seats;and these activities are concentrated and located round the edge so that they generate continual criss-crossing.

2.The seats should be largely in the form of tables for four to eight set in open alcoves—that is,tables that are defined for small groups,with walls,columns,and curtains—but open at both ends.This form helps sustain the life of the group and lets people come in and out freely.Also,when the tables are large,they invite people to sit down with a stranger or another group.

宽敞的凹室吸引着来往的人流
The open alcove—supports the fluidity of the scene

这种形式有助于维持居民的生活。他们可以自由进出。如果酒桌大，他们请客人就座时，就会有陌生人或其他一些人同时就座。

因此：

在社区某处至少有一个可供数百人聚集的大场所。他们在那里可以喝点啤酒和葡萄酒，听听音乐，或进行一些活动，结果，他们就形成连续不断的人流交叉现象。

纵横交错的人行道

频繁的活动

畅通的凹室

酒桌置于两端畅通的凹室内，桌间留有足够的空间，以便顾客能在各种活动之间穿行——**凹室**（179）；提供一堆炉火，作为活动中心之一——**炉火熊熊**（181）；天花须有不同的高度，以便满足不同的社会集团的需要——**天花高度变化**（190）。至于啤酒馆的外形、花园、停车场和周围环境的处理，均要以**建筑群体**（95）为出发点……

Therefore:

Somewhere in the community at least one big place where a few hundred people can gather,with beer and wine,music,and perhaps a half-dozen activities,so that people are continuously criss-crossing from one to another.

☙❧

Put the tables in two-ended alcoves,roomy enough for people to pass through on their way between activities—ALCOVES(179);provide a fire,as the hub of one activity—THE FIRE(181);and a variety of ceiling heights to correspond to different social groupings—CEILING HEIGHT VARIETY(190). For the shape of the building, gardens,parking,and surroundings, begin with BUILDING COMPLEX (95)...

模式91 旅游客栈*

　　……任何大小城镇旅游者都会去观光游览，他们会很自然地向城镇的活动中心聚集——**城市的魅力**（10）、**活动中心**（30）、**散步场所**（31）、**夜生活**（33）和**工作社区**（41）。本模式说明那些迎合旅客需要的旅馆怎样才能最有效地维持这些中心的生命力。

୨୦൭ଓ

　　一个在陌生的地方投宿过夜的人总归是人类社会一名成员，他仍需要结伴交际。要他爬进一个洞穴或单独一人看电视，像在路旁的汽车旅馆中那样，是毫无道理的。

　　从前的客栈，除了我们本国的以外，都是绝妙的地方。素不相识的人在客栈萍水相逢，相遇只此一夜，大家吃吃

91 TRAVELER'S INN*

...any town or city has visitors and travelers passing through, and these visitors will naturally tend to congregate around the cen-ters of activity—MAGIC OF THE CITY(10), ACTIVITY NODES(30), PROMENADE (31), NIGHT LIFE(33), WORK COMMUNITY(41). This pattern shows how the hotels which cater to these visitors can most effectively help to sustain the life of these centers.

❦

A man who stays the night in a strange place is still a member of the human community, and still needs company. There is no reason why he should creep into a hole, and watch TV alone, the way he does in a roadside motel.

At all times, except our own, the inn was a wonderful place, where strangers met for a night, to eat, and drink, play cards, tell stories, and experience extraordinary adventures. But in a moder motel every ounce of this adventure has been lost. The motel owner assumes that strangers are afraid of one another, so he caters to their fear by making each room utterly self-contained and self-sufficient.

But behind the fear, there is a deep need: the need for company—for stories, and adventures, and encounters. It is the business of an inn to create an atmosphere where people can experience and satisfy this need. The most extreme version is the Indian pilgrim's inn, or the Persian caravanserai. There people eat, and meet, and sleep, and talk, and smoke, and drink

喝喝，玩玩纸牌，讲讲故事，谈笑风生，体验一番不同寻常的奇遇。可是，在现代的汽车旅馆里，这种奇遇的影子一点也找不到了。汽车旅馆的老板认为，陌生人都相互害怕，存有戒心；所以为了投合游客的恐惧心理，他就把客房布置成彻底的自给自足式。

但在这恐惧的背后有一种根深蒂固的需要：社交的需要——大家讲述趣闻轶事、历险始末和意外的遭遇。一个客栈最要紧的事就是创造出一种使游客感到满意的气氛。这方面最极端的例子就是印度朝圣者的客栈，或波斯商队的旅馆了。投宿的旅客进餐、会见、睡觉、交谈、抽烟、饮酒都在一个宽敞的地方进行。他们相互交际而避免了危险，他们相互叙述趣闻逸事而心中感到万分愉悦。

本模式从吉他·沙关于印度朝圣者客栈的描写中获得了灵感，在《建筑的永恒之道》一书中我们有如下的记载：

在印度各地，有许多这样的客栈。客栈内有一个庭院供旅客相互见面用。在庭院的一侧有一个共同进餐的地方，同时也就在这一侧有一人照料客栈。在庭院的另外三侧都是客房——在客房的前面有一条拱廊，也许有一步台阶，约 10in 高，另有一步台阶通往各室。傍晚时分，游客在庭院内相互见面，边吃边谈，气氛热烈——这真是别开生面——继而大家一起睡在拱廊内，就在庭院内过夜。

当然，客栈的规模极为重要。客栈的热烈而融洽的气氛取决于经营管理人员居住在客栈内这一事实。他们把整个客栈视为他们自己的住宅。一个家庭是无法管理多于 30 间客房的。

因此：

使游人客栈成为旅游者既可用来住宿过夜、又可——不像绝大多数的旅馆和汽车旅馆那样——用来进行社交活动的场所。客栈的规模要小。每一客栈接纳 30 ~ 40 位旅客；共同进餐；床位都布置在凹室内，环绕中央大空间环状排列。

in one great space,protected from danger by their mutual company,and given entertainment by one another's escapades and stories.

The inspiration for this pattern came from Gita Shah's description of the Indian pilgrim's inn,in *The Timeless Way of Building*:

In India,there are many of these inns.There is a courtyard where the people meet,and a place to one side of the courtyard where they eat,and also on this side there is the person who looks after the Inn,and on the other three sides of the courtyard there are the rooms—in front of the rooms is an arcade,maybe one step up from the courtyard,and about ten feet deep,with another step leading into the rooms. During the evening everyone meets in the courtyard,and they talk and eat together—it is very special—and then at night they all sleep in the arcade,so they are all sleeping together,round the courtyard.

And of course,the size is crucial.The atmosphere comes mainly from the fact that the people who run the place themselves live there and treat the entire inn as their household. A family can't handle more than 30 rooms.

Therefore:

Make the traveler's inn a place where travelers can take rooms for the night,but where—unlike most hotels and motels—the inn draws all its energy from the community of travelers that are there any given evening.The scale is small—30 or 40 guests to an inn;meals are offered communally;there is even a large space ringed round with beds in alcoves.

卧室和凹室　　　　　欢乐的交际活动

共同进餐

&❦&

　　客栈的交际中心区就是旅客相互见面、交谈、跳舞和饮酒作乐的地方——**中心公共区**（129）、**街头舞会**（63）和**啤酒馆**（90）。提供共同进餐的机会，但不是在餐厅宴请宾客，而是大家围坐在同一桌旁，同吃普通的饭菜——**共同进餐**（147）；而且，在私人客房之外，至少有某些地区，人们可以躺下，在公共场所睡觉而不必提心吊胆——**在公共场所打盹**（94）和**共宿**（186）。至于客栈的整体外观、它的花园、停车场和周围的环境，一律按**建筑群体**（95）统一考虑……

❧❦

The heart of the conviviality is the central area,where everyone can meet and talk and dance and drink—COMMON AREAS AT THE HEART(129),DANCING IN THE STREET (65),and BEER HALL(90).Provide the opportunity for communal eating,not a restaurant,but common food around a common table—COMMUNAL EATING(147);and,over and above the individual rooms there are at least some areas where people can lie down and sleep in public unafraid— SLEEPING IN PUBLIC(94),COMMUNAL SLEEPING(186). For the overall shape of the inn,its gardens,parking,and surroundings,begin with BUILDING COMPLEX(95)...

模式92　公共汽车站*

92   BUS STOP*

...within a town whose public transportation is based on MINI-BUSES(20),genuinely able to serve people,almost door to door,for a low price,and very fast,there need to be bus stops within a few hundred feet of every house and workplace.This pattern gives the form of the bus stops.

❧❧❧

Bus stops must be easy to recognize,and pleasant,with enough activity around them to make people comfortable and safe.

Bus stops are often dreary because they are set down independently,with very little thought given to the experience of waiting there,to the relationship between the bus stop and its surroundings.They are places to stand idly,perhaps anxiously,waiting for the bus,always watching for the bus.It is a shabby experience;nothing that would encourage people to use public transportation.

The secret lies in the web of relationships that are present in the tiny system around the bus stop.If they knit together,and reinforce each other,adding choice and shape to the experience,the system is a good one:but the relationships that make up such a system are extremely subtle.For example,a system as simple as a traffic light,a curb,and street corner can be enhanced by viewing it as a distinct node of public life:people wait for the light to change,their eyes wander,perhaps they are not in such a hurry.Place a newsstand and a flower wagon at the corner and the experience becomes more coherent.

……在以**小公共汽车** (20) 为公共交通基础的城镇内，为了真正做到几乎从门到门地为居民低费快速服务，必须设置公共汽车站。这些公共汽车站离开每一住宅和每一工作点都不超过几百英尺。本模式介绍这种公共汽车站的形式。

❧❧❧

公共汽车站必须是易于辨认的和令人愉快的，在它的周围有足够的活动空间，致使旅客感到舒适和安全。

有些公共汽车站常常是令人沉闷的，因为它们总是孤零零地竖立着一个站牌而已。很少有人考虑到公共汽车站和它的周围环境之间的关系，以及等车人的心理状态。旅客站着等车浪费时间，兴许焦急不安，但总是站着等个没完。旅客会感到这种公共汽车站实在蹩脚透顶；公共汽车站没有能吸引旅客利用公共交通的任何设施。

解决上述问题的秘诀就在于关系网。而这些关系目前存在于公共汽车站周围的微系统之中。如果这些关系密切联系相互加强，因地制宜地增设辅助设施，以提供多种体验的可能性，则此微系统就是良好的。但是构成这一系统的各种关系是极为微妙的。例如，人们可以把像交通信号灯、街道镶边石和马路拐角那样简单的系统视为公共生活的一个鲜明的活动中心而加以提高和强化：人们在等待信号灯变换灯光时，常有眼花缭乱之感，也许并不那么急着匆匆忙忙过街。若在街道拐角处有报亭和流动售花车，他们的感受就不同了，就会觉得这一切更加一致了。

街道镶边石、交通信号灯、报亭、鲜花、街道拐角处商店的遮篷、在一小块地区内人群的变化——所有这一切形成了一个相互支持的关系网。

每个公共汽车站变成为这样一个关系网的一部分的可能性是各不相同的：在一种情况下，公共汽车站恰好形成一个吸引旅客进入私密境地的系统：一棵古树使他们浮想

TOWNS
城　镇
883

The curb and the light,the paperstand and the flowers,the awning over the shop on the corner,the change in people's pockets—all this forms a web of mutually sustaining relationships.

The possibilities for each bus stop to become part of such a web are different—in some cases it will be right to make a system that will draw people into a private reverie—an old tree;another time one that will do the opposite—give shape to the social possibilities—a coffee stand,a canvas roof,a decent place to sit for people who are not waiting for the bus.

Therefore:

Build bus stops so that they form tiny centers of public life.Build them as part of the gateways into neighborhoods, work communities,parts of town.Locate them so that they work together with several other activities,at least a newsstand,maps,outdoor shelter,seats,and in various combinations,corner groceries,smoke shops,coffee bar, tree places,special road crossings,public bathrooms,squares....

ഔരു

Make a full gateway to the neighborhood next to the bus stop,or place the bus stop where the best gateway is already—MAIN GATEWAY(53);treat the physical arrangement according to the patterns for PUBLIC OUTDOOR ROOM(69),PATH SHAPE(121),and A PLACE TO WAIT(150);provide a FOOD STAND(93):place the seats according to sun,wind protection,and view—SEAT SPOTS(241)...

联翩；在另一种情况下，恰恰与此相反，要建造一些社会服务设施：一个咖啡商亭、一个帆布顶篷，一个像样的休息场所，供非候车人坐在那里歇息。

两个公共汽车站
Two bus stops

因此：

建造公共汽车站，以便它们形成公共生活的小中心。把它们建成为通往邻里、工作社区和城镇各地的门道的一部分。选择好站址，以便和其他活动联系起来，共同发挥作用，至少有一个报亭、几幅行车路线图和一个室外遮阳篷以及若干坐位，并和其他各种设施，如街角杂货店、吸烟室、咖啡店、树丛、特殊人行横道、公共浴室、广场……组合在一起。

热咖啡店
报亭
公共汽车站
长椅
门道

⊱✿⊰

构筑一个完整的通往公共汽车站附近的邻里的门道，或使公共汽车站位于早就有最好门道的地方——**主门道**（53）；根据模式**户外亭榭**（69）、**小路的形状**（121）和**等候场所**（150）来处理外界环境的布置；提供一个**饮食商亭**（93）；根据向阳、防风和观景的原则，布置坐位——**户外设座位置**（241）……

模式93　饮食商亭*

　　……在邻里中有许多天然的公共集合场所——**活动中心**（30）、**人行横道**（54）、**高出路面的便道**（55）、小广场（61）和**公共汽车站**（92）。从某种程度上说，饮食商亭、小摊和自动售货机的食品香味使这些地方充满生气。

୫୦ଓଃ

　　我们的许多习惯和机构来自下列事实：我们能在街上、在去购物、去工作和找朋友的途中买到普通的便宜食品。

93 FOOD STANDS*

...throughout the neighborhood there are natural public gathering places—ACTIVITY NODES(30),ROAD CROSSINGS(54),RAISED WALKS (55),SMALL PUBLIC SQUARES(61),BUS STOPS(92).All draw their life,to some extent,from the food stands,the hawkers,and the vendors who fill the street with the smell of food.

ꙮ

Many of our habits and institutions are bolstered by the fact that we can get simple,inexpensive food on the street,on the way to shopping,work,and friends.

The food stands which make the best food,and which contribute most to city life,are the smallest shacks and carts from which individual vendors sell their wares.Everyone has memories of them.

But in their place we now have shining hamburger kitchens,fried chicken shops,and pancake houses.They are chain operations,with no roots in the local community. They sell "plastic," mass-produced frozen food,and they generate a shabby quality of life around them.They are built to attract the eye of a person driving:the signs are huge;the light is bright neon.They are insensitive to the fabric of the community.Their parking lots around them kill the public open space.

If we want food in our streets contributing to the social life of the streets,not helping to destroy it,the food stands must

饮食商亭能制作最佳食品。对城市居民的生活贡献最大的饮食商亭是那些最小的棚屋和流动货车。它们由个体商贩经营，出售他们自制的风味小吃。大家对此印象很深。

但现在取而代之的是闪光眩目的汉堡包厨房、炸鸡店、薄煎饼店。这些店在地方社区是没有根底的。它们连续作业，生产食品。它们出售"塑料口袋包装的"、成批生产的冷冻食品，给它们周围造成一种低劣的生活质量。这些店的开设是为了吸引驱车而来的顾客：巨大的广告牌远处可见，夜间的霓红灯光彩夺目。但是它们对社区的结构反应迟钝。它们周围的许多小停车场破坏了公共室外空间。

如果我们在街上购买食品有助于活跃街道的社会生活而不是破坏它，则饮食商亭就必须符合社区要求并相应选好地址。

现在我们建议如下 4 条规定：

1. 饮食商亭应集中在**小路网络和汽车**（52）的**人行横道**（54）附近。有可能从汽车内能看见它们，并希望它们位于某种交叉路口处。但在它们的周围不要有许多专用的小停车场——参阅**停车场不超过用地的 9%**（22）。

2. 饮食商亭应具有与邻里一致的性格。它们可能是自由停放的流动货车，也可能是很小的棚屋，建于街道的拐角处或建筑物的空当内，成为街道建筑的一部分。

3. 这种食品真是"十里飘香"啊。在饮食商亭的周围可以设置有遮阳篷的客座、坐墙、靠背的地方和咖啡店，成为社区较大景观的一部分，而不是密封在平板玻璃结构内，被汽车团团围住。饮食商亭散发出来的香味越浓越好。

4. 这些饮食商亭决不是代销经营的，而是个体商贩自己经营的。最佳的食品总是由家庭餐馆供应的；同样，只有当食品经营者根据他们自己的奇思妙想、自己的烹饪方法、自己的选择而自己制作并销售时，居民才能买到味美可口的最佳食品。

be made and placed accordingly.

We propose four rules:

1.The food stands are concentrated at ROAD CROSSINGS (54)of the NETWORK OF PATHS AND CARS(52).It is possible to see them from cars and to expect them at certain kinds of intersections,but they do not have special parking lots around them—see NINE PER CENT PARKING(22).

2.The food stands are free to take on a character that is compatible with the neighborhood around them.They can be freestanding carts,or built into the corners and crevices of existing buildings;they can be small huts,part of the fabric of the street.

3.The smell of the food is out in the street;the place can be surrounded with covered seats,sitting walls,places to lean and sip coffee,part of the larger scene,not sealed away in a plate glass structure,surrounded by cars.The more they smell,the better.

4.They are never franchises,but always operated by their owners.The best food always comes from family restaurants;and the best food in a foodstand always comes when people prepare the food and sell it themselves,according to their own ideas,their own recipes, their own choice.

Therefore:

Concentrate food stands where cars and paths meet— either portable stands or small huts,or built into the fronts of buildings,half—open to the street.

ଔଔଔ

Treat these food stands as ACTIVITY POCKETS(124) when they are part of a square;Use canvas roofs to make a

因此：

将饮食商亭集中在汽车和小路交汇的地方——或是流动货车，或是小的棚屋，或将饮食商亭建造在建筑物的正面，向街道半开敞。

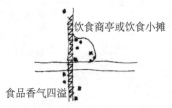

饮食商亭或饮食小摊

食品香气四溢

⁂

当饮食商亭成为广场的一部分时，将它们作为**活动中心**（124）来处理；利用帆布顶篷做它们的简易遮篷——**帆布顶篷**（244）；使它们遵守**个体商店**（87）的经营原则；最佳的食品总是由个体经营者供应的，他们自己经营、自己采购食品原料，并按照自己的风格精工制作美味食品……

simple shelter over them—CANVAS ROOF(244);and keep them in line with the precepts of INDIVIDUALLY OWNED SHOPS(87):the best food always comes from people who are in business for themselves,who buy the raw food,and prepare it in their own style...

模式94　在公共场所打盹

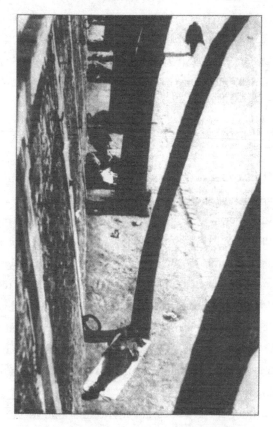

……本模式有助于使**换乘站**（34）、**小广场**（61）、**户外亭榭**（69）、**临街咖啡座**（88）、**步行街**（100）、**有顶街道**（101）和**等候场所**（150）等成为完全的公共场所。

8003

当游人能在公园、公共门厅或门廊内悠然进入梦乡时，这就是这些公共场所取得成功的标志。

94　SLEEPING IN PUBLIC

...this pattern helps to make places like the INTERCHANGE(34),SMALL PUBLIC SQUARES(61), PUBLIC OUTDOOR ROOMS(69),STREET CAFE(88),PEDESTRIAN STREET(100),BUILDING THOROUGHFARE (101),A PLACE TO WAIT(150) completely public.

※※※

It is a mark of success in a park,public lobby or a porch,when people can come there and fall asleep.

In a society which nurtures people and fosters trust,the fact that people sometimes want to sleep in public is the most natural thing in the world.If someone lies down on a pavement or a bench and falls asleep,it is possible to treat it seriously as a need.If he has no place to go—then,we,the people of the town,can be happy that he can at least sleep on the public paths and benches;and,of course,it may also be someone who does have a place to go,but happens to like napping in the street.

But our society does not invite this kind of behavior. In our society,sleeping in public,like loitering,is thought of as an act for criminals and destitutes.In our world,when homeless people start sleeping on public benches or in public buildings,upright citizens get nervous,and the police soon restore "public order."

Thus we cleared these difficult straits,my bicycle and I,together.But a little further on I heard myself hailed.I raised

在培养人民具有高尚品德和信任感的社会中，有时人们想在公共场所睡一觉是人世间最合乎情理的事了。如果有人躺在人行道上或长椅上做一美梦，大家可能把这种行为当作一种生活需要而加以认真对待。如果他无处可去，则我们城镇居民感到高兴的是他至少能在公共的便道上或长椅上甜蜜地睡一觉；当然，也许他有可去之处，但也不妨偶而在街上打个盹。

　　可是，我们的社会现在却视这种行为为粗野的举止，就像闲荡一样，被认为是犯人和贫民干的勾当，是不能容忍的。当我们社会上无家可归的流浪者刚要开始在长椅上或公共建筑内睡眠时，衣冠笔挺的市民就会立刻感到神经紧张不安，警察就会马上过来干涉，恢复"公共秩序"。

　　于是，我和我的自行车总算摆脱了困境。我推车往前走了几步，就听到有人招呼我。我抬头一看，原来是一名警察。简单扼要地说，只是在后来，根据归纳法或演绎法，我已忘记到底是哪一法，我才恍然大悟这是怎么一回事。"你在那里干什么？"他问道。我现在已习惯于这种提问了，我立即答道："休息。""是休息吗？"他继续追问。"是休息。"我理直气壮地回答。"你愿意回答我的问题吗？"他吼叫起来。谈话就这样谈崩的是家常便饭了。我认为我问心无愧地回答了警察的提问，而且我实在什么坏事也没有干。我现在不想再来重述这次谈话曲折的过程了。在这次谈话结束时，我才如梦初醒明白了我休息的方式和我坐在自行车上休息的姿势——我跨坐在我的自行车上，双臂搁在车把上，头搁在手臂上——这违反了我一无所知的公共秩序和公共礼仪……

　　从那以后，我确实再也不那样休息了：我极其不雅地趴在自行车上，双脚着地，双臂搁在车把上，头搁在双臂上摇摇晃晃，东倒西歪。这对于那些痛苦劳累的人，对那些需要鼓励的人来说确是一幅悲惨的景象，一个可悲的例子。他们

my head and saw a policeman.Elliptically speaking,for it was only later,by way of induction,or deduction,I forget which,that I knew what it was.What are you doing there?he said.I'm used to that question,I understood it immediately. Resting,I said.Resting,he said.Resting,I said.Will you answer my question?he cried.So it always is when I'm reduced to confabulation.I honestly believe I have answered the question I am asked and in reality I do nothing of the kind.I won't reconstruct the conversation in all its meanderings.It ended in my understanding that my way of resting,my attitude when at rest,astride my bicycle,my arms on the handlebars,my head on my arms,was a violation of I don't know what,public order,public decency....

What is certain is this,that I never rested in that way again,my feet obscenely resting on the earth,my arms on the handlebars and on my arms my head,rocking and abandoned. It is indeed a deplorable sight,a deplorable example,for the people,who so need to be encouraged,in their bitter toil,and to have before their eyes manifestations of strength only,of courage and joy,without which they might collapse,at the end of the day,and roll on the ground.(Samuel Beckett, *Molloy*.)

It seems,at first,as though this is purely a social problem and that it can only be changed by changing people's attitudes. But the fact is,that these attitudes are largely shaped by the environment itself.In an environment where there are very few places to lie down and sleep people who sleep in public seem unnatural,because it is so rare.

多么需要看见显示出力量、勇气和欢乐的场面，如果没有这一切，在一天行将结束之时，他们的整个身心会垮下来的，会在地上不停地打滚的（Samuel Beckett,*Molloy.*）

乍看起来，这仿佛是一个社会问题，而且解决这一问题只能靠不断改变人们的态度。其实，态度在很大程度上是由环境本身所形成的。在那种只有寥寥数处可供游人躺下睡觉的环境里，无论谁在公共场所打盹，似乎总是不自然的，因为这太罕见了。

因此：

务使在周围环境内有足够数量的长椅板凳，有可席地而坐的舒适地方和角落或可供人们舒展地躺着的沙地。使这些地方成为相对隐蔽的，不受来往人流的影响，也许有一步台阶和若干坐位；还有一片草坪，可供游人躺卧在那里读报和打瞌睡。

遮蔽物

软长椅　　　　　　离开交通线

                            ഉറ

首先，把可供人们打盹的地方辟在沿**建筑物边缘**（160）；在那里放置坐位，甚至也许在公共场所设一两个床龛，可能是很有意思的场所——**床龛**（188）、**户外设座位置**（241）；可是，这首先取决于人们现持的态度——全力以赴创造信任的气氛，既要使去公共场所打盹的人不感到恐惧，又要使看到这种现象的人不大惊小怪。

Therefore:

Keep the environment filled with ample benches, comfortable places,corners to sit on the ground,or lie in comfort in the sand.Make these places relatively sheltered, protected from circulation,perhaps up a step,with seats and grass to slump down upon,read the paper and doze off.

ഗരരജ

Above all,put the places for sleeping along BUILDING EDGES (160);make seats there,and perhaps even a bed alcove or two in public might be a nice touch—BED ALCOVE(188),SEAT SPOTS(241);but above all,it will hinge on the attitudes which people have—do anything you can to create trust,so that people feel no fear in going to sleep in public and so that other people feel no fear of people sleeping in the street.